AQA
GCSE Mathematics

Glyn Payne   Ron Holt   Mavis Rayment   Ian Robinson

Foundation

www.heinemann.co.uk

Heinemann
Inspiring generations

Heinemann Educational Publishers
Halley Court, Jordan Hill, Oxford OX2 8EJ
Part of Harcourt Education

Heinneman is the registered trademark of Harcourt Education Limited
© Harcourt Education Ltd 2006

First published 2006

10 09 08 07 06
10 9 8 7 6 5 4 3 2 1

British Library Cataloguing in Publication Data is available from the British Library on request.

10-digit ISBN: 0 435210 44 0
13-digit ISBN: 978 0 435210 44 1

Edited by Katherine Pate
Designed by Phil Leafe
Typeset by Tech-Set Ltd, Gateshead, Tyne and Wear

Original illustrations © Harcourt Education Limited, 2006

Illustrated by Adrian Barclay and Mark Ruffle

Cover design by mccdesign

Printed in the United Kingdom by Pindar Graphics

Cover photo: Alamy Images ©

Consultants
Andrew Darbourne, Jackie Fairchild, Margaret Hayhurst

Acknowledgements
Harcourt Education Ltd would like to thank those schools who helped in the development and trialling of this course.

The author and publisher would like to thank the following individuals and organisations for permission to reproduce photographs:

iStockPhoto/Matjaz Slanic pp**1**; Corbis pp**3**, **8** first four, **15, 19, 49** bottom, **159, 389, 394, 404, 412, 537**; iStockPhoto/ Ian Francis pp**8** right, Getty Images/PhotoDisc pp**12,36, 78, 431**; Harcourt Education Ltd/Tudor Photography pp**16, 335**; Digital Vision pp**26**; iStockPhoto/James Boulette pp**38**; Getty Images pp**42, 415**; iStockPhoto/Kimberley Girton pp**49** top; iStockPhoto/Ran Plett pp**53**; Science Photo Library pp**58**; iStockPhoto/Alexey Klementiev pp**71**; iStockPhoto/Hannes Nimpuno pp**75**; Photo Library Wales **77**; Alamy Images pp**93, 161, 310, 330, 413**; Empics pp**183, 286**; Harcourt Education Ltd/Debbie Rowe pp**207**; iStockPhoto/Galina Barskaya pp**275**; iStockPhoto pp**331, 335**; iStockPhoto/Gloria-Leigh Logan pp**338**; Photos.com pp**351**; iStockPhoto/Robert St Coeur pp**395**; iStockPhoto/Steven Allen pp**534**

Every effort has been made to contact copyright holders of material reproduced in this book. Any omissions will be rectified in subsequent printings if notice is given to the publishers.

There are links to relevant websites in this book. In order to ensure that the links are up-to-date, that the links work, and that sites are not inadvertently linked to sites that could be considered offensive, we have made the links available on the Heinemann website at www.heinemann.co.uk/hotlinks. When you access the site, the express code is 0440P.

Tel: 01865 888058   www.heinemann.co.uk   www.tigermaths.co.uk

About this book

AQA GCSE Mathematics has been written to meet the requirements of the National Curriculum and provides full coverage of the new two-tier AQA Syllabus which is to be first examined in June 2008.

The book is geared to examination success and is suitable for students in Years 10 and 11. It can be used as a classroom textbook with teacher input, or as a self-help guide for students, using the comment boxes and hints that are a key feature of the book.

The chapters on Number are headed in blue, Algebra chapters in yellow, Shape, Space and Measures chapters in green and Handling Data chapters in orange. This will help you to find related chapters quickly.

Each chapter consists of an introduction outlining the scope of the chapter followed by examples with explanatory notes. Helpful comments and hints feature throughout and key words are highlighted.

  Exercises give students plenty of practice and are structured to provide a clear progression path. Icons indicate whether use of a calculator is advised.

At the end of each chapter exam style questions are included, followed by a summary of key points. Both of these sections include a guide to the examination grade. In addition there are two examination papers at the end of the book, designed to reflect the current AQA papers.

Using and Applying Mathematics (UAM) questions test the ability to solve problems and explain and justify methods of solution. They place extra demands on students. These questions are flagged by the UAM icon.

Proof questions, either in the form of proof by counter example or by more formal methods, now occur on all examination papers. Chapter 22 shows students exactly how to tackle this type of question.

The authors hope that this textbook will make Mathematics more accessible for a wide range of students and will lead to a greater understanding and enjoyment of the subject!

AQA material is reproduced by permission of the Assessment and Qualifications Alliance. AQA take no responsibility for the accuracy of questions or answers in this book. Please note that the following questions are NOT from the live examinations for the current specification. New specifications for GCSE were introduced in 2003.
Ch1 q4, ch8 q5, 9, ch10 q1, 3, ch11, q1, 3, ch13 q1, 2, 3, ch14 q2, ch20 q1, 3, 4, 6

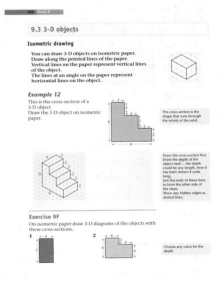

You can find a guide to help you link the specification objectives to this student book at www.tigermaths.co.uk.

Contents

1 Basic rules of number

This chapter will show you how to:

✔ understand place value for whole numbers and decimal numbers
✔ order whole numbers and decimals
✔ understand the four rules for positive and negative whole numbers and decimals
✔ use the correct order of operations
✔ multiply and divide by 10, 100, 1000

If you think you can answer questions on the topics in this chapter, try the revision exercise at the end of the chapter. If you need some help with any of the questions, go to the relevant section of the chapter and look at the examples.

1.1 Whole numbers

Place value for whole numbers

There were 60 479 people at a music festival.

60 479 is a number in the **denary** system, the number system you use every day.

In the denary system, the value of each digit is 10 times the value of the digit on its right-hand side.

The position of a digit in a number tells you its **place value** .

Key words:
place value
place holder

You can write numbers in a place value table:

	Ten millions 10 000 000	Millions 1 000 000	Hundred thousands 100 000	Ten thousands 10 000	Thousands 1000	Hundreds 100	Tens 10	Units 1	
(a)						1	0	4	**(a)** is a three digit or three figure number.
(b)					8	3	4	0	**(b)** is a four digit or four figure number.
(c)				6	0	4	7	9	**(c)** is a five digit or five figure number.

In **(a)** the **4** represents 4 units and has value 4.
In **(b)** the **4** represents 4 tens and has value 40.
In **(c)** the **4** represents 4 hundreds and has value 400.

Notice that **0** is a **place holder** in all of these numbers.

In **(a)** there are no 'tens', in **(b)** there are no 'units' and in **(c)** there are no 'thousands'. You need to write the zero in each of these places to show this.

Example 1

Write down the value of the 6 in each of these numbers:
(a) 46 003   **(b)** 211 068   **(c)** 6 978 123   **(d)** 5006

	TM	M	HT	TT	Th	H	T	U
(a)				4	6	0	0	3
(b)			2	1	1	0	6	8
(c)		6	9	7	8	1	2	3
(d)					5	0	0	6

A place value table makes it easy to see.

(a) The **6** represents 6 thousands, value 6000.
(b) 6 tens, value 60.
(c) 6 millions, value 6 000 000.
(d) 6 units, value 6.

Exercise 1A

1 Write down the value of the 4 in these numbers:

(a) 14

(b) 423

(c) 64 128

(d) 745 000

(e) 400 555

2 John's new motorbike costs £1376.
What is the value of the 3 in this number?

3 The 1996 Olympic Games in Atlanta had 10 750
competitors.
What is the value of the 1 in this number?

4 Write down the value of the red digit in these
numbers:

(a) 15 632

(b) 729

(c) 11 854

(d) 162 759

(e) 500 403

(f) 159 077

(g) 4259

(h) 741 963

(i) 789 002

(j) 40 000

(k) 29 876

(l) 103 301

Writing numbers in words and in figures

You read a number from left to right.

	TM	M	HT	TT	Th	H	T	U
(a)						1	0	4
(b)					8	3	4	0
(c)				6	0	4	7	9

(a) 104 is 1 hundred, no tens and 4 units. You say 'one
hundred and four'.

(b) 8340 is 8 thousands, 3 hundreds, 4 tens and no units.
You say 'eight thousand, three hundred and forty'.

(c) 60 479 is 6 ten-thousands, no thousands, 4 hundreds,
7 tens and 9 units. You say 'sixty thousand, four
hundred and seventy nine'.

You read larger numbers in the same way:

TM	M	HT	TT	Th	H	T	U
	4	3	7	6	5	8	1
2	3	7	0	5	0	0	9

3 hundred thousands 7 ten thousands and 6 thousands makes 376 thousands.

2 ten millions and 3 millions makes 23 million.

For 4 376 581 you say 'four million, three hundred and seventy six thousand, five hundred and eighty one'.

For 23 705 009 you say 'twenty three million, seven hundred and five thousand, and nine'.

You can use the same ideas in reverse when you are given a number in words to write in figures.

Example 2

Write these numbers in figures:

(a) five thousand and forty two

(b) two million, six hundred and fifty thousand and three

(c) eighteen million, one hundred and ninety seven thousand, six hundred and sixty

(d) seventy thousand, nine hundred and twenty nine

	TM	M	HT	TT	Th	H	T	U
(a)					5	0	4	2
(b)		2	6	5	0	0	0	3
(c)	1	8	1	9	7	6	6	0
(d)				7	0	9	2	9

Put them into the correct columns in a place value table.

Remember the 0 as place holder.

Exercise 1B

1 Write these numbers in words:

(a) 1625   **(b)** 21 800   **(c)** 4004

(d) 3 000 303   **(e)** 101 010

2 Write these numbers in figures:
   (a) one thousand, one hundred and forty five
   (b) seven million, seven hundred thousand
   (c) fifteen million, five hundred and five
   (d) sixty thousand and six
   (e) two hundred and two thousand, two hundred and two.

3 Pritesh writes a cheque for £8450 for a new car.
Write this number in words.

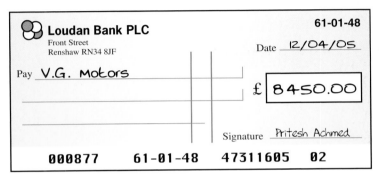

4 According to the local newspaper, there were thirty four thousand, six hundred and three people at the football match this week. Write this number in figures.

5 An estate agent worked out the average price for different types of house in the area. Write each value in words.

Detached 5 bedroom house	£465 000
Detached 3 bedroom bungalow	£280 500
Semi-detached 3 bedroom house	£165 250
Terraced 2 bedroom house	£79 995
1 bedroom flat	£53 750

6 A factory made these numbers of crackers one month:

1st November	Twenty three thousand, four hundred and eight
8th November	Thirty nine thousand and sixty two
15th November	Eighty thousand, three hundred and four
22nd November	One hundred thousand and nine

Write each number in figures.

Ordering whole numbers

The size of a whole number depends on how many digits it has; the more digits, the bigger the number. For example 1234 is bigger than 567.

Sometimes you can easily write numbers in order.

For example   1, 9, 6, 8, 3, 0, 4

written in ascending order is   0, 1, 3, 4, 6, 8, 9

written in descending order is   9, 8, 6, 4, 3, 1, 0

> ascending: going up
> descending: going down

To order large numbers, it helps to write them in a place value table. For two numbers with the same number of digits, look at the digit in the highest place value column. The largest digit belongs to the biggest number.

TM	M	HT	TT	Th	H	T	U
				5	2	3	4
				1	2	3	4

> 5 is bigger than 1.
> So 5234 is bigger than 1234.

If the digits in the highest place value column are equal, compare the digits in the next highest place value column, and so on.

TM	M	HT	TT	Th	H	T	U
			1	3	8	2	0
			1	3	1	7	5
		6	0	2	4	5	1
		6	0	2	4	2	8

> 8 is bigger than 1.
> So 13 820 is bigger than 13 175.
> 5 is bigger than 2.
> So 602 451 is bigger than 602 428.

Example 3

The crowds at four Premiership football matches were:

Arsenal   44 059   Chelsea   45 904

Everton   44 095   Tottenham   45 094

Write them in order of size, biggest first.

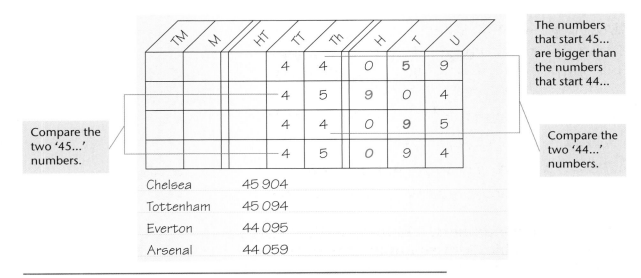

TM	M		HT	TT	Th		H	T	U
				4	4		0	5	9
				4	5		9	0	4
				4	4		0	9	5
				4	5		0	9	4

The numbers that start 45... are bigger than the numbers that start 44...

Compare the two '45...' numbers.

Compare the two '44...' numbers.

Chelsea	45 904
Tottenham	45 094
Everton	44 095
Arsenal	44 059

Exercise 1C

1 Which number in each pair is the biggest?

(a) 1456   1546

(b) 125 462   123 654

(c) 5642   5604

(d) 19 305   19 350

(e) 723   7203

(f) 1 203 000   1 202 999

2 Write these numbers in order from smallest to largest:

(a) 79   93   112   87   108   205   145

(b) 230   203   231   213

(c) 16 324   17 939   16 432   17 940

(d) 236   2345   2349   2136

(e) 1 567 324   1 634 324   563 454

3 The crowds at five rock concerts were:

32 486   32 598   25 654   27 921   25 645

Write these in order, starting with the biggest.

4 The prices of different holidays for a family of four are shown in the table. Write them in order, cheapest first.

Hotel in Tenerife	£2130
Holiday camp in France	£1325
Touring Canada by coach	£2360
Cruise in the Caribbean	£2340
Diving in Australia	£3280
Self-catering in Turkey	£1327

5 The number of grains found in different soil samples were:

Sample	A	B	C	D
Number	156 300 000	98 700 000	160 983 000	159 999 999

Which sample has the largest number of grains?

6 List these five mountains in order, starting with the lowest.

Snowdon	Sawell	Scafell Pike	Ben Nevis	Yes Tor
1085 m	683 m	979 m	1344 m	619 m

Using the signs $<$ and $>$

When you order numbers, you can use special signs for 'less than' and 'greater than'.

$8 < 11$	means	'8 is less than eleven'
$56 < 61$	means	'56 is less than 61'
$40 > 35$	means	'40 is greater than 35'
$102 > 86$	means	'102 is greater than 86'

This also means '11 is greater than 8'.

The **open** (bigger) end of the sign is next to the **bigger** number.

The **pointed** (smaller) end of the sign is next to the **smaller** number.

Example 4

Use $<$ or $>$ to describe the crowds at the football matches in Example 3.

$$45\,904 > 45\,094 > 44\,095 > 44\,059$$

⊞ Exercise 1D

1 Copy the pairs of numbers, using the correct sign, $<$ or $>$.

    (a) 6 9   **(b)** 34 41

    (c) 53 19   **(d)** 159 143

2 Write 'true' or 'false' for each statement.

    (a) $7 > 3$ **(b)** $15 < 9$

    (c) $168 > 98$ **(d)** $3452 < 3544$

3 Write down a value for n (any whole number) in the following. There may be more than one possible answer.

> $486 < n < 488$ means n is a number *between* 486 and 488. n must be bigger than 486 and smaller than 488.

    (a) $486 < n < 488$ **(b)** $159 > n > 157$

    (c) $27\,368 < n < 27\,386$ **(d)** $6001 > n > 5990$

4 True or false?

    (a) $4 < 7 < 11$ **(b)** $28 > 16 > 9$

    (c) $5632 > 3321 > 4892$ **(d)** $62 < 195 < 950$

    (e) $15\,821 > 5811 > 11\,263$

1.2 The four rules for positive whole numbers

Addition

You can use different words for addition.

Find the sum of 21 and 14. Work out 21 plus 14.
What is the total of 21 and 14?

Ed has £21 and Ben has £14, how much do they have altogether?

These all mean add: $21 + 14$.

> sum
> plus
> find the total of   all
> how much   mean
> altogether?   *add*

Example 5

Three buses are taking children on a school trip.

One bus takes 43 children, one takes 39 and the other takes 32.

How many children go on the trip altogether?

```
    4 3
 +  3 9
    3 2
  ─────
  1 1 4      114 children go on the trip altogether.
  1   1
```

✎ Exercise 1E

1 Add these pairs of numbers:
  (a) 16 and 23   **(b)** 75 and 86   **(c)** 183 and 138

2 What is 2349 plus 86?

3 Find the sum of 164, 79 and 288.

4 Work out:
  (a) 162 + 73 + 439 + 5   **(b)** 16 325 + 438 + 7 + 2821
  (c) 1687 + 8 + 63 + 125   **(d)** 17 + 259 + 2 + 1237

5 A shop sold 23 magazines on Monday, 103 on Tuesday, 257 on Wednesday and 87 on Thursday. How many magazines did it sell altogether?

6 Linda's class are collecting vouchers for school equipment. These are the numbers of vouchers the pupils brought in.

  Table A – 6, 3, 5, 9, 17
  Table B – 11, 15, 9, 2, 12
  Table C – 4, 19, 14, 5
  Table D – 12, 23, 7, 9, 18
  Table E – 26, 11, 15, 4, 8

  (a) Work out the number of vouchers for each table.
  (b) What is the total for the whole class?

7 In this addition pyramid, you add the numbers in two bricks together and write the answer in the brick above. The first one has been done for you. Copy and complete the pyramid.

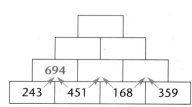

8 In a company, there are 187 people in Sales and 204 in Accounts. How many people is this altogether?

9 On a journey to London, Angus walked 2 miles to the bus station, travelled 27 miles by bus to the railway station and then 434 miles by train.
How far did he travel altogether?

Multiplication

You can use different words for multiplication.

Multiply 36 by 12. Work out 36 times 12.
Find the product of 36 and 12.

These all mean multiply: 36×12.

times
product } mean *multiply*

Example 6

A shopkeeper buys 12 boxes of crisps each containing 36 packets.

How many packets of crisps are there altogether?

Method A

```
      3 6
  ×   1 2
      7 2    ← 36 × 2
  + 3 6 0    ← 36 × 10
    4 3 2
    ₁
```

Method B

	10	2
30	300	60
6	60	12

```
  3 6 0
    7 2
  4 3 2
```

You could work out 36 + 36 + 36 + ... (12 times), but it is much quicker to multiply.

Method C

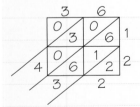

For method C, add along the diagonals.
Start from the bottom right.

 $6 \times 2 = 12$

$6 + 1 + 6 = 13$, so carry 1.

There are 432 packets of crisps.

Exercise 1F

1 What is 7 times 94?

2 Multiply 231 by 15.

3 Copy and complete the multiplication square on the right.

×	8	14	28	123
5				
9				
11				

4 Find the product of 24 and 352.

5 Work out:
   (a) 24×13 (b) 62×38
   (c) 167×41 (d) 327×54

6 A railway carriage holds 48 passengers. A train has 9 full carriages. How many passengers are there altogether?

7 A box contains 180 tins of beans. How many tins are there in 17 boxes?

8 Find the product of 6, 5 and 11.

Multiply 6 and 5 first.

9 The space shuttle burns 3785 litres of fuel every second. How much fuel is burnt in the first minute of its launch?

10 It costs £540 for a holiday in Florida and £107 for a 3-day pass into the theme parks.
   (a) How much would a holiday in Florida cost for 7 people?
   (b) How much would it cost 7 people to go to the theme parks?
   (c) What is the total cost for the holiday and the 3-day pass for 7 people?

Subtraction

You can use different words for subtraction.

Work out 46 minus 32. Find the difference between 46 and 32. Take 32 from 46.

How much more than 32 is 46? How much less than 46 is 32? 46 take away 3.

These all mean subtract: $46 - 32$.

minus
difference
take away
take ... from ...
how much
 more than?
how much
 less than?
} all mean subtract

Example 7

A train travels from London to Edinburgh via Newcastle.

London to Edinburgh is 411 miles.
London to Newcastle is 285 miles.
How far is it from Newcastle to Edinburgh?

$$\overset{3\,\overset{10}{\cancel{1}}}{\cancel{4}11}$$
$$-285$$
$$\overline{126}$$

It is 126 miles from Newcastle to Edinburgh.

> You need to think of this as **how much more than** 285 is 411?

> To check your answer of 126,
> work out 126 + 285   285
> You should get 411   + 126
>   ‾‾‾‾
>   411

Exercise 1G

1 What is 79 minus 24?

2 What is the difference between 83 and 27?

3 What is left if I take 207 from 532?

4 Work out:
  (a) 783 − 126
  (b) 1825 − 532
  (c) 402 − 137
  (d) 3000 − 1562

5 John has 103 CDs and Ali has 59. How many more CDs does John have than Ali?

6 A sports car costs £23 500 and an off-road car costs £18 700. What is the difference in price between the two cars?

7 To get a free DVD Susie needs to collect 500 vouchers. She has collected 392 so far. How many more does she need?

8 Work out 162 minus 43 minus 29.

> Work out 162 minus 43 first.

9 To qualify for the County sports you need 250 points. Lloyd has 187, Fatima has 208 and Laura has 196. How many more points does each one need to reach 250?

10 It is 3423 miles to my holiday destination. I have travelled 987 miles so far. How many more miles do I have to travel?

Division

You can use different words and phrases for division.

How many times does 7 to into 28? Share 28 by 7.

Work out $\frac{28}{7}$. How many 7s are there in 28?

These all mean divide: 28 ÷ 7.

share
how many
  times does
  ... go into ... ?   all
how many   mean
  ... are there *divide*
  in ... ?

Example 8

Eight friends share a lottery win of £5136 equally between them.

How much do they each receive?

8 goes 6 times into 51, remainder 3.

$$8 \overline{) 51^{3}36}$$

$$642$$

They each receive £642.

Example 9

Work out 3358 ÷ 23.

```
        1 4 6
  23)3 3 5 8
      2 3        ←— 23 × 1   23 goes once into 33
    1 0 5        ←— Remainder 10, bring down 5
      9 2        ←— 23 × 4   23 goes 4 times into 105
    1 3 8        ←— Remainder 13, bring down 8
    1 3 8        ←— 23 × 6
        0        ←— No remainder
```

This method is called 'long division'.

The answer is 146.

 Exercise 1H

1 Work out:

   (a) 18 ÷ 3   **(b)** 256 ÷ 4   **(c)** 665 ÷ 5

2 Share 56 sweets equally between 4 people.

3 Work out:

    (a) 105 ÷ 15   **(b)** 168 ÷ 21   **(c)** 154 ÷ 11

4 Eight friends won £40 in a raffle. How much will each receive?

5 A box contains 12 eggs. How many boxes can I fill with 156 eggs?

6 Share 1860 bolts equally between 5 containers. How many are in each container?

7 A pack of writing pads contains equal numbers of sheets of 6 colours. There are 186 sheets. How many is this of each colour?

8 In a supermarket 938 tins of fruit are stacked on 7 identical shelves. How many tins are on each shelf?

9 A parrot has 161 nuts to last her 23 days. If she eats the same number each day, how many nuts can she have each day?

10 230 people are going on a trip. A bus holds 41 people. How many buses do they need?

Mixed problems

To solve a problem, you need to decide whether to add, subtract, multiply or divide to find the answer. In some of the questions you might need more than one **operation** .

> **Key words:**
> operation

> add +   subtract −
> multiply ×   divide ÷
> are *operations*

Example 10

A holiday brochure gives the cost of a holiday in Majorca.

	Number of days	
	7	**14**
Adult	£227	£425
Child aged 3–16	£148	£263
Child under 3	Free	Free

Mr and Mrs Frost and their four children, Adam (11 yrs), Alice (8 yrs), Josh (4 yrs) and Julie (2 yrs), decide to go for two weeks.

How much will it cost them?

Cost for Mr & Mrs Frost
$$= £425 \times 2 = £850$$

Cost for Adam, Alice and Josh
$$= £263 \times 3 = £789$$

Cost for Julie   =   £0   (children under 3 go free)

Total cost   = £1639

14 days = 2 weeks.

Exercise 1I

1 A newsagent has 347 copies of the *Daily News*, 659 copies of the *Free Print Paper* and 68 copies of *Life and Times*. How many papers is this altogether?

2 John has 103 CDs in his collection and Tom has 87. How many more CDs does John have than Tom?

3 How many weeks are there in 91 days?

4 I get £50 for my birthday. I buy a jumper for £17 and some shoes for £24.
How much do I have left?

5 Ester wins £168 on the Lottery. She shares it equally between herself and her two sisters. How much will each get?

6 How many hours are there in 4 days?

7 580 mince pies are to be packed into packets of 12.
   (a) How many full packets will I get?
   (b) How many mince pies will be left over?

8 At Jen's yard there are 15 horses. A vet is inspecting their ears, hooves and tails.
   (a) How many ears does the vet inspect?
   (b) How many hooves does the vet inspect?
   (c) How many hooves, ears and tails does the vet inspect altogether?

 9 A box of raisins holds 12 packets and a box of sultanas holds 20 packets. If I have 6 boxes of raisins and 5 boxes of sultanas, how many packets of fruit do I have altogether?

UAM **10** 'Easy Seat' has sold 1160 sofas. The numbers sold each
week are:

Week	Week 1	Week 2	Week 3	Week 4	Week 5
Sofas sold	204	198	211	187	

How many sofas did they sell in Week 5?

UAM **11** A company sells Easter eggs in packs of 3 and packs of
6. They produce 531 eggs one day. They decide to
make 60 packs containing 6 eggs.
   (a) How many eggs will this use?
   (b) How many packs containing 3 eggs can they
     make with the eggs left over?

UAM **12** At a car boot sale Les had 84 books, 129 CDs and
75 DVDs to sell.
He sold 68 books, 98 CDs and 43 DVDs.
   (a) How many items altogether did he have to take
     home?
   (b) He sold books for 20p, CDs for 75p and DVDs for 50p.
     How much money did he make?

1.3 Order of operations

Key words:
index
indices

Brackets. You must work out the value of any
brackets **first**.

Indices are 'powers' and
you will meet these later.
For now we will use the
index ^{2} for 'squared'.
$5^2 = 5 \times 5 = 25$.

Index or **Indices** . You must work out any
indices **second**.

Divide

Multiply
    You must work out any
    divide and multiply
    calculations **third**.

Divide and Multiply can be
done in either order. You
usually work from left to
right.

Add

Subtract
    You must work out any
    add and subtract
    calculations **last**.

Add and Subtract can be
done in either order. You
usually work from left to
right.

Example 11

Work out:

(a) (i) $2 \times 8 - 4 \times 3$ **(ii)** $2 \times (8 - 4) \times 3$

(b) (i) $18 + 4 \div 2 - 11$ **(ii)** $(18 + 4) \div 2 - 11$

(c) (i) $2 + 3^2 - 1 \times 5$ **(ii)** $(2 + 3)^2 - 1 \times 5$

(a) (i) $2 \times 8 - 4 \times 3 = 16 - 12 = 4$

     (Mult, Mult, Sub)

     (ii) $2 \times (8 - 4) \times 3 = 2 \times 4 \times 3 = 24$

     (Br, Mult, Mult)

(b) (i) $18 + 4 \div 2 - 11 = 18 + 2 - 11 = 9$

     (Div, Add, Sub)

     (ii) $(18 + 4) \div 2 - 11 = 22 \div 2 - 11 = 11 - 11 = 0$

     (Br, Div, Sub)

(c) (i) $2 + 3^2 - 1 \times 5 = 2 + 9 - 1 \times 5 = 2 + 9 - 5 = 6$

     (Ind, Mult, Add, Sub)

     (ii) $(2 + 3)^2 - 1 \times 5 = 5^2 - 1 \times 5 = 25 - 1 \times 5 = 25 - 5 = 20$

     (Br, Ind, Mult, Sub)

> After each line of calculation make a note of the operation you used. This helps you to keep the operations in the correct order.

> Use short forms of the operations e.g. 'mult' for multiplication.

Exercise 1J

1 Work out:

  (a) $3 + 2 \times 7$ **(b)** $5 \times 8 - 3$ **(c)** $9 + 4 \times 5 - 2$

  (d) $9 \times 5 + 3 \times 6$ **(e)** $4 - 3 + 7 \times 2$ **(f)** $3 + 3 \times 5$

2 Find the value of:

  (a) $4 \times (3 + 4)$ **(b)** $16 - (2 \times 4)$

  (c) $5(13 - 9)$ **(d)** $(2 + 8) \times (7 - 4)$

  (e) $7 + (6 - 4) - 3$ **(f)** $(13 + 5) \div 3 + 2$

> $5(13 - 9)$ is another way of writing $5 \times (13 - 9)$.

3 Work out:

  (a) $6 \times 3 + 5^2$ **(b)** $(4 + 5) \times 2^2$

  (c) $4 + (7 - 3) \times 3$ **(d)** $110 - (3 + 2)^2$

  (e) $3 + 6^2 \div 2$ **(f)** $4(15 - 13) \div 2^2$

> Remember:
> $2^2 = 2 \times 2$
> $5^2 = 5 \times 5$ etc.

4 Copy these and put in the correct signs to make the statements true.

(a) 3 ☐ 2 ☐ 4 = 10

(b) 3 ☐ 2 ☐ 4 = 18

(c) 5 ☐ 1 ☐ 7 ☐ 3 = 60

(d) 10 ☐ 6 ☐ 2 ☐ 3 = 43

(e) 2 ☐ 7 ☐ 6 ☐ 2 = 11

(f) 9 ☐ 2 ☐ 3 ☐ 20 = 25

5 Work out each calculation and match the answer to a letter. Now rearrange your letters to make a word. The starting letter has been done for you. Which letter has not been used?

Answer	7	8	1	4	14	9	5	6
Letter	A	C	I	F	N	S	T	E

(a) $25 - 3 \times (4 + 3) = 25 - 3 \times 7 = 25 - 21 = 4$ $4 \to F$

(b) $7 \times 8 \div (19 - 15)$ (c) $19 - 4 \times 3$

(d) $(18 - 6) \div (24 - 12)$ (e) $81 - (8 \times 6) - (4 \times 7)$

(f) $(8 \times 7 - 36) \div 2^2$ (g) $(15 \times 4) - (17 \times 3)$

(h) $6 \times 4 \div 3 - 1$ (i) $5 \times 4 - 6 \times 2$

1.4 The four rules for positive and negative numbers

Negative numbers

The temperature in Moscow yesterday was $-25\,°C$. This means that the temperature was 25° **below** $0\,°C$.

The Himalayas (the mountain range which includes Mount Everest) rise almost 9 km above sea level but have their base about 60 km **below** sea level. This can be written as -60 km.

Numbers can be shown on a number line.

The number line can be horizontal or vertical.

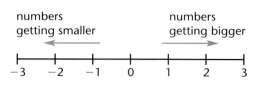

numbers getting smaller   numbers getting bigger

$-3 \quad -2 \quad -1 \quad 0 \quad 1 \quad 2 \quad 3$

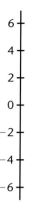

Example 12

Use the thermometer to answer the questions.

(a) The temperature in Manchester at 0600 is $-2\,°C$ and at noon it is $7\,°C$.
What is the rise in temperature between these times?

(b) The temperature in my garden shed this morning was $-4\,°C$.
By the afternoon it had risen by $10\,°C$.
By the evening it had fallen to $7\,°C$ lower than it was in the afternoon.

  (i) What was the temperature in the afternoon?

  (ii) What was the temperature in the evening?

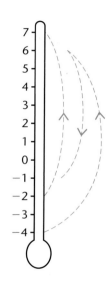

Remember to count 0° as one of the steps because 0 is a place holder.

(a) The rise in temperature is $9\,°C$.

Start at $-2\,°C$ and count the steps up to $7\,°C$.

(b) (i) The temperature in the afternoon was $6\,°C$.

Start at $-4\,°C$ and count up 10 steps.

   (ii) The temperature in the evening was $-1\,°C$.

From $6\,°C$, count down 7 steps.

Exercise 1K

1 In Helsinki, the temperature at midnight is $-6\,°C$. At noon the next day the temperature is $4\,°C$.
What is the rise in temperature between these times?

2 On a spring day the temperature at 9 am was $-1\,°C$. By 2 pm the temperature had risen by $12\,°C$. By 9 pm the temperature had dropped $14\,°C$ from what it was at 2 pm.

   (a) What was the temperature at 2 pm?

   (b) What was the temperature at 9 pm?

3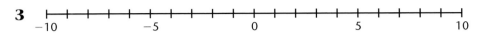

   (a) Copy the number line and mark on it:

      $A = 0$ $B = 7$ $C = -6$ $D = 1$ $E = -2$

   (b) Write the numbers in part **(a)** in order from largest to smallest.

> Use the number line to help.

4 The temperatures in five cities one December day were:

   London $-2\,°C$ Rome $3\,°C$ New York $-5\,°C$
   Moscow $-23\,°C$ Madrid $-14\,°C$

   Write these temperatures in order, starting with the coldest.

5 **(a)** Use the scale to find the height of each item compared to sea level.

   (b) At what height will the bird be if it dives 20 m?

   (c) The diver swims to the bottom of the sea. How far has he dropped?

   (d) The fish swims up 15 m and then down 10 m. How far below the surface is the fish now?

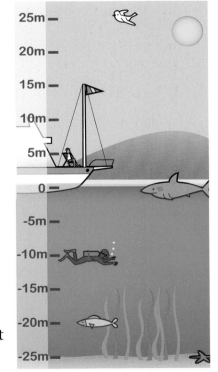

6 The freezer units in a laboratory are kept at different temperatures. They read:

   $-8\,°C$ $-2\,°C$ $-15\,°C$ $-6\,°C$ $-4\,°C$
   $-22\,°C$ $-25\,°C$ $-18\,°C$

   Write these temperatures in order from coldest to warmest.

Adding and subtracting positive and negative numbers

You can add and subtract using a number line.

To **add** a **positive** number, count **up** the number line. $5 + 8 = 13$

> You could write
> $5 + (+8) = 13$

$5 - 8 = -3$ can also be written as
$5 - (+8) = -3$

> You could write
> $5 - (+8) = -3$

To **subtract** a **positive** number, count **down** the number line.

To **add** a **negative** number, count **down**.

So $5 + (-8) = -3$ is the same as $5 - 8 = -3$.

To **subtract** a **negative** number, count **up**.

So $5 - (-8) = 13$ is the same as $5 + 8 = 13$.

In this example, 5 is the starting value.

You can use any number for the starting value – the rules are the same.

For example
$$-4 + (+6) = -4 + 6 = 2$$
$$-4 - (-6) = -4 + 6 = 2$$
$$-4 - (+6) = -4 - 6 = -10$$
$$-4 + (-6) = -4 - 6 = -10$$

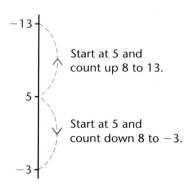

Start at 5 and count up 8 to 13.

Start at 5 and count down 8 to -3.

> Putting ice in a drink makes it cooler.

> Taking ice out of a drink makes it warmer.

The 'rules of signs' are:

$+\ +$ is the same as $+$
$-\ -$ is the same as $+$
$-\ +$ is the same as $-$
$+\ -$ is the same as $-$

Example 13

Work out:

(a) $12 + (-5)$ (b) $-2 + (+8)$

(c) $-3 - (+10)$ (d) $-11 - (-7)$

(a) $12 + (-5) = 12 - 5 = 7$

(b) $-2 + (+8) = -2 + 8 = 6$

(c) $-3 - (+10) = -3 - 10 = -13$

(d) $-11 - (-7) = -11 + 7 = -4$

Exercise 1L

1 Work out:
    (a) $+7 + (-2)$ **(b)** $-5 + (+3)$
    (c) $-8 + (-3)$ **(d)** $3 + (-6) + (-8)$
    (e) $-5 + (+9) + (-8)$ **(f)** $-4 + (-3)$

2 Complete these subtractions:
    (a) $+8 - (+5)$ **(b)** $-7 - (+4)$
    (c) $-9 - (-3)$ **(d)** $-6 - (+12)$
    (e) $-2 - (+3) - (-4)$ **(f)** $3 - (-2)$
    (g) $4 - 5 + 2$ **(h)** $10 - 3 + 4$

3 The recorded evening temperature in 5 cities in England was $-3\,°C$, $1\,°C$, $-4\,°C$, $2\,°C$ and $-1\,°C$.
    (a) Write these in order from coldest to warmest.
    (b) What was the difference in temperature between the coldest and the warmest city?

4 **(a)** The temperature at noon was $8\,°C$. By midnight it was $-3\,°C$. By how many degrees had the temperature fallen?
    (b) In July the temperature in Aberdeen was $15\,°C$. In Cardiff it was $11°$ higher. What was the temperature in Cardiff?

5 Copy and complete these addition and subtraction tables.

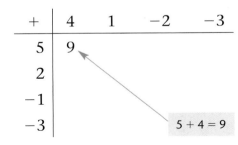
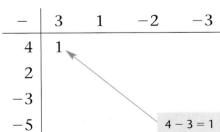

+	4	1	−2	−3
5	9			
2				
−1				
−3				

$5 + 4 = 9$

−	3	1	−2	−3
4	1			
2				
−3				
−5				

$4 - 3 = 1$

6 Copy these calculations and write in the missing numbers.
    (a) $-4 + \square = 3$ **(b)** $8 - \square = 10$
    (c) $7 + \square = -2$ **(d)** $\square - (-3) = 1$

Multiplying and dividing positive and negative numbers

There are similar 'rules of signs' for multiplying and dividing positive and negative numbers.

Multiplying

+	×	+	=	+
+	×	−	=	−
−	×	+	=	−
−	×	−	=	+

Dividing

+	÷	+	=	+
+	÷	−	=	−
−	÷	+	=	−
−	÷	−	=	+

All these rules come from following patterns in multiplication tables.

$2 \times 2 = 4$
$1 \times 2 = 2$ ⟩ -2
$0 \times 2 = 0$ ⟩ -2
$-1 \times 2 = -2$ ⟩ -2
$-2 \times 2 = -4 \longrightarrow 2 \times -2 = -4$ ⟩ $+2$
$\qquad 1 \times -2 = -2$ ⟩ $+2$
$\qquad 0 \times -2 = 0$ ⟩ $+2$
$\qquad -1 \times -2 = 2$
$\qquad -2 \times -2 = 4 \longrightarrow 4 \div -2 = -2$ etc.
$\qquad -3 \times -2 = 6 \longrightarrow 6 \div -3 = -2$

Example 14

Work out:

(a) $(-8) \times (+4)$ **(b)** $(-5) \times (-3)$
(c) $(+20) \div (-2)$ **(d)** $(+36) \div (+9)$

> Work out the sign of the answer first using the 'rules of signs' tables. Then work out the number.

(a) $(-8) \times (+4) = -32$ (b) $(-5) \times (-3) = 15$
(c) $(+20) \div (-2) = -10$ (d) $(+36) \div (+9) = 4$

> When the answer is positive you need not write the + sign.

Exercise 1M

1 Complete these multiplications:

(a) $(-7) \times (+3)$ **(b)** $(-5) \times (-4)$
(c) $4 \times (-3)$ **(d)** $(-2) \times 5$
(e) $(-3) \times (-3)$ **(f)** $(-4) \times (-2) \times (-3)$

> A number without a sign is always positive.

2 Work out these divisions:

   (a) $(+15) \div (-3)$ **(b)** $(-27) \div (-3)$

   (c) $(+16) \div (+4)$ **(d)** $\dfrac{-16}{-4}$

   (e) $\dfrac{-15}{5}$ **(f)** $(-36) \div (-6)$

> $\dfrac{-16}{-4}$ means $-16 \div -4$

3 Copy these calculations and write in the missing numbers.

   (a) $(-9) \div \square = +3$ **(b)** $21 \div \square = -3$

   (c) $(-40) \div \square = -10$ **(d)** $\square \div (-6) = -6$

   (e) $\square \div 2 = 17$ **(f)** $\dfrac{-120}{\square} = 10$

4 Write the missing value for each calculation.

   (a) $3 \times \square = -21$ **(b)** $\square \times (-6) = 24$

   (c) $(-3) \times \square = -9$ **(d)** $6 \times \square = -48$

   (e) $(-5) \times \square = 25$ **(f)** $3 \times \square = 120$

5 In this multiplication pyramid you multiply the numbers in two bricks and write the answer in the brick above. The first one has been done for you.

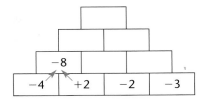

   Copy and complete the multiplication pyramid.

6 In these division pyramids you divide the number on left by the number on its right and write the answer in the brick above.

   Copy and complete the division pyramids.

(a)

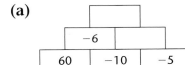

(b)

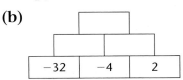

Exercise 1N Mixed questions

1 Work out:

(a) $2 - 4 =$ (b) $-1 - 3 =$

(c) $-8 + 3 =$ (d) $4 + (-2) =$

(e) $(-10) + 12 =$ (f) $31 - 45 =$

(g) $(-2) + (-4) =$ (h) $2 + 5 =$

(i) $5 \times (-4) =$ (j) $(-16) \div 4 =$

(k) $(-3) \times 7 =$ (l) $18 \div (-3) =$

(m) $(-5) \times (-6) =$ (n) $0 \times (-9) =$

(o) $(-24) \div (-8)$ (p) $110 \div 10 =$

2 The temperature at 9 p.m. was 5 °C. Find the temperature if it dropped by:

(a) 1 °C   (b) 5 °C   (c) 9 °C

(d) 12 °C   (e) 20 °C

3 The temperature in London is 4 °C. The temperature in Moscow is −8 °C. What is the difference between these temperatures?

4 Copy and complete these calculations:

(a) $-6 + \square = -9$ (b) $5 + \square = -1$

(c) $4 - \square = 1$ (d) $8 - \square = 10$

(e) $5 \times \square = -30$ (f) $\square \times -3 = -12$

(g) $\square \div (-4) = -3$ (h) $\square - (-8) = 6$

5 The depth of the sea can be written as a negative number. For example, 4 m below sea level is −4 m. A cornetfish swims at −4 m. A bream swims at seven times this depth. How deep is this?

6 One morning the temperature went up from −4 °C to 7 °C. By how many degrees did the temperature rise?

7 The temperature on a rock was 15 °C. At night the temperature fell by 18 °C. What was the temperature on the rock at night?

1.5 Decimals

Place value for decimal numbers

In a place value table the **decimal point** is to the right of the 'units' column.

Key words:
decimal point   tenths
hundredths   thousandths
decimal place

The columns to the right of the decimal point represent parts of a whole number.

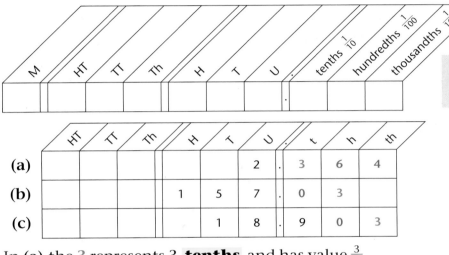

The value of each digit is 10 times the value of the digit on its right-hand side.

HT	TT	Th	H	T	U	.	t	h	th
(a)					2	.	3	6	4
(b)			1	5	7	.	0	3	
(c)				1	8	.	9	0	3

In **(a)** the 3 represents 3 **tenths** and has value $\frac{3}{10}$

the 6 represents 6 **hundredths** and has value $\frac{6}{100}$

the 4 represents 4 **thousandths** and has value $\frac{4}{1000}$

In **(b)** the 3 represents 3 hundredths and has value $\frac{3}{100}$

In **(c)** the 3 represents 3 thousandths and has value $\frac{3}{1000}$

0 is a place holder in **(b)** and **(c)**, there are no 'tenths' in **(b)** and no 'hundredths' in **(c)**.

You put a zero in each of these places to show this.

(a) 2.364   you say 'two point three six four'

(b) 157.03   you say 'one hundred and fifty seven point nought three'

(c) 18.903   you say 'eighteen point nine nought three'

2.364 has 3 **decimal places** .

157.03 has 2 decimal places.

Example 15

Write down the value of the 9 in each of these numbers:

(a) 34.097 (b) 108.94 (c) 92.506 (d) 0.129

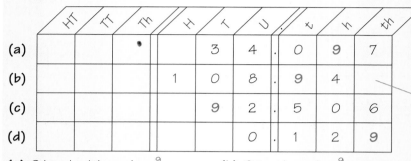

Write the numbers in a place value table.

	HT	TT	Th		H	T	U	.	t	h	th
(a)			•			3	4	.	0	9	7
(b)				1	0	8	.	9	4		
(c)				9	2	.	5	0	6		
(d)					0	.	1	2	9		

You do not need zero here.

(a) 9 hundredths, value $\frac{9}{100}$ (b) 9 tenths, value $\frac{9}{10}$

(c) 9 tens, value 90   (d) 9 thousandths, value $\frac{9}{1000}$

Exercise 10

1 Write the value of the 5 in each of these numbers:

(a) 0.005 (b) 0.05 (c) 0.5

2 Write in words:

(a) 0.8 (b) 0.407 (c) 1.256

Write them in a place value table.

3 Write these decimal numbers in figures:

(a) four tenths (b) one point four
(c) two thousandths (d) six point one zero three

4 Write down the value of the 4 in these numbers:

(a) 132.4 (b) 5.24 (c) 7.426
(d) 0.154 (e) 4.23

5 Write as a decimal number in figures:

(a) $\frac{1}{10}$ (b) $\frac{3}{100}$ (c) $\frac{4}{10} + \frac{3}{100}$
(d) $\frac{27}{100}$ (e) $\frac{3}{10} + \frac{4}{1000}$

6 Statistics show that British families
have an average of 2.4 children.
Write this number in words.

7 The makers of a timing device claim that it is
accurate to within one thousandth of a second.
Write this number as a decimal.

Multiplication and division by 10, 100, 1000 …

Look at these numbers written in a place value table:

	HT	TT	Th	H	T	U	.	t	h	th
(a)						6	.			
(b)					6	0	.			

In **(a)** there is a 6 in the units column so **(a)** has value 6.

In **(b)** the 6 has moved one place to the left and now represents 6 tens, so **(b)** has value 60.

> You need 0 as a place holder here.

Moving digits **one** place to the **left** is the same as **multiplying by 10.**

This idea can be extended.

	HT	TT	Th	H	T	U	.	t	h	th
(a)					1	9	.	0	7	
(a) × 10				1	9	0	.	7		
(a) × 100			1	9	0	7	.			
(a) × 1000		1	9	0	7	0	.			

$19.07 \times 10 = 190.7$

$19.07 \times 100 = 1907$

$19.07 \times 1000 = 19\,070$

> You need 0 as a place holder here.

> Moving digits *one* place to the *left* is the same as *multiplying by 10.*
> Moving digits *two* places to the *left* is the same as *multiplying by 100.*
> Moving digits *three* places to the *left* is the same as *multiplying by 1000.*

If you move digits to the right, you divide.

> Right is the 'opposite' of left. Divide is the 'opposite' of multiply.

> Moving digits *one* place to the *right* is the same as *dividing by 10.*
> Moving digits *two* places to the *right* is the same as *dividing by 100.*
> Moving digits *three* places to the *right* is the same as *dividing by 1000.*

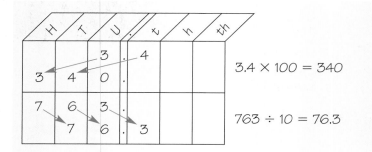

	HT	TT	Th		H	T	U	.	t	h	th
(a)					4	0	8	.			
(a) ÷ 10						4	0	.	8		
(a) ÷ 100							4	.	0	8	
(a) ÷ 1000							0	.	4	0	8

$408 \div 10 = 40.8$

$408 \div 100 = 4.08$

$408 \div 1000 = 0.408$

You need 0 as a place holder here to show there are no units.

Place holders are needed to fill spaces at the *right* of whole numbers, or at the *left* of decimal numbers.

For example, in 0.000 18 place holders are needed; in 0.18 000 place holders are not needed.

$0.18\,000 = 0.18.$

Example 16

Work out: **(a)** 3.4 × 100   **(b)** 763 ÷ 10

H	T	U	.	t	h	th
		3	.	4		
3	4	0	.			
7	6	3	.			
	7	6	.	3		

$3.4 \times 100 = 340$

$763 \div 10 = 76.3$

 Exercise 1P

1 Work out the following. Use a place value table to help you.
- **(a)** 5.2 × 10
- **(b)** 5.2 × 100
- **(c)** 5.2 × 1000
- **(d)** 0.48 × 10
- **(e)** 64 × 100
- **(f)** 3.85 × 10
- **(g)** 0.3 × 1000
- **(h)** 0.006 × 100
- **(i)** 0.066 × 100

2 The cost of one jumper is £9.99. What is the cost of:
- **(a)** 10 jumpers
- **(b)** 100 jumpers?

3 Work out the following. Use a place value table to help you.
- **(a)** 6.4 ÷ 10
- **(b)** 6.4 ÷ 100
- **(c)** 9 ÷ 10
- **(d)** 8.5 ÷ 100
- **(e)** 13.2 ÷ 10
- **(f)** 0.2 ÷ 100
- **(g)** 14 ÷ 1000
- **(h)** 0.04 ÷ 10
- **(i)** 215 ÷ 100

4   The mass of 10 identical robots is 53.4 kg.
What is the mass of 1 robot?

5   Find the value of:
   (a) 3.6 × 100   **(b)** 4.9 ÷ 10   **(c)** 7.32 × 10
   (d) 12.8 ÷ 100   **(e)** 0.005 × 100   **(f)** 7 ÷ 10
   (g) 15.3 ÷ 10   **(h)** 6.02 × 100   **(i)** 5 ÷ 100

6   A paperclip has a mass of 0.8 grams.
What is the mass of 1000 paperclips?

7   100 pupils each pay £6.75 to visit an animal park.
How much is this altogether?

8   Ten tickets for the cinema cost a total of £23.50.
How much will 1 ticket cost?

9   **(a)** 170 000 apples are put into bags of 10.
How many bags are there?
   (b) 100 of these bags are put into a box.
How many boxes are there?
   (c) A van carries 10 boxes each journey. How many
journeys does the van need to deliver all the boxes?

10   True or false? If the statement is false, say why.
   (a) 50 is ten times as big as 5.
   (b) 4 000 is ten times bigger than 40.
   (c) 70 000 is one hundred times bigger than 700.
   (d) 0.9 is the same as 9 ÷ 10.
   (e) 60 is equal to 6 000 ÷ 10.

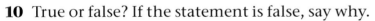

Ordering decimals

Use the same method for ordering decimal numbers as for
whole numbers.

Put the numbers in a place value table.

Examine the numbers from the left. The one with the
largest digit in the highest place value column is the
biggest.

If two or more numbers have equal digits in the highest
place value column, compare the digits in the next
highest place value column, and so on.

Example 17

Write these in order, starting with the biggest:

6.504   6.800   7.290   6.510

	H	T	U	.	t	h	th
(a)			6	.	5	0	4
(b)			6	.	8	0	0
(c)			7	.	2	9	0
(d)			6	.	5	1	0

7.290, 6.8, 6.51, 6.504

(c) has the largest units digit, so it must be the biggest.

(b) has a larger 'tenths' digit than either (a) or (d), so (b) is second biggest.

(d) has a larger 'hundredths' digit than (a), so (d) is the third biggest.

You can write $7.29 > 6.8 > 6.51 > 6.504$ using the signs for 'greater than'.

See Section 1.1.

In (b) and (c) there are some zeros as place holders (in red). These are not needed because they come after the last non-zero decimal digit. In this example they have been included so that you can clearly see the value of the digits in the decimal places.

A common mistake is to look at the decimal parts of the three numbers 6.504, 6.8 and 6.51 and think that because 504 takes up more space than 51 and they are both longer than 8, then the order must be $6.504 > 6.51 > 6.8$. This takes no notice of the place value of the digits and gives the *wrong* order for these three numbers.

Exercise 1Q

1 Which number in each pair is the smaller?

    (a) 0.5 and 0.05   **(b)** 1.23 and 1.3

    (c) 6.509 and 6.5   **(d)** 4.023 and 4.032

2 Write these numbers in order from smallest to largest.

    (a) 0.4   0.6   0.41

    (b) 2.36   2.345   2.349

    (c) 0.23   0.203   0.231   0.2

    (d) 9.503   9.305   9.5   9.53

3 Copy the number line and place the following numbers on it.

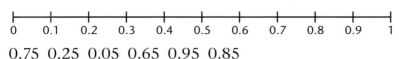

0.75 0.25 0.05 0.65 0.95 0.85

4 Copy and complete each statement using the < or > sign.

(a) 6.4 ☐ 6.04   (b) 0.305 ☐ 0.3

(c) 7.23 ☐ 7.203   (d) 11.206 ☐ 11.026

5 Write down a value for *n* in the following. (There is more than one possible answer.)

(a) $6 > n > 5$ (b) $6.1 < n < 6.2$

(c) $115.2 > n > 115$ (d) $3.13 < n < 3.14$

> $6 > n > 5$ means a number between 6 and 5. *n* is smaller than 6 and more than 5.

6 The times in seconds for a heat of the 100 m race were:

10.32   10.302   10.4   10.325   10.3

Write these in order from the quickest to the slowest.

> The quickest is the one with the lowest time.

7 The table shows the average height in centimetres of some flowers. Write them in order from the shortest to the tallest.

Flower	Height in cm
Bluebell	23.4
Primula	33.04
Daffodil	33.4
Tulip	33.35
Iris	33.38

8 What value can *n* have in the following? Give all possible values.

(a) $25.3 \leqslant n \leqslant 25.5$ if *n* has 1 decimal place

(b) $3.69 \geqslant n \geqslant 3.67$ if *n* has 2 decimal places

(c) $4.5 \leqslant n \leqslant 4.6$ if *n* has 2 decimal places

> ⩽ greater than or equal to
> ⩾ less than or equal to

> 1 decimal place means 1 digit after the decimal point.

Adding decimals

> **To add decimal numbers, line up the numbers so that digits of equal place value are under each other and the decimal points are in line.**

You can then use the same method as for whole numbers, 'carrying' where necessary.

Example 18

Find the total of 26.08 and 174.369.
Do not use a calculator and show all your working.

```
    2 6.0 8 0
 +1 7 4.3 6 9
  2 0 0.4 4 9
      1 1    1
```

Decimal points in line.

The **0** place holder helps to line up the numbers correctly.

Digits written in columns as in a place value table.

Decimal point in line in the answer.

Example 19

Alison spent £14.83 in the supermarket, 76 pence in the newsagents and £3.05 on her bus fare. How much did she spend altogether?

```
   1 4.8 3
 +  0.7 6
    3.0 5
   1 8.6 4
     1 1
```

Write 76 pence in £ as £0.76 so that digits of equal place value are under each other.

Alison spent £18.64 altogether.

Exercise 1R

1 Find the total of each of the following:
   (a) 0.5 + 0.36   (b) 5.29 + 7.3 + 8.245
   (c) 3.5 + 2.46 + 0.723 + 0.28   (d) 4.56 + 6 + 7.2 + 0.384

2 At the cinema, Ranjit spent £2.45 on a ticket, £1.15 on sweets, £1.10 on popcorn and £1.29 on a drink.
   How much did Ranjit spend altogether?

3 Mr Robinson has some fencing wire 2.3 m long, another piece 0.87 m long and a piece 1.125 m long. What is the total length of wire?

4 Sita is comparing costs of bottled water at different shops. The cost, in pounds, is given for 8 shops.

Out-of-town	1.59	1.46	1.72	1.62
Town centre	1.67	1.42	1.80	1.58

    (a) Find the total cost for the out-of-town shops.

    (b) Find the total cost for the town centre shops.

    (c) Sita bought all 8 bottles.
       How much did she spend in total?

5 The mass of 3 different dogs was 11.2 kg, 15.65 kg and 8.2 kg. What is the total mass of the dogs?

6 On a school trip Sara spent £2.38 on drinks, £0.98 on an ice cream, £1.24 on a pack of transfers for her sister and £14.99 on a DVD.
How much did she spend altogether?

Subtracting decimals

> **To subtract decimal numbers line up the digits, keeping the decimal points in line.**
> **Make sure the number you are subtracting *from* goes on the top.**

Always put in any zeros as place holders.

You can use the same method as for whole numbers, 'borrowing' where necessary.

Example 20

Work out 11.8 − 2.647. Do not use a calculator and show all your working.

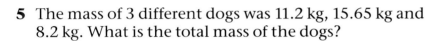

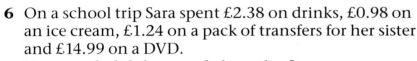

11.8 must go on the top.

You *must* put in the **0** place holders.

```
   ¹1 ⁷7 ⁹9 ¹1
   1̸1.8̸ 0̸ 0
 −  2.6 4 7
   ─────────
    9.1 5 3
```

Example 21

Grace earns £183.65 a week. £46.09 is deducted for income tax and National Insurance.

How much pay does she take home?

'Deduct' means take away.

$$
\begin{array}{r}
1\,\overset{7}{\cancel{8}}\,\overset{1}{3}.\,\overset{5}{\cancel{6}}\,\overset{1}{5} \\
-\ \ 4\,6.\,0\,9 \\
\hline
1\,3\,7.\,5\,6 \\
\end{array}
$$

Grace takes home £137.56 per week.

You can check your subtraction by adding the bottom two lines to see if you get the number on the top line. See Example 7.

 ## Exercise 1S

1 Work out:

   (a) $6.2 - 1.7$ **(b)** $3.27 - 1.53$

   (c) $9.01 - 4.68$ **(d)** $9 - 1.42$

2 Rosy is 42.5 kg and Matt is 48.3 kg. How much heavier is Matt than Rosy?

3 What is the difference between 1.634 and 2.58?

4 If I cut 35.6 cm from a piece of string 90.2 cm long, how much will I have left?

5 The times, in seconds, of the five fastest goals scored are given in this table.

Goal	A	B	C	D	E
Time	12.58	34.6	59.84	64.07	72.29

What is the difference between the quickest and the slowest time?

6 If I buy a pack of pens for £1.39 and a pair of compasses for £1.25, how much change will I get from £5?

Multiplying decimals

You multiply decimals in the same way as whole numbers. You need to be able to put the decimal point in the correct place in your answer.

To multiply decimals:
1 **Ignore the decimal points and multiply the numbers as if they were whole numbers.**
2 **Count the total number of decimal places in the numbers you are multiplying together.**
3 **Put the decimal point in your answer so that it has the same total number of decimal places.**

> These are the digits to the *right* of the decimal point.

Example 22

Work out 4.16×0.2.

> Notice, *no* decimal points.

```
    416
  ×   2
  ─────
    832        The answer is 0.832.
```

> 4.16 has **2** decimal places and 0.2 has **1** decimal place. The answer must have **2 + 1 = 3** decimal places.

> You can check your answer by using your common sense!
> 4.16×0.2 is approximately the same as 4×0.2, so the answer should be about 0.8 (which it is!).

Example 23

Ann needs some carpet for her bedroom.
The bedroom is a rectangle, 3.8 metres long and 2.75 metres wide.
How many square metres of carpet does she need?

> Area of a rectangle
> = length × width.

```
You need to multiply 3.8 by 2.75
        275
  ×      38
  ──────────
       2200     ← 275 × 8
  +    8250     ← 275 × 30
  ──────────
      10450
The answer is 10.450 square metres.
```

> Ignore the decimal points and multiply 38 by 275. Put the larger whole number on the top.

> 2.75 has **2** decimal places and 3.8 has **1** decimal place. The answer must have **2 + 1 = 3** decimal places.

> Ann will probably need to buy 12 square metres because carpet is usually in a given width, often 3 or 4 metres.

You do not need a zero place holder at the right of decimal numbers.

You could write the answer as 10.45 square metres.

Exercise 1T

1 Work out:
   (a) 32×0.3 (b) 67×0.4
   (c) 142×0.05 (d) 53×0.12

2 Work out:
   (a) 4.3×0.7 (b) 6.2×1.3
   (c) 15.4×0.08 (d) 16.4×3.5

3 I want to make 21 shelves, each 0.6 m long. How much wood do I need?

4 A rectangular picture is 6.4 cm wide and 9.5 cm long. What is the area of the picture?

> Area of rectangle = length × width.

5 A square lawn has side lengths of 4.7 m. What is the area of the lawn?

6 Fishing line costs £1.15 per metre. If I buy 4.5 m, how much will it cost?

Dividing decimals

> **Key words:**
> divisor

The rules for dividing whole numbers can be used for decimal division.

You need to take care placing the decimal point in your answer.

Example 24

Share £343.80 equally between six people.

$$6\overline{)3\overset{4}{4}3.\overset{1}{8}0}$$
$$57.30$$

Each person receives £57.30.

> When you divide by a whole number, the decimal point in the answer is in line with the one in the original number.

> This is approximately £300 ÷ 6 = £50 so £57.30 is likely to be the correct answer.

To divide by a number which is *not* a whole number, first multiply top and bottom by the same number so that you get a whole number divisor.

> The **divisor** is the number you are dividing by.

For example, in the calculation

$$\frac{2.832}{0.3} = 9.44$$

You can avoid dividing by a decimal if you multiply 0.3 by 10.

> $0.3 \times 10 = 3.$

Remember to multiply 2.832 by 10 as well.

Then use the method shown in Example 24.

Example 25

Work out:

(a) $2.832 \div 0.3$ **(b)** $14.215 \div 0.05$ **(c)** $0.000\,764 \div 0.004$

(a) $\dfrac{2.832}{0.3} = \dfrac{28.32}{3}$

$$3\overline{)28.\overset{1}{3}\overset{1}{2}}$$
$$9.44$$

Answer $= 9.44$

> Multiply top *and* bottom by 10 (the answer will still be the same). This makes the divisor a *whole number*.

> Making the number on the *bottom* a *whole number* is the secret to doing correct division.

(b) $\dfrac{14.215}{0.05} = \dfrac{1421.5}{5}$

$$5\overline{)1\overset{4}{4}2\overset{2}{1}.\overset{1}{5}}$$
$$284.3$$

Answer $= 284.3$

> $0.05 \times 100 = 5.$
> Multiply top *and* bottom by 100.

(c) $\dfrac{0.000\,764}{0.004} = \dfrac{0.764}{4}$

$$4\overline{)0.7\overset{3}{6}4}$$
$$0.191$$

Answer $= 0.191$

> $0.004 \times 1000 = 4.$
> Multiply top *and* bottom by 1000.

⊞ Exercise 1U

1 Work out:

    (a) $16.35 \div 3$ **(b)** $31.4 \div 5$ **(c)** $43.5 \div 3$

    (d) $1.284 \div 4$ **(e)** $0.34 \div 8$

2 Convert these calculations to give a whole number divisor:

(a) $\dfrac{1.64}{0.5}$ (b) $\dfrac{10.432}{0.001}$ (c) $\dfrac{7.35}{0.07}$ (d) $\dfrac{13.36}{0.15}$

3 Work out:
(a) $6.3 \div 0.3$ (b) $13.4 \div 0.02$ (c) $72.5 \div 0.5$
(d) $39 \div 1.3$ (e) $25 \div 0.1$

4 Work out:
(a) $6.8 \div 0.2$ (b) $3.9 \div 0.03$ (c) $1.32 \div 0.4$
(d) $12.5 \div 0.25$ (e) $0.03 \div 0.2$

5 Theatre tickets for three friends cost £28.20. If all three seats cost the same amount, what was the cost of one ticket?

6 The bill at a café comes to £11.60. Four people share it equally. How much will each person pay?

7 I have a 4 m length of rope. I want to cut it into 0.6 m lengths. How many whole lengths will I get? What length of rope will be left over?

8 The area of a rectangular garden is 51.66 m². If the length is 6.3 m, what is the width?

> Area of rectangle
> = length × width.
> Area ÷ length = width.

Exercise 1V Mixed questions

1 How much change will I get if I pay for a shirt costing £17.75 with a £20 note?

2 The total cost of four tickets for the cinema is £15.40. What is the cost of 1 ticket?

3 A teacher buys 74 text books each costing £6.45. What is the total cost of the books?

4 An apple costs 35p.
(a) How many apples can I buy for £5?
(b) How much change will I have?

5 Harry buys a jumper for £15.99, a DVD for £11.75, a pair of trainers for £35.44 and a baseball cap for £5.79.

(a) How much did he spend?

(b) How much change did he have from £80?

6 A coach company is taking 400 people on a trip. Each coach seats 46 people. The cost is £120 for each coach.

(a) How many coaches do they need for everyone?

(b) What is the total cost for the coaches?

(c) The total cost is shared equally between the people on the trip. How much does each person pay?

7 Flora earns £5.25 per hour. One week she worked 15 hours and had £18.75 deducted for tax and National Insurance. How much did she take home?

8 The cost of electricity is 9.768p per unit for the first 100 units, then 5.9p per unit for the rest. If you use 235 units, how much will you have to pay?

9 A garden centre buys 375 trays of plants. There are 44 plants in a tray.

(a) How many plants is this altogether?

(b) They paid £1500 for all the plants. They sell small trays of 10 plants for £1.24. How much profit will they make?

10 Copy the cross-number grid below. Calculate the answers and write them in the squares on the grid. Put in the decimal points on the lines where necessary. The answer to 1 across has been done as an example.

Clues:

Across

1 $3.29 + 1.96$
3 $24.5 \div 0.7$
5 $31.54 - 7.37$
6 $2.52 - 0.8$
7 1.6×4
8 $8.5 \div 0.05$
10 $101\,000 \div 1000$
11 $3843 \div 7$
12 $104.6 - 95.9$
13 11.1×0.5
16 3.046×1000
19 $9 \times 0.9 \times 0.9$
20 27.5×2.4

Down

1 $32.80 + 19.38$
2 $64\,720 - 12\,510$
3 3.1×1000
4 $12.84 \div 6$
7 $8.12 - 1.28$
9 6.5×1.1
10 $1.36 \div 0.8$
11 0.25×2196
12 $0.7 + 2.8 + 23.3 + 60.9$
14 $7.82 - 2.47$
15 0.07×700
17 $4.14 \div 0.9$
18 165×0.04

Exercise 1W   Revision exercise

1 Write down the value of the red digit in each number.
Section 1.1

(a) 25 632   (b) 729   (c) 11 854

(d) 162 759   (e) 500 403   (f) 17 005 637

(g) 259 000 000   (h) 741 963 248

2 The local newspaper reported that there were thirty five thousand and thirty people at the carnival this week.
Section 1.1

(a) Write this number in figures.

(b) The following week the paper reported a lottery win of £108 108. Write this number in words.

3 The crowds at five rock concerts were:
Section 1.1

35 456   31 545   35 654   27 427   29 248

Write these in order from smallest to biggest.

4 Write down a value for n (any whole number) in the following. There may be more than one possible answer.
Section 1.1

(a) $25 < n < 30$ (b) $4352 > n > 4289$

(c) $89\,523 > n$ (d) $618 < n$

5 The number of errors made by 5 data entry clerks were 162, 73, 439, 8 and 64. What is the total number of errors?
Section 1.2

6 A carton holds 128 boxes of matches. Each box contains 43 matches. How many matches are there altogether in the carton?
Section 1.2

7 The distance from Amsterdam to New York is 3420 miles. If I have already travelled 1856 miles, how much further do I have to go?
Section 1.2

8 A piece of rope 1113 metres long is cut into 7 equal pieces. What is the length of each piece?
Section 1.2

9 Work out:
Section 1.3

(a) $5 + 4 \times 2$ (b) $3 \times (7 - 4)$

(c) $6 \times 5 + 2 \times 3$ (d) $4 \times (9 - 4) - 3$

(e) $9 + 21 \div 3$ (f) $5 + (24 \div 3) \times 7$

(g) $2 + 5^2 - (10 - 4)$ (h) $6 + (11 - 8)^2 - 4$

10 In an experiment the temperature of a liquid started at 3 °C then dropped to −6 °C.
How many degrees did the temperature fall?

Section 1.4

11 Work out:
  (a) $(-3) + (+7)$ **(b)** $(-8) - (+2)$
  (c) $7 - (-4)$ **(d)** $(-5) + (-7)$

Section 1.4

12 Work out:
  (a) $(-8) \times (+4)$ **(b)** $(-5) \times (-3)$
  (c) $49 \div (-7)$ **(d)** $(-64) \div (-8)$

Section 1.4

13 Write down the value of the 9 in each of these numbers:
  (a) 52.9   **(b)** 3.29   **(c)** 7.957
  (d) 0.159   **(e)** 9.56

Section 1.5

14 Work out:
  (a) 1.62×10 **(b)** $3.42 \div 10$ **(c)** 14.6×100
  (d) $32.5 \div 100$ **(e)** 0.58×1000 **(f)** $9 \div 10$
  (g) $54 \div 1000$ **(h)** 0.006×100

Section 1.5

15 **(a)** The mass of 10 identical footballs is 15.4 kg.
      What is the mass of 1 ball?
  (b) A staple has a mass of 0.06 grams.
      What is the mass of 1000 staples?

Section 1.5

16 The times in seconds for a 100 m hurdles race were:
  12.34   12.304   12.4   12.405   12.3
  Write these in order from the quickest to the slowest.

Section 1.5

The quickest is the one with the lowest time.

17 Meg bought a T-shirt for £6.85, a pair of jeans for £25.99, a belt for 99p and a jumper for £9.70. How much did she spend altogether?

Section 1.5

18 Assam weighs 53.5 kg with his boots and pack. If the boots and pack weigh 13.75 kg, how much does Assam weigh?

Section 1.5

19 To find the area of a rectangular lawn you multiply length by width. What is the area of this lawn?

Section 1.5

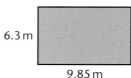

6.3 m

9.85 m

20 To find the diameter of a wheel you do the following calculation:

Section 1.5

20.15 ÷ 3.1

What is the diameter of the wheel?

Examination style questions

1 (a) Copy and complete the shopping bill by Asif.

Leeks	3 kg at £1.60 per kg	
Bananas	2 kg at £1.15 per kg	
3 bottles of water at £0.75 each		
	Total	£

(b) The shop gives Asif 1 discount point for every £2 spent.
How many discount points is Asif given?

(c) Asif buys 7 pens at 63p each.
He pays with a £20 note.
How much change does he receive?

(7 marks)
AQA, Spec A, F, June 2003

2 (a) The entry prices at a theme park are

     Adults   £5.25 each
     Children   £3.40 each

Find the cost for 2 adults and 4 children to visit the theme park.

(b) The entry prices for a group are

     Adults   £5.00 each
     Children   £3.00 each

   (i) A group of adults and children goes to the theme park for a cost of £44.
There are 4 adults in the group.
How many children are in the group?

  (ii) Another group also pays £44.
Find a different answer for the number of adults and children.

(c) The temperature, in °C, at midday at the theme park on six
winter days was recorded.

Day	Monday	Tuesday	Wednesday	Thursday	Friday	Saturday
Temperature	-3	-2	0	-4	-1	-1

     (i) Which day was the warmest at midday?
     (ii) Which day was the coldest at midday?   *(6 marks)*
                                          AQA, Spec A, F, June 2003

3 Magazines are stored in piles of 100.
Each magazine is 0.4 cm thick.
Calculate the height of one pile of magazines.   *(2 marks)*
                                          AQA, Spec A, I, June 2003

4 (a) In the number 2784 what does the digit 7 represent?
  (b) The number 2784 is multiplied by 10.
       In the new number, what does the digit 2 represent?   *(2 marks)*
                                 AQA, Spec B, 3F, November 2002

5 (a) (i) Write 7483 in words.
       (ii) Write 7483 to the nearest 1000.
  (b) From the list of numbers: 2   11   17   29   39   58   71
      Write down
      (i) two numbers which have a sum of 60
      (ii) two numbers which have a difference of 10.   *(4 marks)*
                                 AQA, Spec B, 3F, June 2004

6 Chloe orders the following items at a burger bar.

     3 burgers at £1.49 each   4 large fries at £1.19 each
     2 milkshakes at £1.10 each   3 colas at 69p each

Copy and complete the bill below.

BEST BURGERS		
	£	Pence
3 burgers at £1.49		
4 large fries at £1.19		
2 milkshakes at £1.10		
3 colas at 69p		
Total		

                                       (4 marks)
                               AQA, Spec B, 3F, June 2004

7 (a) Work out $12 - (3 + 7)$

(b) Put brackets in each of these calculations to make them correct.

(i) $18 - 4 - 2 = 16$

(ii) $3 + 4 \times 5 = 35$

(iii) $20 \div 5 - 3 = 10$

(4 marks)
AQA, Spec B, 3I, November 2003

8 (a) (i) Write 3748 in words.

(ii) Write 3748 to the nearest 100.

(iii) Write 3748 to the nearest 10.

(b) Here is a list of numbers

17   21   24   27   35   43   51

What is the largest number you can make when you multiply together two of the numbers on the list? *(5 marks)*
AQA, Spec B, 3F, June 2003

9 The table shows the midday temperatures of five cities on a Sunday in January.

City	Temperature (°C)
Copenhagen	-2
Geneva	0
London	3
Moscow	-4
Paris	-1

(a) Which is the warmest city?

(b) Which is the coldest city?

(c) How much warmer was Paris than Moscow?

(d) On Monday, the midday temperature in Copenhagen was 3 °C higher than it was on Sunday.
What was the midday temperature in Copenhagen on Monday? *(4 marks)*
AQA, Spec B, 3F, March 2003

Summary of key points

Place value (grade G/F)

The value of each digit in a number depends on its position. To see the value, put the number in a place value table.

TM	M	HT	TT	Th	H	T	U	.	t	h	th
			4	7	0	3	8	.	1	6	

The **7** in 47 038.16 represents 7 thousands and has value 7000.
The **6** in 47 038.16 represents 6 hundredths and has value $\frac{6}{100}$.

Ordering numbers (grade G/F)

Write the numbers in a place value table. Read from the left. The one with the largest digit in the highest place value column is the biggest.

Th	H	T	U	.	t	h	th
1	4	6	3	.	5		
1	4	0	8	.	6		
1	4	0	8	.	9		

In this example $1463.5 < 1408.9 > 1408.6$

Order of operations

You must follow this order to get the correct answer to calculations.

$$\text{Brackets} \rightarrow \text{Indices} \rightarrow \begin{array}{c}\text{Divide}\\\text{Multiply}\end{array} \rightarrow \begin{array}{c}\text{Add}\\\text{Subtract}\end{array}$$

Rules of signs for positive and negative numbers (grade F/E)

| Addition and subtraction | Multiplication | Division |

Addition and subtraction
+ + is the same as +
− − is the same as +
− + is the same as −
+ − is the same as −

+	×	+	=	+
+	×	−	=	−
−	×	+	=	−
−	×	−	=	+

+	÷	+	=	+
+	÷	−	=	−
−	÷	+	=	−
−	÷	−	=	+

Multiplying and dividing by 10, 100, 1000 (grade G/F)

To *multiply* by 10 move the digits 1 place to the *left*.
To *divide* by 10 move the digits 1 place to the *right*.

To *multiply* by 100 move the digits 2 places to the *left*.
To *divide* by 100 move the digits 2 places to the *right*.

To *multiply* by 1000 move the digits 3 places to the *left*.
To *divide* by 1000 move the digits 3 places to the *right*.

Remember to add place holders.

The four rules for decimals (grades G to C)

In decimal addition and subtraction, always line up the decimal points under each other.

When multiplying decimals, count the *total* number of decimal places in the numbers being multiplied. The answer must have this number of decimal places.

When dividing decimals, multiply top *and* bottom by 10 (or 100 or 1000 ...) to make the number you are dividing by (the divisor) into a whole number.

2 Approximations and estimation

This chapter will show you how to:
- ✔ understand why you round numbers and use approximations
- ✔ round numbers to the nearest 10, 100, 1000
- ✔ round numbers to a given number of significant figures or decimal places
- ✔ estimate answers to calculations by using approximations

2.1 Rounding

Key words:
rounding
approximation

Sometimes you don't need to know an exact answer.

The land area of the UK is 94 251 square miles.

How can the land area of the UK be measured so accurately? Do you really need to know the exact area?

The population of Italy is 58 138 394

As people are born or die, the population of Italy will be changing all the time.

You don't need to know the amount down to the last penny.

You could use **rounded** or **approximate** values.

The land area of the UK is *about* 90 000 square miles.	The population of Italy is *approximately* 58 million.	York lady wins *nearly* £8 million.

All of these statements give enough information.

An approximation should be easy to read and understand.

Rounding to the nearest 10, 100, 1000

You can use a place value table to round numbers.
The land area of the UK is 94 251 square miles.

	HT	TT	Th	H	T	U	.	t	h	th
	9	4	2	5	1		.			
(a)	9	4	2	5	0		.			
(b)	9	4	3	0	0		.			
(c)	9	4	0	0	0		.			
(d)	9	0	0	0	0		.			

> When you round to the nearest 10, the last non-zero digit is in the Tens column.

(a) shows 94 251 rounded to the nearest 10 since 51 is nearer 50 than 60

(b) shows 94 251 rounded to the nearest 100 since 251 is nearer to 300 than 200 (just!)

(c) shows 94 251 rounded to the nearest 1000 since 4251 is nearer to 4000 than 5000

(d) shows 94 251 rounded to the nearest 10 000 since it is nearer to 90 000 than 100 000

> A number rounds to different answers, depending on the accuracy required.

You usually round a large number to the size of the place value of its largest digit. The most important thing is to give an answer which is easy to understand and makes sense.

Example 1

Round these numbers to the degree of accuracy indicated:
(a) 273 to the nearest 10
(b) 27 350 to the nearest 100
(c) 273 589 to the nearest 1000

(a)

- ← 280

- ← 273
- ← 270

73 is nearer to 70 than 80
273 is 270, to the nearest 10

> A number line will help you round.

continued ▼

(b)

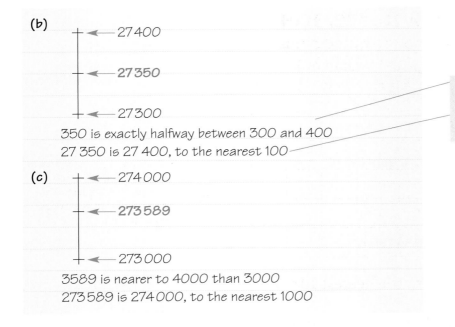

350 is exactly halfway between 300 and 400
27 350 is 27 400, to the nearest 100

When the number falls *exactly* halfway between two limits you always round *upwards*.

(c)

3589 is nearer to 4000 than 3000
273 589 is 274 000, to the nearest 1000

Example 2

Round 8550:
(a) to the nearest 100
(b) to the nearest 1000

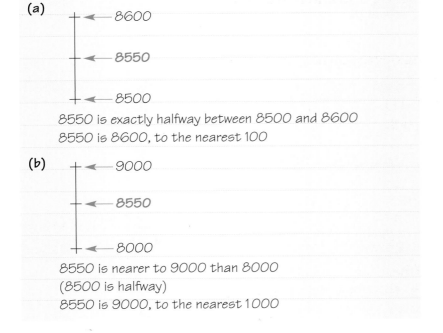

8550 is exactly halfway between 8500 and 8600
8550 is 8600, to the nearest 100

Halfway, so round upwards.

8550 is nearer to 9000 than 8000
(8500 is halfway)
8550 is 9000, to the nearest 1000

Exercise 2A

1 Round these numbers to the nearest 10:
  (a) 38   (b) 592   (c) 245
  (d) 3192   (e) 24 385

2 Round these numbers to the nearest 100:
  (a) 634   (b) 4271   (c) 6850
  (d) 25 351   (e) 165 387

3 Round these numbers to the nearest 1000:
  (a) 8734   (b) 15 397   (c) 243 591
  (d) 423 500   (e) 400 526

4 As you drive into Hartley and Eastpool you see these signs.

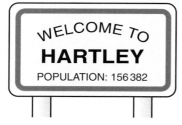 

  (a) Round each population to the nearest:
    (i) ten   (ii) hundred   (iii) thousand
    (iv) ten thousand   (v) hundred thousand
  (b) Do you think it is sensible to put the exact number
      of the population on the signs? Give a reason for
      your answer.

5 A warehouse has the following stock:

Item	Number in stock
$\frac{1}{2}$ inch washers	2 354 816
3 m copper pipe	14 535
90° elbow bends	8783
20 mm straight connectors	12 494

  (a) Round each number in stock to the nearest
      thousand.
  (b) Would it be sensible for the stock manager to
      round the numbers to the nearest thousand? Give
      a reason for your answer.

6 A newspaper quoted the number of people attending a demonstration as 34 500 to the nearest hundred.

Which of the statements could be true and which must be false?

(a) There were 34 526 at the demo.

(b) There were 34 406 at the demo.

(c) The attendance was 34 450.

(d) A total of 33 951 people were there.

(e) There were 34 549 demonstrators.

Rounding sensibly

In some problems where the answer is not a whole number, you need to decide whether to round up or down.

Example 3

Jenny needs 63 tiles to finish tiling her bathroom.
The tiles are sold in boxes of 12.
How many boxes will she need?

Number of boxes = 63 ÷ 12 = 5.25

She needs to buy 6 boxes. (She will have 9 tiles left.)

Remember that
$5 \times 12 = 60$ and
$6 \times 12 = 72$

You must round up or she will not have enough tiles to finish the job.

Example 4

Fay is packing eggs into boxes. Each box holds 6 eggs.
She has 125 eggs. How many boxes can she fill?

Number of boxes = 125 ÷ 6 = 20 remainder 5

She can fill 20 boxes, with 5 eggs left over.

Round down as only whole boxes are filled.

Exercise 2B

1 185 pupils are going on a school trip. Each bus can take 38 pupils. How many buses are needed?

2 Mince pies are packed into boxes of 6. I have 74 pies. How many boxes can I fill?

3 A school sells 108 tickets for a play. The audience sits in rows of 14 chairs. How many full rows of chairs are needed?

4 A room has a wall area of 270 m². Amina wants to paint the walls. She knows that 1 litre of paint covers 35 m² of wall.
   (a) How many litres of paint does Amina need?
   (b) The paint comes in 3 litre tubs. How many tubs will Amina need?

5 At Easter, a factory makes 5 million small chocolate eggs.
   (a) Each packet contains 24 eggs. How many packets will they have?
   (b) 40 packets are packed into boxes. How many full boxes will they have?
   (c) 18 boxes are packed into a carton. How many cartons will they have?
   (d) A truck can carry 54 cartons. How many trucks are needed to deliver all the cartons at the same time?

Rounding to a given number of decimal places

Some calculations do not give an exact answer.

The question may ask you to give the answer to a number of **decimal places (d.p.)** .

> **Key words:**
> decimal place (d.p.)

> Your final answer must have only as many decimal places as the question asks for, no more and no less.

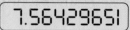

An answer may fill the whole calculator display.

> **To round to a number of decimal places:**
> **1** **Count the number of decimal places you want, to the right of the decimal point.**
> **2** **Look at the *next* digit.**
>   **If it is 5 *or more* you *round up* the digit in the previous decimal place,**
>   **If it is 4 *or less* you *leave* the previous decimal digit *as it is*.**

Example 5

Round each of these numbers to 2 decimal places (2 d.p.).
   (a) 7.2056   **(b)** 0.32491   **(c)** 12.698

   (a) 7.20|56   = 7.21   (2 d.p.)
   (b) 0.32|491 = 0.32   (2 d.p.)
   (c) 12.69|8   = 12.70   (2 d.p.)

> The *next* digit is 5, so round the 0 up to 1.

> The *next* digit is 4, so leave the 2 *as it is*.

> The *next* digit is 8, so round the 9 up. This makes 10, so 'carry' to the next column.

> You need to write the zero as the question asks for 2 d.p. 6 changes to 7

Example 6

Write 8.149 73 to   **(a)** 1 d.p.   **(b)** 2 d.p.   **(c)** 3 d.p.

(a) 8.1|4973   = 8.1   (1 d.p.)

(b) 8.14|973   = 8.15   (2 d.p.)

(c) 8.149|73   = 8.150   (3 d.p.)

> The *next* digit is **4**, so leave 1 *as it is*.

> The *next* digit is **9**, so round the **4** *up* to 5.

> The *next* digit is **7**, so round the **9** *up*. This makes 10, so 'carry' 1 to the next column. 4 changes to 5

> Write the 0 because you must have 3 d.p. The answers of 8.15 in (b) and 8.150 in (c) are *not* the same, since 8.15 is accurate to 2 d.p. and 8.150 is accurate to 3 d.p.

Exercise 2C

1 Round these numbers to 1 decimal place:

  (a) 5.83   **(b)** 7.39   **(c)** 2.15   **(d)** 5.681

  (e) 4.332   **(f)** 15.829   **(g)** 11.264   **(h)** 17.155

  (i) 145.077 **(j)** 521.999

2 Round to 2 decimal places:

  (a) 3.259   **(b)** 6.542   **(c)** 0.877   **(d)** 0.031

  (e) 11.055   **(f)** 4.007   **(g)** 3.899   **(h)** 2.3093

  (i) 0.0009   **(j)** 5.1299

3 Round to 3 d.p.:

  (a) 1.2546   **(b)** 5.2934   **(c)** 4.1265   **(d)** 0.0007

  (e) 0.000 08

4 Round 15.1529 to:

  (a) 1 decimal place   **(b)** 2 d.p.   **(c)** 3 d.p.

  (d) nearest whole number

5 Tom bought a new CD for £14.75.

  What is this:

  (a) to the nearest pound?   **(b)** to the nearest 10p?

6 The mass of a dog is given as 5.625 kg.

  Write this to:

  (a) the nearest kg   **(b)** 1 d.p.   **(c)** 2 d.p.

7 In a race, the winner's time was given as 15.629 seconds.

  (a) Write this to the nearest second.

  (b) Write this to 1 d.p.

  (c) Write this to 2 d.p.

  The time for second place was 15.634 seconds.

  (d) Round this as you did for **(a)**, **(b)** and **(c)** above.

    What do you notice?

  (e) Write a comment about rounding results of races.

Rounding to a given number of significant figures

Sometimes you are asked to round a value to a number of **significant figures (s.f.)** .

Significant figures are 'important' figures. The most significant figure in a number is the one with the greatest place value.

The most significant figure is the *first non-zero* figure reading from the left.

To round to a number of significant figures:
1 **Start from the most significant figure and count the number of figures that you want.**
2 **Look at the *next* digit.**
   If it is *5 or more* you *round up* the previous digit.
   If it is *4 or less* you *leave* the previous digit *as it is*.
3 **Use zero place holders to locate the decimal point and indicate place value.**

This is *very* important.

Example 7

Round **(a)** 240 to 1 significant figure (1 s.f.)
        (b) 192 to 1 s.f.
        (c) 152 to 2 s.f.

(a) $2|40$ $= 200$ (1 s.f.)
(b) $1|92$ $= 200$ (1 s.f.)
(c) $1|5|2$ $= 150$ (2 s.f.)

The most significant figure is 2. The *next* digit is **4**, so leave the **2** as it is.
Use zeros as place holders.

The most significant figure is 1. The *next* digit is **9**, so round **1** *up* to 2.
Use zeros as place holders.

The *next* digit is **2** so leave the **5** *as it is*.
Use zeros as place holders.

Example 8

Round 26 818 to **(a)** 1 s.f.   **(b)** 2 s.f.   **(c)** 3 s.f.

(a) $2|6\,818$ $= 30\,000$ (1 s.f.)
(b) $26|818$ $= 27\,000$ (2 s.f.)
(c) $26|8|18$ $= 26\,800$ (3 s.f.)

The most significant figure is 2.
The *next* digit is **6**, so you round **2** *up* to 3.
Use zeros as place holders.

The *next* digit is **8** so round the **6** *up* to 7.
Use zeros as place holders.

The *next* digit is **1**, so leave the **8** *as it is*.
Use zeros as place holders.

Any of 30 000, 27 000 or 26 800 would be a sensible approximation of 26 818.

Example 9

Write 0.007 038 6 to: **(a)** 1 s.f. **(b)** 2 s.f. **(c)** 3 s.f.

(a) *0.007|038 6* = 0.007
(b) *0.0070|386* = 0.0070
(c) *0.007 03|86* = 0.007 04

In (a), (b) and (c) the zeros after the decimal point but before the 7 locate the decimal point and indicate the place value of the 7.

> The most significant figure is the *first non-zero* digit reading from the left. Here is is 7.
> The *next* digit is 0, so leave the 7 *as it is.*

> The *next* digit is 3, so leave the 0 *as it is.* Write the 0 after the 7 because it is the second significant figure.

> The *next* digit is 8, so round the 3 *up* to 4.

Exercise 2D

1 Round these numbers to 1 significant figure:
   (a) 384   **(b)** 2862   **(c)** 19 473
   (d) 257 394   **(e)** 84 693   **(f)** 4.86

2 Round to 2 s.f.:
   (a) 612   **(b)** 6842   **(c)** 32 841
   (d) 153 945   **(e)** 267 149   **(f)** 5.32

3 Round these numbers to 1 s.f.:
   (a) 0.037   **(b)** 0.042   **(c)** 0.0059
   (d) 0.0035   **(e)** 0.509

4 Round to 2 s.f.
   (a) 0.574   **(b)** 0.000 382   **(c)** 0.001 92
   (d) 0.203   **(e)** 0.001 99

5 The population of Harpool was given as 293 465. What is this rounded to:
   (a) 1 s.f.   **(b)** 2 s.f.?

6 The attendance at some football matches were:

Man Utd	Coventry	Oxford	Luton	Carlisle
48 354	21 924	11 589	5937	2051

   (a) Write each number rounded to 1 s.f.
   (b) Round each number to 3 s.f.
   (c) Which attendance figures, **(a)** or **(b)**, do you think a newspaper would give for each team? Give a reason for your answer.

7 The size of a pollen grain is given as 0.0385 mm. What is this to 2 s.f.?

8 Round 15.045 correct to:

(a) 1 s.f.   (b) 2 s.f.   (c) 3 s.f.

2.2 Estimating using approximations

Key words:
approximation
estimate

When you do a calculation on a calculator, do you always believe the answer?

What if you press a wrong button? Would you realise what you had done?

You can check your answer using **approximations** .

> In the exam you might be asked to find estimates to calculations by using approximations.

> To **estimate** an approximate answer:
> 1 **Round all of the numbers in the calculation to 1 significant figure.**
> 2 **Do the calculation using these approximations.**

Example 10

Use approximations to estimate the answers to these calculations:

(a) $\dfrac{19 \times 59}{31}$ (b) 6.89×3.375 (c) $38.45 \div 7.56$

(d) $\dfrac{181.03 \times 0.48}{8.641}$ (e) $\dfrac{6.85^2}{2.19 \times 11.7}$

(a) $\dfrac{19 \times 59}{31} \approx \dfrac{20 \times 60}{30} = \dfrac{1200}{30} = 40$

(b) $6.89 \times 3.375 \approx 7 \times 3 = 21$

(c) $38.45 \div 7.56 \approx 40 \div 8 = 5$

(d) $\dfrac{181.03 \times 0.48}{8.641} \approx \dfrac{200 \times 0.5}{9} = \dfrac{100}{9} \approx 11$

(e) $\dfrac{6.85^2}{2.19 \times 11.7} \approx \dfrac{7^2}{2 \times 10} = \dfrac{49}{20} \approx \dfrac{50}{20} = 2.5$

$\approx$ means 'approximately equal to'

All 1 s.f. approximations and the estimated answers are shown in blue.

$11 \times 9 = 99$, which is close enough.

49 is almost 50, and it makes the calculation easier.

If you work out the calculations from Example 10, you will see that they are close to the approximate answers.

(a) $\dfrac{19 \times 59}{31} = 36.161\ 29...$ approximate answer = 40

(b) $6.89 \times 3.375 = 23.253\ 75$ approximate answer = 21

(c) $38.45 \div 7.56 = 5.0859...$ approximate answer = 5

(d) $\dfrac{181.03 \times 0.48}{8.641} = 10.0560...$ approximate answer = 11

(e) $\dfrac{6.85^2}{2.19 \times 11.7} = 1.8312...$ approximate answer = 2.5

For **(d)** you could use different approximations.
These would give different approximate answers.

$$\frac{200 \times 0.5}{10} = \frac{100}{10} = 10$$ Using $8.641 \approx 10$

$$\frac{180 \times 0.5}{9} = \frac{90}{9} = 10$$ Using $181.03 \approx 180$ and $8.641 \approx 9$

$$\frac{180 \times 0.5}{10} = \frac{90}{10} = 9$$ Using $181.03 \approx 180$ and $8.641 \approx 10$

For **(e)** you could use 12 (instead of 10) as an approximation for 11.7

$$\frac{7^2}{2 \times 12} = \frac{49}{24} \approx 2$$

It doesn't matter which approximation you use, as long as they are *sensible* and *make the calculations easy*.

Exercise 2E

1 Use approximation to estimate the answers to these calculations:

(a) 67×53

(b) $11 \times 38 \times 12$

(c) $157 \div 47$

(d) $\dfrac{103 \times 32}{12 \times 11}$

(e) $\dfrac{4236 \times 4982}{38 \times 51}$

(f) $\dfrac{276 \times 36}{114 \times 3}$

2   Calculate the actual answers to question 1. Check to see how close your estimates are.

 3   Estimate the answers to these:

    (a) 4.38×5.12 **(b)** $17.3 \div 3.925$

    (c) 148.2×9.604 **(d)** $\dfrac{6.432 \times 3.618}{3.97 + 4.2}$

    (e) $\dfrac{462.3 \times 0.48}{(4.852)^2}$ **(f)** $\dfrac{(5.3)^2 \times 7.759}{192 \times 2.095}$

 4   Calculate the answers to question 3. Check to see how close your estimates are.

 5   A milkman travels 32.5 miles each day delivering milk. He works 27 days each month. Estimate how many miles he travels each month.

 6   On a holiday tour there were 23 coaches. Each coach held 48 people. How many people went on the tour?

 7   For a project a workshop needs a total of 73.23 m of metal rod. They have 24 pieces of metal rod, each 3.75 m long. Use estimation to decide if they have enough rod for the project.

 8   A theatre has 43 rows of 38 seats and 22 rows of 11 seats. Estimate the number of seats in the theatre.

 9   There are 1025 paperclips in a packet. There are 144 packets in a box. 36 boxes are packed in a carton. Estimate the number of paperclips in a carton.

10   A car can travel 1.6 km in 58.6 seconds. Estimate how far it can travel in 1 hour.

Examination style questions

1 Find an approximate value of $\dfrac{41 \times 197}{78}$.

You must show all your working. *(2 marks)*

AQA, Spec A, F, June 2003

2 Kim buys 71 stamps which cost 19 pence each.
By using suitable approximations, estimate the total cost
of the stamps.
You must show your working. *(2 marks)*

AQA, Spec B, 3I, June 2004

3 Find an approximate value of $\dfrac{296 \times 8.13}{0.39}$.

You must show all your working. *(3 marks)*

AQA, Spec B, 3I, June 2004

4 **(a)** Work out 394×52.

(b) **(i)** Write 34.2497 to 1 decimal place.

(ii) Write 34.2497 to 3 decimal places. *(5 marks)*

AQA, Spec B, 3F, June 2004

5 Eileen is making some four-digit numbers.
Each number contains all the digits 7, 4, 9 and 1.

(a) Write down the smallest four-digit number Eillen can make.

(b) Write down the largest four-digit even number Eileen
can make.

(c) Round your answer to part **(b)** to the nearest 100. *(4 marks)*

AQA, Spec B, 3F, June 2004

6 **(a)** Work out $3 \div 0.72$.

(i) Write down the full calculator display.

(ii) Give your answer to the nearest whole number.

(b) **(i)** Calculate $\dfrac{9.8}{6.7 - 1.2}$.

(ii) Give your answer to an appropriate degree of accuracy. *(4 marks)*

AQA, Spec A, I, June 2004

7 The populations of three towns are given below.

Arton 15 748   Barton 9683   Carton 12 403

(a) Write the number 15 748 to the nearest thousand.

(b) Put the towns in order of their population, with the smallest first.

(3 marks)

AQA, Spec A, F, June 2004

8 Calculate the value of $\dfrac{17.32 + 14.29}{4.18 - 1.97}$.

Give your answer to three significant figures.

(3 marks)

AQA, Spec B, 3I, November 2003

Summary of key points

Rounding to the nearest 10, 100, 1000 (grade G)

For example,
41 625 = 41 630 (to the nearest 10)
41 625 = 41 600 (to the nearest 100)
41 625 = 42 000 (to the nearest 1000)

If a number is halfway between two limits, round up.

For example,
3450 is halfway between 3400 and 3500.
3450 is 3500 to the nearest hundred.

Sensible rounding (grade F)

When problems involving division calculations do not give whole number answers, you need to decide whether to round up or down.

For example,
Pens are packed in boxes of 12. How many boxes can you fill with 116 pens?

Number of boxes = $\dfrac{116}{12}$ = 9 remainder 8

You can fill 9 boxes (with 8 pens left over).

Rounding to a given number of decimal places (grade F)

Count the number of decimal places you want.
If the *next* digit is *5 or more*, round *up*,
If the *next* digit is *4 or less*, leave it *as it is*.

For example,   8.427 = 8.43 (to 2 d.p.)
6.014 = 6.01 (to 2 d.p.)

Rounding to a given number of significant figures (grade E)

1 Start your count from the first non-zero digit from the left, this is the first significant figure.
2 Count the number of significant figures you want.
   If the *next* digit is *5 or more*, round *up*.
   If the *next* digit is *4 or less*, leave it *as it is*.
3 Put in zeros to locate the decimal point and indicate place value.

For example,   132.89 = 130 (to 2 s.f.)
                 0.0719 = 0.072 (to 2 s.f.)

Estimating using approximations (grade C)

To estimate an approximate answer:
1 Round all of the numbers to 1 s.f.
2 Work out the answer using these rounded numbers.

For example,

$$\frac{7.73 \times 11.52}{3.87} \approx \frac{8 \times 10}{4} = \frac{80}{4} = 20$$

(exact answer = 23.010 2...)

$$56.25 \times 0.431 \approx 60 \times 0.4 = 24$$

(exact answer = 24.243 75)

3 Fractions

> **This chapter will show you how to:**
> ✔ find equivalent fractions and write a fraction in its simplest form
> ✔ put fractions in order of size
> ✔ find a fraction of a quantity
> ✔ use improper fractions and mixed numbers
> ✔ use the four rules for fractions
> ✔ convert between fractions, decimals and percentages

3.1 Understanding fractions

> **Key words:**
> numerator
> denominator
> proper fraction

This birthday cake has been cut into six equal pieces.

One of the pieces has been eaten.

You say that $\frac{1}{6}$ (one sixth) of the cake has been eaten.

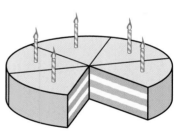

> A fraction has a top number called the **numerator** and a bottom number called the **denominator** .

numerator — 1 piece eaten

$\frac{1}{6}$

denominator — The cake is divided into 6

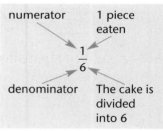

> When the numerator is smaller than the denominator the fraction is a **proper fraction** .
> A proper fraction is always part of a whole.

You can use pictures to show proper fractions.

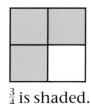

4 equal parts
3 parts shaded

$\frac{3}{4}$ is shaded.

5 equal parts
2 parts shaded

$\frac{2}{5}$ is shaded.

Example 1

Write the fraction shaded in this diagram:

8 equal parts
5 parts are shaded

You can use fractions to represent
information given in words.

$\frac{5}{8}$ is shaded.

Example 2

Tom has 45 DVDs.
10 are sport DVDs, 15 are music DVDs and the rest are
film DVDs.
Write each of these as fractions of his collection.

Sport $= \frac{10}{45}$ Music $= \frac{15}{45}$ Film $= \frac{20}{45}$

Numerator = number of
'Sport' or 'Music' or 'Film'.
Denominator = total
number of DVDs.

Exercise 3A

1 For each diagram write:
  (i) the fraction shaded
  (ii) the fraction unshaded

(a) **(b)** **(c)** **(d)** **(e)**

2 For each of the following, draw a diagram to show the
fraction shaded:
  (a) $\frac{1}{2}$ **(b)** $\frac{3}{4}$ **(c)** $\frac{1}{3}$
  (d) $\frac{3}{5}$ **(e)** $\frac{7}{10}$

3 There are 17 animals in the vet's waiting room. There
are 9 dogs, 6 cats and 2 rabbits. What fraction of the
animals are:
  (a) dogs **(b)** cats **(c)** rabbits?

4 In a vehicle survey there were 12 cars and 9 lorries.
  (a) How many vehicles were surveyed altogether?
  (b) What fraction of the vehicles were cars?
  (c) What fraction were lorries?

5 This is how Joti spent his day.

Activity	Sleeping	School	Eating	TV	Football	Homework
Hours	10	7	1	2	3	1
Fraction						

(a) Copy and complete the table to show the fraction of the day he spent on each activity.

(b) Make a similar table for the way you spend your day.

6 A box of chocolates contains 6 soft centres, 4 hard centres and 3 nut chocolates. What fraction of the chocolates are:

(a) soft centres   (b) hard centres

(c) nut chocolates?

Equivalent fractions

Key words:
equivalent fractions
simplest form

These three diagrams are exactly the same size and have exactly the same amount shaded.

$\frac{3}{4}$ is shaded

$\frac{6}{8}$ is shaded

$\frac{12}{16}$ is shaded

The diagrams show that $\frac{3}{4} = \frac{6}{8} = \frac{12}{16}$.

They are called **equivalent fractions**.

There is a connection between these three fractions:

$$\overset{\times 2}{\underset{\times 2}{\frac{3}{4} = \frac{6}{8}}} \qquad \overset{\times 2}{\underset{\times 2}{\frac{6}{8} = \frac{12}{16}}} \qquad \overset{\times 4}{\underset{\times 4}{\frac{3}{4} = \frac{12}{16}}}$$

You could also write:

$$\overset{\div 2}{\underset{\div 2}{\frac{6}{8} = \frac{3}{4}}} \qquad \overset{\div 2}{\underset{\div 2}{\frac{12}{16} = \frac{6}{8}}} \qquad \overset{\div 4}{\underset{\div 4}{\frac{12}{16} = \frac{3}{4}}}$$

To find equivalent fractions, multiply or divide the numerator and denominator by the same number.

When there is no number that divides exactly into the numerator and denominator, a fraction is in its **simplest form** .

No number divides exactly into 3 and 4, so $\frac{3}{4}$ is the simplest form of this fraction.

Example 3

Find three more fractions that are equivalent to $\frac{2}{9}$.

(i) Multiply numerator and denominator by 2 to get

$$\overset{\times 2}{\underset{\times 2}{\frac{2}{9} = \frac{4}{18}}}$$

(ii) Multiply numerator and denominator by 3 to get

$$\overset{\times 3}{\underset{\times 3}{\frac{2}{9} = \frac{6}{27}}}$$

(iii) Multiply numerator and denominator by 10 to get

$$\overset{\times 10}{\underset{\times 10}{\frac{2}{9} = \frac{20}{90}}}$$

You can choose *any* multiplier to give an equivalent fraction, as long as you multiply the numerator *and* the denominator.

Example 4

Fill in the missing numbers to make equivalent fractions.

$$\frac{2}{\square} = \frac{6}{9} = \frac{\square}{30} = \frac{\square}{90} = \frac{14}{\square}$$

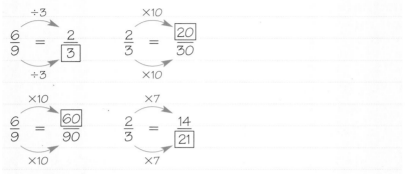

> Start with the fraction you know, $\frac{6}{9}$ in this example.

The equivalent fractions are: $\frac{2}{3} = \frac{6}{9} = \frac{20}{30} = \frac{60}{90} = \frac{14}{21}$

> You can use the fraction you were given or any of the answers you have worked out.

Example 5

Write these fractions in their simplest form:

(a) $\frac{8}{10}$ **(b)** $\frac{36}{48}$ **(c)** $\frac{40}{72}$

(a)

$$\frac{8}{10} \overset{\div 2}{\underset{\div 2}{=}} \frac{4}{5} \qquad \text{Answer} = \frac{4}{5}$$

(b)

$$\frac{36}{48} \overset{\div 2}{\underset{\div 2}{=}} \frac{18}{24} \overset{\div 2}{\underset{\div 2}{=}} \frac{9}{12} \overset{\div 3}{\underset{\div 3}{=}} \frac{3}{4} \qquad \text{Answer} = \frac{3}{4}$$

> You could work out (b) more quickly if you spotted that both 36 and 48 divide exactly by 4 or 6 or 12.

(c)

$$\frac{40}{72} \overset{\div 2}{\underset{\div 2}{=}} \frac{20}{36} \overset{\div 2}{\underset{\div 2}{=}} \frac{10}{18} \overset{\div 2}{\underset{\div 2}{=}} \frac{5}{9} \qquad \text{Answer} = \frac{5}{9}$$

> In (c), you might spot that both 40 and 72 divide exactly by 4 or 8.

 Even if you do not spot the quickest way, you will get the correct answer if you follow the rules.

Exercise 3B

1 Find 2 equivalent fractions for each of the following:

    (a) $\frac{1}{5}$ **(b)** $\frac{2}{3}$ **(c)** $\frac{5}{7}$ **(d)** $\frac{3}{8}$

2 Which of these fractions are equivalent to $\frac{2}{5}$?

$\frac{8}{20}$ $\frac{15}{25}$ $\frac{18}{45}$ $\frac{16}{30}$ $\frac{14}{35}$

Write each in its simplest form.

3 Copy and complete these equivalent fractions.

(a) $\frac{3}{5} = \frac{6}{\square} = \frac{\square}{15} = \frac{\square}{50} = \frac{36}{\square}$

(b) $\frac{\square}{7} = \frac{8}{14} = \frac{12}{\square} = \frac{\square}{35} = \frac{40}{\square}$

(c) $\frac{3}{\square} = \frac{\square}{8} = \frac{15}{20} = \frac{21}{\square} = \frac{\square}{100}$

4 Write each fraction in its simplest form:

(a) $\frac{6}{8}$ (b) $\frac{12}{20}$ (c) $\frac{10}{15}$ (d) $\frac{9}{30}$

(e) $\frac{40}{48}$ (f) $\frac{16}{24}$ (g) $\frac{60}{72}$ (h) $\frac{27}{45}$

5 Luke ate $\frac{15}{30}$ of a pie and Katy ate $\frac{8}{32}$. Write these fractions in their simplest form.

6 In a garden $\frac{2}{12}$ is lawn, $\frac{3}{9}$ is flower beds and $\frac{9}{18}$ is a vegetable plot.

(a) Write each fraction in its simplest form.

(b) Draw a rectangle split into 6 equal parts to represent the garden. Shade in and label the fractions for the lawn, flower beds and vegetable plot.

Ordering fractions

You can see that $\frac{5}{8}$ is bigger than $\frac{3}{8}$.
When fractions have the *same* denominator, the one with the bigger numerator is the bigger fraction.

Key words:
lowest common multiple

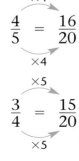

$\frac{5}{8}$ $\frac{3}{8}$

$\frac{5}{8} > \frac{3}{8}$

> means 'is greater than'

Which fraction is bigger, $\frac{4}{5}$ or $\frac{3}{4}$?
These fractions do *not* have the same denominator.
It is difficult to compare them.

You can use equivalent fractions to help you to decide which is bigger.

$$\overset{\times 4}{\frac{4}{5} = \frac{16}{20}} \underset{\times 4}{}$$

$$\overset{\times 5}{\frac{3}{4} = \frac{15}{20}} \underset{\times 5}{}$$

Look for a number that is in the 'times-tables' of both denominators. 20 is the smallest number in both the 4-times and the 5-times tables. It is called the **lowest common multiple (LCM)**. You will meet this again in Chapter 14.

$\frac{16}{20}$ is bigger than $\frac{15}{20}$, so $\frac{4}{5}$ is bigger than $\frac{3}{4}$.
You can write $\frac{4}{5} > \frac{3}{4}$.

> The quickest way to find a common denominator is to multiply the two original denominators together.

> **To order fractions, use equivalent fractions to write them with the same denominator, then compare them.**

Example 6

Put these fractions in order, smallest first: $\frac{1}{3}, \frac{7}{20}, \frac{3}{10}$.

The smallest number that 3, 20 and 10 all divide into is 60.

> 60 is the smallest number you can use.
> *Any* number in the 3×, 10× and 20× table would work.

$$\overset{\times 20}{\underset{\times 20}{\frac{1}{3}}} = \frac{20}{60}$$

$$\overset{\times 3}{\underset{\times 3}{\frac{7}{20}}} = \frac{21}{60}$$

> Multiply both top and bottom by the same number.

$$\overset{\times 6}{\underset{\times 6}{\frac{3}{10}}} = \frac{18}{60}$$

The order is $\frac{18}{10}, \frac{20}{60}, \frac{21}{60}$ or $\frac{3}{10}, \frac{1}{3}, \frac{7}{20}$

> You can write this as
> $\frac{3}{10} < \frac{1}{3} < \frac{7}{20}$

Exercise 3C

1 Put these fractions in order, smallest first:

(a) $\frac{1}{4}, \frac{1}{2}, \frac{3}{8}$ (b) $\frac{3}{5}, \frac{1}{2}, \frac{4}{10}$ (c) $\frac{5}{6}, \frac{1}{3}, \frac{3}{4}$

2 Change these to equivalent fractions, then write them using > or <:

> < means less than
> > means more than

(a) $\frac{3}{5}$ $\frac{5}{8}$ (b) $\frac{1}{3}$ $\frac{3}{10}$ (c) $\frac{5}{6}$ $\frac{4}{9}$

(d) $\frac{3}{4}$ $\frac{5}{6}$ (e) $\frac{3}{5}$ $\frac{5}{7}$ (f) $\frac{1}{4}$ $\frac{2}{5}$

3 Put in order from smallest to largest:

(a) $\frac{1}{4}, \frac{3}{5}, \frac{7}{10}$

(b) $\frac{5}{6}, \frac{2}{3}, \frac{7}{12}$

(c) $\frac{17}{40}, \frac{5}{8}, \frac{7}{10}$

(d) $\frac{57}{100}, \frac{13}{25}, \frac{14}{20}$

Put each fraction into its simplest form first.

4 Ali has $\frac{9}{40}$ of a bowl of cherries. Meg has $\frac{18}{72}$ of the cherries and Bill has $\frac{7}{30}$. Place them in order from smallest to largest to find who has the biggest fraction of cherries.

3.2 Finding a fraction of a quantity

In mathematics, 'of' means multiply so

$$\frac{3}{5} \text{ of } 80 = \frac{3}{5} \times 80$$

$$= \frac{3 \times 80}{5}$$

$$= \frac{240}{5}$$

$$= 48$$

Sharing by 2 is the same as finding $\frac{1}{2}$ of the quantity. $\div 4$ is equivalent to $\frac{1}{4}$.

Another way to do this is to work out $\frac{1}{5}$ of 80 first, then multiply by 3 because $\frac{3}{5} = 3 \times \frac{1}{5}$.

$$\frac{1}{5} \text{ of } 80 = \frac{1}{5} \times 80$$

$$= \frac{80}{5}$$

$$= 16$$

So $\frac{3}{5}$ of $80 = 3 \times 16 = 48$

You can see that both methods give 48 for the answer. You can choose whichever one you prefer.

Example 7

A factory employs 216 people.

$\frac{5}{12}$ of them are men.

(a) How many men work in the factory?

(b) How many women work in the factory?

(a) $\frac{5}{12}$ of 216 = $\frac{5}{12} \times 216$

$\phantom{\frac{5}{12} \text{ of } 216} = \dfrac{5 \times 216}{12}$

$\phantom{\frac{5}{12} \text{ of } 216} = \dfrac{1080}{12}$

$\phantom{\frac{5}{12} \text{ of } 216} = 90$

90 men work in the factory.

(b) So the number of women $= 216 - 90$

$\phantom{\text{So the number of women }} = 126$

Exercise 3D

 1 Work out:

(a) $\frac{1}{4}$ of 24 metres   (b) $\frac{2}{3}$ of £36

(c) $\frac{3}{5}$ of 60 kg   (d) $\frac{7}{9}$ of 36 litres

(e) $\frac{7}{10}$ of 40 minutes   (f) $\frac{5}{8}$ of 48 pages

> Diagrams can help to answer these questions e.g.
>
>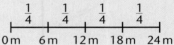

2 Work out the following. You may use a calculator to help.

(a) $\frac{5}{6}$ of 162 km   (b) $\frac{7}{11}$ of 121 days

(c) $\frac{13}{15}$ of £630   (d) $\frac{8}{9}$ of 153 euros

(e) $\frac{3}{4}$ of 450 g   (f) $\frac{2}{3}$ of 5 litres

 3 Which is bigger:

(a) $\frac{4}{5}$ of 15 or $\frac{2}{3}$ of 24   (b) $\frac{3}{10}$ of 200 or $\frac{7}{8}$ of 96

(c) $\frac{1}{3}$ of 240 or $\frac{1}{4}$ of 440   (d) $\frac{3}{8}$ of 104 or $\frac{2}{3}$ of 69?

4 Tom is awake for $\frac{5}{8}$ of a day. How many hours is this?

> Change a day into hours first.

5 Hannah spends $\frac{3}{4}$ of her pocket money and saves the rest. One week she gets £10 pocket money.

(a) How much does she spend?

(b) How much does she save?

6 Amit's grandfather says he can either have $\frac{2}{3}$ of £75 or $\frac{3}{8}$ of £120 for his birthday. Which one should he choose?

 7 A man earns £420.80 per week. Out of this he has to pay $\frac{3}{20}$ in tax. What will his take home pay be?

3.3 Improper fractions and mixed numbers

> **Key words:**
> improper fractions
> mixed number

This is a **proper** fraction.

The numerator is smaller than the denominator.

$$\frac{3}{8}$$

numerator

denominator

$\frac{9}{4}$ is an **improper fraction** . The numerator is greater than the denominator.

It can be represented in picture form as

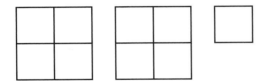

You can see that this is the same as $2\frac{1}{4}$ squares.
$2\frac{1}{4}$ is a **mixed number** because it contains a whole number part (2) and a fractional part ($\frac{1}{4}$).

You need to be able to write an improper fraction as a mixed number and vice versa.

Writing an improper fraction as a mixed number

For example, to change $\frac{9}{4}$ to $2\frac{1}{4}$:

1 Divide the numerator by the denominator and write down the whole number part of your answer.

$9 \div 4 = 2$ remainder 1

2 The remainder gives you the fractional part. The denominator is the same as the denominator of the improper fraction.

$2\frac{1}{4}$

remainder 1

denominator of mixed number

Example 8

Write these improper fractions as mixed numbers:

(a) $\frac{7}{2}$ (b) $\frac{20}{3}$ (c) $\frac{29}{6}$

(a) $7 \div 2 = 3$ remainder 1, so $\frac{7}{2} = 3\frac{1}{2}$

(b) $20 \div 3 = 6$ remainder 2, so $\frac{20}{3} = 6\frac{2}{3}$

(c) $29 \div 6 = 4$ remainder 5, so $\frac{29}{6} = 4\frac{5}{6}$

Exercise 3E

1 Write these improper fractions as mixed numbers:

(a) $\frac{9}{2}$ (b) $\frac{8}{3}$ (c) $\frac{24}{5}$

(d) $\frac{15}{4}$ (e) $\frac{23}{10}$ (f) $\frac{253}{100}$

(g) $\frac{23}{7}$ (h) $\frac{47}{9}$ (i) $\frac{27}{3}$

2 Change the improper fractions into mixed numbers then write them using > or <.

(a) $\frac{17}{5}$ $\frac{11}{2}$ (b) $\frac{10}{3}$ $\frac{23}{9}$

(c) $\frac{15}{4}$ $\frac{25}{6}$ (d) $\frac{13}{2}$ $\frac{43}{8}$

(e) $\frac{83}{10}$ $\frac{153}{20}$ (f) $\frac{24}{4}$ $\frac{125}{25}$

3 Jody ate $\frac{7}{3}$ apples and Megan ate $\frac{13}{4}$ apples. Who ate the most apples?

Writing a mixed number as an improper fraction

For example, to change $2\frac{1}{4}$ to $\frac{9}{4}$:

1 Multiply the whole number part by the denominator of the fractional part, 2 is the same as $\frac{8}{4}$.

2 Add the fractional part of your answer to the result of step 1.

$\frac{8}{4} + \frac{1}{4} = \frac{9}{4}$

Example 9

Write these mixed numbers as improper fractions:

(a) $1\frac{3}{8}$ (b) $4\frac{2}{7}$ (c) $10\frac{4}{9}$

(a) $1 \times 8 = 8$ $\frac{8}{8} + \frac{3}{8} = \frac{11}{8}$

(b) $4 \times 7 = 28$ $\frac{28}{7} + \frac{2}{7} = \frac{30}{7}$

(c) $10 \times 9 = 90$ $\frac{90}{9} + \frac{4}{9} = \frac{94}{9}$

Exercise 3F

1 Write these mixed numbers as improper fractions.

(a) $2\frac{1}{2}$ (b) $4\frac{1}{5}$ (c) $6\frac{3}{4}$

(d) $5\frac{1}{3}$ (e) $2\frac{2}{9}$ (f) $3\frac{5}{8}$

(g) $4\frac{5}{6}$ (h) $3\frac{7}{10}$ (i) $2\frac{2}{7}$

2 Molly ate $2\frac{1}{4}$ oranges. Shona ate $1\frac{3}{4}$ oranges.
How many orange quarters did each girl eat?

3 Copy and complete with the correct sign, $<$, $>$ or $=$

(a) $3\frac{1}{2} \square \frac{7}{2}$ (b) $\frac{17}{5} \square 2\frac{4}{5}$ (c) $\frac{11}{4} \square 3\frac{1}{4}$

(d) $1\frac{8}{9} \square \frac{16}{9}$ (e) $2\frac{4}{11} \square \frac{28}{11}$ (f) $\frac{300}{50} \square 6$

> Change both numbers into improper fractions or mixed numbers to compare them.

3.4 The four rules for fractions

You can add, subtract, multiply and divide fractions.

Adding fractions

When fractions have the *same* denominator it is easy to add them. You simply add the numerators.

For example: $\dfrac{2}{7} + \dfrac{3}{7} = \dfrac{5}{7}$

Notice that you do *not* add the denominators.

When fractions do *not* have the same denominator you need to use equivalent fractions to change one or both fractions so that the denominators are the same.

For example: $\frac{2}{5} + \frac{3}{10}$

$\frac{2}{5}$ can be written as $\frac{4}{10}$

So the addition becomes $\frac{4}{10} + \frac{3}{10} = \frac{7}{10}$

> **To add fractions, you need to change them to equivalent fractions with the same denominator.**

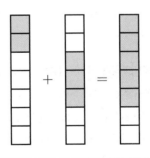

> Equivalent fractions are in Section 3.1

> Multiplying both numerator and denominator by 2. $\dfrac{2}{5} = \dfrac{4}{10}$ $\times 2$

Example 10

Work out $\frac{2}{3} + \frac{4}{5}$

$$\frac{2}{3} + \frac{4}{5} = \frac{10}{15} + \frac{12}{15}$$
$$= \frac{22}{15}$$
$$= 1\frac{7}{15}$$

The denominators are 3 and 5. Look for a number that is in the 3-times and 5-times tables. 15 is in both.

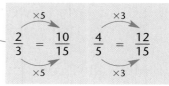

See Section 3.3, Improper fractions and mixed numbers.

Example 11

Work out $3\frac{1}{4} + 1\frac{5}{6}$

Method 1

$$3\frac{1}{4} + 1\frac{5}{6} = \frac{13}{4} + \frac{11}{6}$$
$$= \frac{39}{12} + \frac{22}{12}$$
$$= \frac{61}{12}$$
$$= 5\frac{1}{12}$$

Write mixed numbers as improper fractions first.

Multiplying by 3 and by 2 to get a denominator of 12 in each fraction.

Always give your answer as a mixed number.

Method 2

$$3\frac{1}{4} + 1\frac{5}{6} = 3 + 1 + \frac{1}{4} + \frac{5}{6}$$
$$= 4 + \frac{3}{12} + \frac{10}{12}$$
$$= 4 + \frac{13}{12}$$
$$= 4 + 1\frac{1}{12}$$
$$= 5\frac{1}{12}$$

Add whole numbers together.

Multiply to get a common denominator of 12.

Convert improper fraction to mixed number and add.

Exercise 3G

1 Work out:

    (a) $\frac{3}{8} + \frac{3}{8}$ **(b)** $\frac{4}{9} + \frac{7}{9}$ **(c)** $\frac{3}{11} + \frac{10}{11}$

2 Work out:

    (a) $\frac{3}{4} + \frac{1}{3}$ **(b)** $\frac{2}{5} + \frac{1}{2}$ **(c)** $\frac{2}{3} + \frac{3}{5}$

    (d) $\frac{4}{5} + \frac{3}{4}$ **(e)** $\frac{1}{3} + \frac{5}{6}$ **(f)** $\frac{1}{6} + \frac{3}{8}$

3 Work out the following, giving your answers as mixed numbers:

(a) $1\frac{1}{4} + 2\frac{2}{3}$ (b) $1\frac{1}{6} + 2\frac{1}{3}$ (c) $1\frac{3}{5} + 1\frac{1}{10}$

(d) $2\frac{4}{5} + 1\frac{1}{4}$ (e) $2\frac{3}{10} + 1\frac{5}{6}$ (f) $1\frac{3}{8} + 2\frac{2}{3}$

4 After a party I have $\frac{2}{3}$ of one cake and $\frac{1}{6}$ of another cake left over. How much cake do I have altogether?

5 I have a piece of stair carpet $3\frac{1}{3}$ metres long and another piece $2\frac{1}{4}$ metres long. What length of carpet do I have altogether?

6 Jenny travelled by bus for $2\frac{3}{4}$ hours and then by train for $3\frac{1}{2}$ hours. How long was her journey?

7 A box has a mass of $\frac{1}{8}$ kg. Its contents have a mass of $2\frac{2}{3}$ kg. What is the total mass of the box and its contents?

Subtracting fractions

To subtract fractions, you need to change them to equivalent fractions with the same denominator.

You use the same method as for adding fractions.

Example 12

Work out:

(a) $\frac{9}{11} - \frac{3}{11}$ (b) $\frac{7}{8} - \frac{11}{16}$

(c) $\frac{9}{10} - \frac{2}{3}$ (d) $3\frac{1}{4} - 1\frac{5}{6}$

(a) $\frac{9}{11} - \frac{3}{11} = \frac{6}{11}$

The denominators are the same, so you can subtract.

(b) $\frac{7}{8} - \frac{11}{16} = \frac{14}{16} - \frac{11}{16}$

$= \frac{3}{16}$

Writing $\frac{7}{8}$ as $\frac{14}{16}$

(c) $\frac{9}{10} - \frac{2}{3} = \frac{27}{30} - \frac{20}{30}$

$= \frac{7}{30}$

Using a denominator of 30 and writing equivalent fractions.

(d) $3\frac{1}{4} - 1\frac{5}{6} = \frac{13}{4} - \frac{11}{6}$

$= \frac{39}{12} - \frac{22}{12}$

$= \frac{17}{12}$

$= 1\frac{5}{12}$

Convert to improper fractions first. These are the same fractions as in Example 11.

Write the answer as a mixed number.

Exercise 3H

1 Work out:

(a) $\frac{5}{8} - \frac{3}{8}$ (b) $\frac{5}{6} - \frac{1}{6}$ (c) $1\frac{3}{5} - \frac{4}{5}$

> In (c) change to improper fractions.

2 Work out:

(a) $\frac{7}{10} - \frac{3}{5}$ (b) $\frac{8}{9} - \frac{5}{6}$ (c) $\frac{3}{4} - \frac{2}{5}$

3 Work out:

(a) $4\frac{1}{3} - 2\frac{3}{4}$ (b) $2\frac{1}{3} - 1\frac{4}{5}$ (c) $3\frac{1}{8} - 1\frac{2}{3}$

4 I had $1\frac{2}{3}$ bars of chocolate but ate $\frac{3}{4}$ of a bar after lunch. How much do I have left?

5 I cut $1\frac{3}{4}$ m of copper pipe from a length $3\frac{1}{2}$ m long. How much copper pipe do I have now?

6 At my party we drank $3\frac{2}{3}$ litres of cola. There were $4\frac{1}{2}$ litre bottles to start with. How much is left?

7 $\frac{2}{5}$ of a garden is lawn and $\frac{1}{3}$ is patio.

(a) How much of the garden is lawn and patio together?

(b) How much of the garden is left?

8 Kai spends $\frac{1}{3}$ of his allowance at the cinema and $\frac{1}{6}$ on a magazine. What fraction does he have left?

Multiplying fractions

To multiply two fractions you multiply the numerators, and multiply the denominators. Change mixed numbers to improper fractions before multiplying.

> Writing a fraction in its simplest form is in Example 5.

You need to write the answer in its simplest form.

Example 13

Work out: (a) $\frac{4}{7} \times \frac{3}{5}$ (b) $\frac{2}{3} \times 5$ (c) $1\frac{1}{2} \times \frac{4}{5}$

> Multiply the numerators and the denominators.

(a) $\frac{4}{7} \times \frac{3}{5} = \frac{4 \times 3}{7 \times 5} = \frac{12}{35}$

(b) $\frac{2}{3} \times 5 = \frac{2}{3} \times \frac{5}{1} = \frac{2 \times 5}{3 \times 1} = \frac{10}{3} = 3\frac{1}{3}$

continued ▼

> Write 5 as $\frac{5}{1}$. Then multiply numerators and denominators. The answer is written in its simplest form.

(c) $1\frac{1}{2} \times \frac{4}{5} = \frac{3}{2} \times \frac{4}{5}$

$\qquad = \frac{3 \times 4}{2 \times 5}$

$\qquad = \frac{12}{10}$

$\qquad = 1\frac{2}{10}$

$\qquad = 1\frac{1}{5}$

First write $1\frac{1}{2}$ as an improper fraction.

Write the answer as a mixed number and then in its simplest form.

Example 14

Molly buys $1\frac{1}{2}$ metres of material to make a skirt.
She only uses $\frac{3}{4}$ of the material.
How many metres of material does she use?

$\frac{3}{4}$ of $1\frac{1}{2} = \frac{3}{4} \times \frac{3}{2}$

$\qquad = \frac{3 \times 3}{4 \times 2}$

$\qquad = \frac{9}{8}$

$\qquad = 1\frac{1}{8}$

Molly uses $1\frac{1}{8}$ metres of material.

Remember that 'of' means multiply.

Write $1\frac{1}{2}$ as an improper fraction.

Write the final answer as a mixed number.

Exercise 3I

1 Work out:

(a) $\frac{3}{5} \times \frac{2}{7}$ **(b)** $\frac{1}{4} \times \frac{5}{9}$ **(c)** $\frac{1}{2} \times \frac{1}{2}$

(d) $\frac{1}{4} \times \frac{1}{4}$ **(e)** $\frac{4}{5} \times \frac{5}{7}$ **(f)** $\frac{3}{8} \times \frac{5}{9}$

2 Work out:

(a) $\frac{2}{3}$ of $\frac{3}{8}$ **(b)** $\frac{5}{7}$ of $\frac{14}{15}$ **(c)** $\frac{3}{4}$ of $\frac{4}{5}$

'of' means multiply.

3 Work out:

(a) $1\frac{1}{5} \times 3\frac{1}{3}$ **(b)** $1\frac{1}{3} \times 1\frac{1}{8}$ **(c)** $1\frac{5}{19} \times 1\frac{7}{12}$

(d) $2\frac{1}{7} \times 1\frac{2}{5}$ **(e)** $1\frac{1}{4} \times 1\frac{4}{6}$ **(f)** $2\frac{1}{2} \times 1\frac{3}{4} \times \frac{3}{10}$

Convert the mixed numbers to improper fractions.

4 Find the area of a rectangular photograph $3\frac{1}{4}$ inches long and $3\frac{1}{5}$ inches wide.

Area of rectangle
$\qquad$ = length × width.

5 A pizza has a mass of $\frac{2}{5}$ kg. What is the mass of $2\frac{1}{2}$ pizzas?

6 A rectangular path is $3\frac{1}{8}$ metres long and $\frac{2}{5}$ metres wide. What is its area?

7 It takes $3\frac{1}{3}$ minutes to fill a bag with sand. How long will it take to fill $10\frac{1}{2}$ bags?

Reciprocals

The reciprocal of $\frac{3}{4}$ is $\frac{4}{3}$ and the reciprocal of $\frac{4}{3}$ is $\frac{3}{4}$.

The reciprocal of 6 is $\frac{1}{6}$, and the reciprocal of $\frac{1}{6}$ is 6.

Remember that $6 = \frac{6}{1}$

To find the reciprocal of a fraction, you **invert** it (turn it upside down).

When you multiply a number by its reciprocal, you always get 1.

Note: 0 has no reciprocal.

Example 15

Find the reciprocal of (a) $\frac{5}{7}$ (b) $2\frac{2}{3}$.

(a) $\frac{7}{5}$

Check: $\frac{5}{7} \times \frac{7}{5} = \frac{1}{1} \times \frac{1}{1} = 1$

(b) $2 \times 3 = 6$ $\frac{6}{3} + \frac{2}{3} = \frac{8}{3}$

Reciprocal of $\frac{8}{3}$ is $\frac{3}{8}$

Check: $\frac{8}{3} \times \frac{3}{8} = \frac{1}{1} \times \frac{1}{1}$

Change all mixed numbers into improper fractions before you find the reciprocal.

Exercise 3J

1 Invert the following fractions:

(a) $\frac{2}{3}$ (b) $\frac{4}{5}$ (c) $1\frac{2}{7}$ (d) $2\frac{5}{9}$

2 Find the reciprocals of the following fractions:

(a) $\frac{3}{7}$ (b) $\frac{9}{2}$ (d) 9   (d) $3\frac{4}{5}$

(f) $2\frac{5}{8}$ (g) $\frac{17}{3}$ (h) 18   (i) $10\frac{3}{7}$

Dividing fractions

To divide two fractions:
1 **invert** (turn upside down) the fraction you are dividing by
2 change the division sign to a multiplication sign.

$4 \div \frac{1}{2}$ means how many $\frac{1}{2}$s in 4?

Answer: 8 ⬜⬜⬜⬜

$4 \div \frac{1}{2} = \frac{4}{1} \times \frac{2}{1} = 8$

So $\frac{1}{3} \div \frac{2}{5} = \frac{1}{3} \times \frac{5}{2} = \frac{5}{6}$

You turn the 2nd fraction upside down. This is the fraction you are dividing by.

If the division involves mixed numbers, change these to improper fractions first.

Example 16

Work out: **(a)** $\frac{7}{8} \div 3$ **(b)** $\frac{2}{9} \div \frac{3}{4}$ **(c)** $1\frac{3}{4} \times \frac{5}{6}$

(a) $\frac{7}{8} \div 3 = \frac{7}{8} \div \frac{3}{1}$

$\quad\quad = \frac{7}{8} \times \frac{1}{3}$

$\quad\quad = \frac{7}{24}$

> Write 3 as $\frac{3}{1}$, then use the rule for division.

(b) $\frac{2}{9} \div \frac{3}{4} = \frac{2}{9} \times \frac{4}{3}$

$\quad\quad = \frac{2 \times 4}{9 \times 3}$

$\quad\quad = \frac{8}{27}$

(c) $1\frac{3}{4} \div \frac{5}{6} = \frac{7}{4} \div \frac{5}{6}$

$\quad\quad = \frac{7}{4} \times \frac{6}{5}$

$\quad\quad = \frac{42}{20}$

$\quad\quad = \frac{21}{10}$

$\quad\quad = 2\frac{1}{10}$

> Write $1\frac{3}{4}$ as an improper fraction.

> Write the answer as a mixed number in its simplest form.

Exercise 3K

1 Work out:

(a) $\frac{1}{4} \div 3$ (b) $\frac{2}{5} \div \frac{3}{8}$ (c) $7 \div \frac{2}{5}$

(d) $\frac{3}{4} \div 12$ (e) $\frac{3}{4} \div \frac{3}{8}$ (f) $\frac{5}{9} \div \frac{2}{3}$

2 Work out:

(a) $1\frac{1}{2} \div \frac{3}{4}$ (b) $1\frac{2}{3} \div 1\frac{4}{6}$ (c) $5\frac{1}{4} \div 3\frac{1}{2}$

3 To wrap a parcel takes $\frac{3}{5}$ m of string. How many parcels can I wrap with $7\frac{1}{2}$ m of string?

4 $3\frac{1}{2}$ litres of orange juice is poured into $\frac{1}{3}$ litre glasses. How many full glasses are there?

5 How many pieces of ribbon $\frac{3}{8}$ m long can I cut from a piece $2\frac{1}{4}$ m long?

6 Share $2\frac{1}{4}$ cakes between 6 people. How much will they each receive?
Give your answer in its simplest form.

3.5 Fractions, decimals and percentages

You will need to know:
- about place value
- how to multiply and divide by 10, 100, 1000

Changing a percentage into a fraction or a decimal

Key words:
percentage

Percentage (%) means 'out of a hundred'.

So if you score 60 marks out of 100 in a maths test your mark can be written as $\dfrac{60}{100}$ or as 60%.

There is more on percentages in Chapter 15.

To change a percentage to a fraction, you write it as a fraction with a denominator of 100.

So $78\% = \dfrac{78}{100}$, $43\% = \dfrac{43}{100}$, $7\% = \dfrac{7}{100}$

When you have written a percentage as a fraction with denominator 100, you can change it to its simplest form.

Always use a denominator of 100 to change a percentage into a fraction.

Example 17

Write these percentages as fractions in their simplest form:
(a) 58%   **(b)** 95%   **(c)** 8%   **(d)** $12\frac{1}{2}\%$

(a) $58\% = \dfrac{58}{100} = \dfrac{58}{100} \overset{\div 2}{\underset{\div 2}{=}} \dfrac{29}{50}$

(a), (b) and (c) use the 'simplifying' skills from Section 3.1.

(b) $95\% = \dfrac{95}{100} = \dfrac{95}{100} \overset{\div 5}{\underset{\div 5}{=}} \dfrac{19}{20}$

(c) $8\% = \dfrac{8}{100} = \dfrac{8}{100} \overset{\div 4}{\underset{\div 4}{=}} \dfrac{2}{25}$

(d) $12\frac{1}{2}\% = \dfrac{12\frac{1}{2}}{100} = \dfrac{12\frac{1}{2}}{100} \overset{\times 2}{\underset{\times 2}{=}} \dfrac{25}{200} \overset{\div 25}{\underset{\div 25}{=}} \dfrac{1}{8}$

Multiply by 2 first to make the numerator a whole number. Don't stop at $\frac{25}{200}$, you must simplify as much as possible.

You already know how to divide by 100 using decimals. You move the digits 2 places to the right.

You can use this to change a percentage to a decimal.

$$78\% = \frac{78}{100} = 0.78$$

See Chapter 1, Section 1.5.

$\frac{78}{100}$ means $78 \div 100$

To change a percentage to a decimal:
1 Write it as a fraction with a denominator of 100.
2 Divide the numerator by 100 (move the digits 2 places to the right).

Example 18

Change these percentages to decimals:
(a) 43%   **(b)** 7%

(a) $43\% = \dfrac{43}{100} = 0.43$

(b) $7\% = \dfrac{7}{100} = 0.07$

Exercise 3L

1 Write these percentages as fractions in their simplest form.

    (a) 47%   **(b)** 20%   **(c)** 75%   **(d)** 84%
    (e) 1%   **(f)** 65%   **(g)** 96%   **(h)** $12\frac{1}{2}\%$

2 Write these percentages as decimals:

    (a) 69%   **(b)** 43%   **(c)** 40%   **(d)** 25%
    (e) 70%   **(f)** 33.3%   **(g)** 184%   **(h)** 5%

3 Write the following percentages as (i) fractions in their simplest form (ii) decimals:

    (a) 30%   **(b)** 85%   **(c)** 2%   **(d)** 123%

4 Two friends were comparing test results. Sam got 70% and Tom $\frac{12}{20}$. By changing them both into fractions in their simplest form, find which one scored the highest in the test.

Changing a decimal or a fraction into a percentage

To change a decimal into a percentage is the reverse of changing a percentage to a decimal.

$$85\% = \frac{85}{100} = 0.85$$

$\div 100$

$\times 100$

So $0.85 = (0.85 \times 100)\% = 85\%$

To multiply by 100, move the digits 2 places to the left. See Section 1.5.

Example 19

Change into percentages:

(a) 0.075   **(b)** 1.64

(a) $0.075 = (0.075 \times 100)\% = 7.5\%$

(b) $1.64 = (1.64 \times 100)\% = 164\%$

$1.64 > 1$, so the answer is more than 100%.

To change a decimal to a percentage, multiply by 100.

To change a fraction into a percentage you first need to change the fraction into a decimal.

To do this you divide the numerator of the fraction by the denominator.

$\frac{2}{5} = 2 \div 5 = 0.4$

$\frac{2}{5} = 0.4 = (0.4 \times 100)\% = 40\%$

$\frac{2}{5}$ means $2 \div 5$

$$\begin{array}{r} 0.\,4 \\ 5\overline{)2.^20} \end{array}$$

You can go directly from a fraction to a % like this:
$\frac{2}{5} \times 100 = \frac{200}{5} = 40\%$

To change a fraction to a percentage:
1 **Change the fraction to a decimal by dividing the numerator by the denominator.**
2 **Change the decimal to a percentage by multiplying by 100.**

Example 20

Write these fractions as percentages:

(a) $\frac{9}{10}$ (b) $\frac{5}{8}$ (c) $\frac{31}{40}$

(a) $\dfrac{9}{10} = 0.9 = (0.9 \times 100)\% = 90\%$

To divide by 10 move digits 1 place to the right.

(b) $\dfrac{5}{8} = 8\overline{)5.^50^20^40}$

      0.625

(b) done by 'non-calculator' method. This could be on a non-calculator paper.

$= (0.625 \times 100)\% = 62.5\%$

(c) $\dfrac{31}{40} = 0.775 = (0.775 \times 100)\% = 77.5\%$

$31 \div 40$ done on a calculator. You would *not* have to do this on a non-calculator paper.

Now that you can change between percentages, fractions or decimals, you can make comparisons between all three.

Example 21

Write these in order of size, smallest first,

$$\frac{2}{5} \qquad 0.39 \qquad 45\% \qquad \frac{7}{20}$$

$\dfrac{2}{5} = 0.4$

$45\% = \dfrac{45}{100} = 0.45$

It is usually easiest to change them all to decimals.

$\dfrac{7}{20} = 0.35$

A calculator has been used to change the fractions into decimals.

Since $0.35 < 0.39 < 0.4 < 0.45$

The order is $\dfrac{7}{20} < 0.39 < \dfrac{2}{5} < 45\%$

Exercise 3M

1 Write these decimals as percentages.

     (a) 0.32   (b) 0.79   (c) 2.39   (d) 0.125

2 Write these fractions as decimals and then as percentages.

     (a) $\frac{7}{10}$ (b) $\frac{4}{5}$ (c) $\frac{3}{4}$ (d) $\frac{3}{8}$

3 Write the following as percentages and then put them in order from smallest to largest:

$\frac{1}{10}$ $\frac{7}{100}$ $\frac{2}{5}$ $\frac{8}{25}$

4 Write the following fractions as percentages. You can use a calculator to help.

(a) $\frac{17}{34}$ **(b)** $\frac{13}{52}$ **(c)** $\frac{25}{125}$ **(d)** $\frac{50}{32}$

5 Copy and complete the following table of equivalent fractions, decimals and percentages.

Percentage	Fraction	Decimal
60%		
		0.48
	$\frac{3}{10}$	
		1.75
5%		

 6 Tessa was comparing her test results. She had $\frac{16}{20}$ in History, 75% in Mathematics and $\frac{28}{40}$ in English. By changing all the results into percentages, put her results from lowest to highest.

7 In a sale you can buy a TV with '15% off' or one with '$\frac{1}{4}$ off' the original price. Which one is the best offer?

Change both into percentages to compare them.

8 Write the following in order from smallest to largest:

(a) 64% $\frac{7}{10}$ 0.625 $\frac{14}{25}$

(b) 0.438 $\frac{9}{20}$ 42%   0.4

Changing a decimal into a fraction

To write a decimal as a fraction you need to look at the place value of the *last* significant figure.

For significant figures see Section 2.1. For place value see Section 1.1.

	H	T	U	.	t	h	th
(a)			0	.	4	8	
(b)			0	.	2	2	5
(c)			0	.	0	3	7
(d)			2	.	1	9	

Key words:
terminating decimal

In **(a)** the last significant figure is in the hundredths column.

So $0.48 = \dfrac{48}{100}$

$\dfrac{48}{100}$ simplifies to $\dfrac{12}{25}$

'hundredths' have denominator 100.

$$\dfrac{48}{100} \xrightarrow{\div 4} = \dfrac{12}{25} \xleftarrow{\div 4}$$

In **(b)** the last significant figure is in the thousandths column.

So $0.225 = \dfrac{225}{1000}$

'thousandths' have denominator 1000.

$$\dfrac{225}{1000} \xrightarrow{\div 25} = \dfrac{9}{40} \xleftarrow{\div 25}$$

In **(c)** the last significant figure is in the thousandths column.

So $0.037 = \dfrac{37}{1000}$

$\frac{37}{1000}$ cannot be simplified further.

(d) is a mixed number – it has a whole number part and a fractional (decimal) part. The last significant figure of the decimal part is in the hundredths column, so $2.19 = 2\frac{19}{100}$.

Decimals such as 0.48, 0.225 and 0.037 are **terminating decimals** . They end, or terminate, rather than going on for ever.

🖩 Exercise 3N

1 Change the following decimals into fractions. Give each answer in its simplest form.

(a) 0.3	**(b)** 2.7	**(c)** 0.09
(d) 0.007	**(e)** 0.75	**(f)** 0.025
(g) 0.45	**(h)** 0.625	**(i)** 2.05
(j) 16.125	**(k)** 4.105	**(l)** 23.24

Fractions and recurring decimals

Key words:
recurring decimal

Look at these fractions written as decimals:

$\frac{1}{3} = 0.333\,333...$ $\frac{5}{12} = 0.416\,666...$

$\frac{8}{11} = 0.727\,272...$ $\frac{3}{7} = 0.428\,571\,428\,571\,42...$

When you change these fractions to decimals using a calculator the decimal places fill the whole of the calculator display.

These fractions (and many others too) have decimal answers that never end.

These decimals have either a repeating digit or a repeating pattern of digits. They are called **recurring decimals**.

> **You can write recurring decimals using 'dot' notation to make the repeating pattern clear.**

$\frac{1}{3} = 0.333\,333...$ is written as $0.\dot{3}$

$\frac{5}{12} = 0.416\,666...$ is written as $0.41\dot{6}$

$\frac{8}{11} = 0.727\,272...$ is written as $0.\dot{7}\dot{2}$

$\frac{3}{7} = 0.428\,571\,428\,571\,42...$ is written as $0.\dot{4}2857\dot{1}$

A recurring decimal is an alternative way of writing an exact fraction. You can always convert a recurring decimal to a fraction.

0.428571428

This is what your calculator display will show when you work out $3 \div 7$.

'Recur' means 'repeat'.

When a single digit repeats you put a dot over this digit.

When more than one digit repeats, you put a dot over the *first* digit of the pattern *and* a dot over the *last* digit of the pattern.

Example 22

Write **(a)** 0.777 7...   **(b)** 0.353535... as fractions.

(a) Let $x = 0.777\,7...$ ①

Then $10x = 7.777\,7...$ ②

$9x = 7$ Subtract ① from ②.

$x = \dfrac{7}{9}$

(b) Let $x = 0.353\,535...$ ①

Then $100x = 35.353\,535...$ ②

$99x = 35$ Subtract ① from ②.

$x = \dfrac{35}{99}$

Exercise 3O

1 Change these fractions to decimals. Write **(i)** the full calculator display **(ii)** the recurring decimal using the 'dot' notation.

(a) $\frac{2}{9}$ **(b)** $\frac{1}{6}$ **(c)** $\frac{7}{11}$

(d) $\frac{5}{12}$ **(e)** $\frac{4}{15}$

2 Write these fractions as recurring decimals using the 'dot' notation:

(a) $\frac{1}{3}$ **(b)** $\frac{2}{3}$ **(c)** $\frac{7}{9}$

(d) $\frac{4}{11}$ **(e)** $\frac{5}{6}$ **(f)** $\frac{45}{55}$

(g) $\frac{5}{54}$ **(h)** $1\frac{1}{6}$ **(i)** $3\frac{7}{12}$

(j) $8\frac{3}{18}$ **(k)** $2\frac{16}{36}$ **(l)** $3\frac{1}{7}$

3 Convert these recurring decimals to fractions.

(a) 0.636 363...   **(b)** 0.444 4...

(c) 0.171 717...

Examination style questions

1 Heather is revising fractions for her homework. This is how she answers one of the questions.

$$\frac{1}{2} + \frac{1}{3} = \frac{2}{5}$$

Heather is wrong.
Show the correct way to work out $\frac{1}{2} + \frac{1}{3}$.   *(3 marks)*
AQA, Spec A, F, June 2004

2 (a) Which two of these fractions are equivalent to $\frac{1}{4}$?

     $\frac{2}{8}$ $\frac{5}{16}$ $\frac{6}{24}$ $\frac{11}{40}$

(b) Change $\frac{1}{4}$ to to a decimal.   *(3 marks)*
AQA, Spec A, F, June 2003

3 Tom has £2200.
He gives $\frac{1}{4}$ to his son and $\frac{2}{5}$ to his daughter.
How much does he keep for himself?
You must show all your working.   *(3 marks)*
AQA, Spec A, F, June 2003

4 Work out $4\frac{1}{3} - 1\frac{2}{5}$.   *(3 marks)*
AQA, Spec B, 3I, June 2004

5 Which of these fractions is closest to $\frac{1}{4}$?
You must show your working.

$\frac{1}{5}$ $\frac{3}{10}$ $\frac{7}{20}$ $\frac{7}{30}$

(3 marks)
AQA, Spec B, 5F, June 2004

6 (a) What fraction of the shape is shaded?
Give your answer in its simplest form.

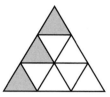

(b) Find $\frac{7}{10}$ of £50.

(c) Work out $\frac{1}{2} + \frac{1}{5}$.

(6 marks)
AQA, Spec A, F, June 2003

7 (a) What fraction of the shape is shaded?
Give your answer in its simplest form.

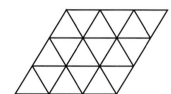

(b) Copy this shape
and shade $\frac{1}{2}$ of it.

(3 marks)
AQA, Spec B, 5F, June 2003

Summary of key points

Fractions

A fraction has a numerator (top number) and a
denominator (bottom number).

In a proper fraction the numerator is smaller than the
denominator. A proper fraction is always part of a
whole. For example, $\frac{3}{4}$.

In an improper fraction the numerator is smaller than
the denominator. For example, $\frac{7}{2}$.

A mixed number has a whole number part and a
fractional part. For example, $3\frac{1}{2}$.

Equivalent fractions

Equivalent fractions are fractions which have the same value and can be found by multiplying or dividing numerator and denominator by the same number.
For example,

$$\overset{\times 5}{\frac{2}{3}} = \frac{10}{15}$$
$$\underset{\times 5}{}$$

Simplifying fractions

The simplest form of a fraction is when there is no number that will divide exactly into the numerator and the denominator.
For example,

$$\overset{\div 4}{\frac{12}{20}} = \frac{3}{5}$$
$$\underset{\div 4}{}$$

$\frac{3}{5}$ is the simplest form, because no number divides exactly into 3 and 5.

Ordering fractions

To put fractions in order they need to have the same denominator.
Use equivalent fractions to do this. For example, $\frac{1}{3} = \frac{4}{12}$ and $\frac{1}{4} = \frac{3}{12}$ so $\frac{1}{3} > \frac{1}{4}$.

The four rules for fractions

You can only add or subtract fractions if they have the *same* denominator. Use equivalent fractions to do this.
For example, $\frac{4}{5} - \frac{3}{10} = \frac{8}{10} - \frac{3}{10} = \frac{5}{10} = \frac{1}{2}$.

To multiply two fractions you multiply the numerators and multiply the denominators.
For example, $\frac{4}{7} \times \frac{2}{3} = \frac{4 \times 2}{7 \times 3} = \frac{8}{21}$.

When you multiply a number by its reciprocal you always get 1.

To divide two fractions, invert the fraction you are dividing by and change the sign to multiply. For example, $\frac{2}{5} \div \frac{1}{4} = \frac{2}{5} \times \frac{4}{1} = \frac{8}{5} = 1\frac{3}{5}$

When you multiply or divide mixed numbers you must first change them to improper fractions.

Converting between fractions, decimals and percentages

To change a percentage to a fraction, write it as a fraction with denominator 100. For example, $36\% = \frac{36}{100}$.

To change a percentage to a decimal:
1 Change the percentage into a fraction with denominator 100.
2 Divide the numerator by 100.
   For example, $43\% = \frac{43}{100} = 0.43$.

To change a decimal to a percentage, multiply by 100.

To change a fraction to a percentage:
1 Change the fraction to a decimal by dividing the numerator by the denominator.
2 Multiply by 100.
   For example, $\frac{2}{5} = 2 \div 5 = 0.4 = (0.4 \times 100)\% = 40\%$.

To change a decimal to a fraction, look at the place value of the *last* significant figure. This will be the denominator of your fraction.
For example,

$$0.24 = \frac{24}{100} \xrightarrow{\div 4} = \frac{6}{25} \text{ (in its simplest form)}$$

You can write recurring decimals using 'dot' notation. Place a dot over the first and last digits of the repeating pattern.
For example, $\frac{7}{9} = 0.7777\ldots = 0.\dot{7}$
$$\frac{23}{111} = 0.207\,207\,207\ldots = 0.\dot{2}0\dot{7}.$$

This work ranges from grade G (changing simple fractions to decimals to %), through grade F (writing simple fractions, decimals, % in order of size), grade E (easy subtraction of fractions), grade D (fractions problems or questions on division of fractions) to grade C (mixed number questions). When fractions are used in the context of a problem, the grade rating usually increases by one grade.

This chapter will show you how to:

✔ use letters to represent numbers
✔ write simple expressions using letters to represent unknown numbers
✔ simplify algebraic expressions by collecting like terms
✔ multiply out brackets (by multiplying a single term over a bracket)
✔ simplify expressions involving brackets

4.1 Using letters to write simple expressions

Key words:
unknown
algebra
expression

To solve problems you often have to use a letter to represent an **unknown** number.

Using letters in mathematics is called **algebra** .

Suppose you have a bag of marbles but you do not know how many marbles are in the bag.

You could use m to represent this unknown quantity.

If you now add 4 marbles to the bag you will have m marbles plus 4 marbles. You can write this as $m + 4$ marbles.

$m + 4$ is called an **expression** in terms of m.

Example 1

Use algebra to write expressions for these:
(a) 3 more than x **(b)** 5 less than w **(c)** a added to b

(a) $x + 3$
(b) $w - 5$
(c) $a + b$

When you multiply two numbers the order in which you multiply them does not matter.

$5 \times 4 = 20$ and $4 \times 5 = 20$

It is the same when you are using algebra.

$3 \times x$ and $x \times 3$ are both the same and can be thought of as 3 lots of x, or $x + x + x$.
You write 3 lots of x as $3x$.

Always put the number first.

In an exam $x3$ would be marked as wrong.

y times y, or $y \times y$, is written as y^2 ('y squared').

$d \div 4$ can be written as $\dfrac{d}{4}$.

You can also multiply two unknowns together.

$g \times h = gh, \qquad x \times y = xy$

You leave out the '$\times$ sign'.

Example 2

Use algebra to write expressions for these:
(a) A bag contains m marbles. 4 are taken out. How many marbles are left?
(b) $a + a + a + a$.
(c) I have 5 boxes of strawberries. Each has n strawberries. How many strawberries do I have altogether?
(d) x oranges are shared equally between three people. How many oranges does each one get?

(a) $m - 4$
(b) $a + a + a + a = 4a$
(c) $5 \times n = 5n$
(d) $x \div 3 = \dfrac{x}{3}$

Remember to put the number first.

Example 3

Write expressions for the area of these shapes:

(a) a rectangle of length l and width w

(b) a square of side b

(a) Area $= l \times w$

$\quad\quad = lw$

(b) Area $= b \times b$

$\quad\quad = b^2$

Area of a rectangle
$\quad$ = length × width.

Exercise 4A

1 Use algebra to write expressions for these:

(a) 5 more than x (b) 3 less than w

(c) 8 more than m (d) 12 less than d

(e) 6 added to x (f) y subtract 2

(g) p added to 4   (h) 1 taken away from a

(i) x added to y (j) r take away t

2 Write these expressions using algebra.

(a) $g + g + g + g$ (b) $r + r + r + r + r$

(c) $h + h + h + h + h + h$ (d) $t + t + t$

Use Example 2(b) to help.

3 Write an algebraic expression for each of these:

(a) 3 lots of y (b) z divided by 3

(c) k divided by 4   (d) f times 8

(e) n multiplied by 10   (f) 12 divided by x

(g) a multiplied by b (h) g multiplied by g

4 Use algebra to write expressions for these.
Use x to stand for the number I choose.

(a) I choose a number and add 2

(b) I choose a number and multiply it by 7

(c) I choose a number and take 5 away

(d) I choose a number and double it

(e) I choose a number and halve it

(f) I choose a number, multiply it by 3, then subtract 2

(g) I choose a number, divide it by 4, then add 5

(h) I choose a number and multiply it by itself.

5 Write expressions for the area of these shapes:

Use Example 3 to help.

(a) square, side a a

(b) square, side x x

(c) rectangle, width 4 and length l

4
l

(d) rectangle, width x and length y

x
y

6 Write an expression for the total cost, in pounds, of:

(a) 4 singles at x pounds each

(b) 6 albums at y pounds each

(c) d singles at £4 each

(d) t albums at £13 each

(e) 5 singles at j pounds each and 8 albums at k pounds each

(f) a singles at £3 each and b albums at £11 each

7 Hayley has 4 bags of beads. Each bag has n beads in it.

(a) How many beads does she have altogether?

(b) Hayley puts 2 more beads in each bag. How many beads are in each bag now?

(c) How many beads does she have altogether now?

(d) Hayley gives 3 beads from one bag to her friend. How many beads are left in this bag?

4.2 Simplifying algebraic expressions

Key words:
simplify

You will need to know:

- **how to add and subtract positive and negative numbers**

You use the same rules for algebra as you do when working with numbers.

3 lots of 6, added to 7 lots of 6, is the same as 10 lots of 6

$(3 \times 6) + (7 \times 6) = 10 \times 6$

because $3 + 7 = 10$

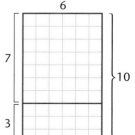

You can do the same when you have a letter standing for an unknown quantity:

$$3n + 7n = 10n$$

$$5x + x = 6x$$

Remember that $3n$ means $3 \times n$.

x means $1x$ (you don't need to write the 1).

The same works when you subtract (take away).

10 sixes, take away 7 sixes, is the same as 3 sixes.

$$(10 \times 6) - (7 \times 6) = 3 \times 6$$

because $10 - 7 = 3$

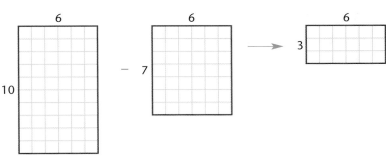

When you have a letter it stands for an unknown quantity:

$$10n - 7n = 3n$$

$$6x - x = 5x$$

These answers are written in a shorter form than the original expression.

To **simplify** an expression, you write it in as short a form as possible.

Example 4

Write each of these expressions in a shorter form.

(a) m   m   m

What is the total length of pipe?

(b) $3h + 6h + 2h$ (c) $6b + 5b - 7b$

(d) $6x - 2x + 8x - 10x$

(e)
   t [4] + t [5] + t [3]

What is the total area?

Remember to write the number first.

(a) $m + m + m = 3m$

(b) $3h + 6h + 2h = 11h$

$3 + 6 + 2 = 11$

(c) $6b + 5b - 7b = 11b - 7b$

$= 4b$

$6b + 5b = 11b$

$11 - 7 = 4$

continued ▼

(d) $6x - 2x + 8x - 10x = 6x + 8x - 2x - 10x$

$$= 14x - 12x$$

$$= 2x$$

(e) $4t + 5t + 3t = 12t$

> Just as with numbers you can add and subtract the terms in any order as long as each term keeps its own sign.
> $6 - 2 + 8 - 10 = 2$
>
> $4 + 5 + 3 = 12$

Exercise 4B

1 Write each of these expressions in a shorter form:

(a) What is the total height? n n n

(b) What is the total length?

(c) $a + a + a + a + a + a$

(d) $g + g$

(e) What is the total area?

(f) What is the total area?

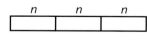

(g) $5c + c$

(h) $4t + 5t + 3t$

(i) $x + 6x + 3x$

(j) $5l + l + 8l + 2l$

c means 1*c*.

2 Write each of these in a longer form:

(a) Length: **(b)** Area:

(c) $8y$ **(d)** $4r$ **(e)** $2w$ **(f)** $6v$

3 Simplify these algebraic expressions:

(a) $10b - 7b$ **(b)** $6y - y$ **(c)** $12z - 8z$

(d) $8t - 2t$ **(e)** $6j + 5j - 7j$ **(f)** $5u + 3u - 6u$

(g) $h + h + h$ **(h)** $7t - 5t + 3t$ **(i)** $6x - x - 3x$

(j) $4r - r + 5r - 2r$

4 In a magic square the sum of the expressions in each row, each column, and in the two diagonals, is the same. Copy and complete each of these magic squares:

(a)

		$6x$
	$5x$	
$4x$	$3x$	

(b)

$5y$	$8y$	$11y$
		$9y$

In this magic square $4x + 5x + 6x = 15x$, so each row, column and diagonal must add up to $15x$.

4.3 Collecting like terms

Key words:
term
like term

$3a + 6b + 5a$ This expression has three **terms** .

$3a$ is one term. $6b$ is the second term, $5a$ is the third term.

Terms which use the same letter are called like terms .

$3a$ and $5a$ are like terms in the expression $3a + 6b + 5a$.

Terms which use the same combination of letters are also called like terms.

$6xy + 5x^2 - 2xy + 3x^2 + 2x$
$6xy$ and $2xy$ are like terms,
$5x^2$ and $3x^2$ are like terms,
$3x^2$ and $2x$ are *not* like terms.

You can simplify algebraic expressions by collecting like terms together.

$$3a + 6b + 5a = 3a + 5a + 6b$$
$$= 8a + 6b$$

For example:

$$6xy + 5x^2 - 2xy + 3x^2 + 2x = 6xy - 2xy + 5x^2 + 3x^2 + 2x$$
$$= 4xy + 8x^2 + 2x$$

Rearrange so that like terms are next to each other.

You keep the $+$ or $-$ sign with the term so the $-$ stays with $2xy$, the term is $-2xy$.

Example 5

Simplify these expressions by collecting like terms:

(a) $2a + 7b + 3a$ **(b)** $4p + 5 + 3p - 2$

(c) $6f + 5g - 4f + g$ **(d)** $4ab - 6a + 3ab - 2a$

(a) $2a + 7b + 3a = 2a + 3a + 7b$
$$= 5a + 7b$$

(b) $4p + 5 + 3p - 2 = 4p + 3p + 5 - 2$
$$= 7p + 3$$

(c) $6f + 5g - 4f + g = 6f - 4f + 5g + g$
$$= 2f + 6g$$

(d) $4ab - 6a + 3ab - 2a = 4ab + 3ab - 6a - 2a$
$$= 7ab - 8a$$

Collect the terms in p, and collect the terms which are just numbers.

Remember to keep the $-$ sign with the $4f$.

$4ab$ and $3ab$ are like terms, $-6a$ and $-2a$ are like terms.

Example 6

Write an expression for the perimeter of this rectangle, in its simplest form.

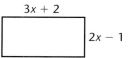

$3x + 2 + 3x + 2 + 2x - 1 + 2x - 1$
$$= 3x + 3x + 2x + 2x + 2 + 2 - 1 - 1$$
$$= 10x + 2$$

To work out the perimeter of a shape you need to add together the lengths of all its sides.

Exercise 4C

1 Simplify these expressions by collecting like terms:

(a) $2c + 7d + 3c$ **(b)** $3m + 4r + 2m$

(c) $6x + 4y + x$ **(d)** $4a + 8b + 6a$

(e) $2q + 6 + 3q + 2$ **(f)** $7p + 2 + 5p + 3$

(g) $5j + 4 + 6j + 1$ **(h)** $6w + 7 + 2w + 3$

x means $1x$.

2 Simplify these expressions by collecting like terms:

(a) $5x + 4y - 2x$ **(b)** $7a + 3b - 5a$

(c) $8k + 3m - 4k$ **(d)** $12h + 7j - 4h$

(e) $4q + 5 + 3q - 2$ **(f)** $6p + 7 + 2p - 3$

(g) $5t + 2 - 3t + 1$ **(h)** $9z + 4 - 8z + 6$

3 Write expressions for the perimeter of these shapes. Write each expression in its simplest form.

(a) a square, side $3a$ **(b)** a square, side $x + 1$

(c) a rectangle, width y and length $2x$

(d) a rectangle, width $2x$ and length $3x + 1$

$2x$ ▭ $3x + 1$

4 Simplify these expressions:
   (a) $6a + 5b + 4a + 2b$ **(b)** $4m + 3r + 2m + 2r$
   (c) $2x + 3y + 3x + 5y$ **(d)** $5q + 8r + 6q + r$ Keep the $-$ sign with the $4k$.
   (e) $6k + 5l - 4k + l$ **(f)** $7v + 2w - 5v - 3w$
   (g) $4y + 4z - 3y - z$ **(h)** $9c - 7d - 8c - 6d$

5 Simplify these expressions:
   (a) $5x + 3x + 2y + 2x + 4y$
   (b) $6p + 7q + 2p + 3p - q$
   (c) $7g + 3h - 4g + g - 2h$
   (d) $7t + 7n - 4t - t + 2n - 3n$
   (e) $4a + 6b - 2a + 3c + 2b - 4c$
   (f) $5j + 6k + j + 4l - 2k + 3l$
   (g) $8d + 5e + 6f - 3d + e - 4f$
   (h) $6x - 2y - 5x + 8z - 3y - 4z$

6 Simplify these expressions:
   (a) $4ab + 6a + 3ab - 2a$
   (b) $6x^2 + 5x + 4x + 2x^2$
   (c) $7t^2 + 4 - 3t^2 + 1$
   (d) $6xy + 5x^2 - 2xy + 3x^2 + 2x$
   (e) $9xy + 5x - 6xy + 6x^2 + 2x$
   (f) $4ab + 6a + 2ab - 5b - 3ab + a - 2b$

7 In a magic square the sum of the expressions in each
   row, column, and the two diagonals, is the same.
   Copy and complete each of these magic squares:

(a)

$6a + 7b$		
$a + 6b$		
$8a + 2b$	$4a + 3b$	

(b)

$3a + 2b$		$8a - 2b$
	$5a + b$	
$2a + 4b$		

(c)

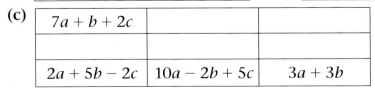

$7a + b + 2c$		
$2a + 5b - 2c$	$10a - 2b + 5c$	$3a + 3b$

4.4 Multiplying algebraic terms

You will need to know:
- **how to multiply positive and negative numbers**

Remember:

$$3 \times x = x \times 3 = 3x \qquad y \times y = y^2 \qquad g \times h = gh$$

You can simplify complicated multiplications. For example:

$$3f \times 4g = 12fg$$

This is because
$$
\begin{aligned}
3f \times 4g &= 3 \times f \times 4 \times g \\
&= 3 \times 4 \times f \times g \\
&= 12 \times fg \\
&= 12fg
\end{aligned}
$$

$f \times 4 = 4 \times f$

> **To multiply algebraic terms:**
> 1 **Multiply the numbers.**
> 2 **Multiply the letters.**

Example 7

Simplify these expressions:

(a) $2 \times 5b$ (b) $3c \times 2d$ (c) $3x \times 4x$

(a)
$$
\begin{aligned}
2 \times 5b &= 2 \times 5 \times b \\
&= 10 \times b \\
&= 10b
\end{aligned}
$$

(b)
$$
\begin{aligned}
3c \times 2d &= 3 \times c \times 2 \times d \\
&= 3 \times 2 \times c \times d \\
&= 6 \times cd \\
&= 6cd
\end{aligned}
$$

Multiply the numbers first, then multiply the letters.

(c)
$$
\begin{aligned}
3x \times 4x &= 3 \times x \times 4 \times x \\
&= 3 \times 4 \times x \times x \\
&= 12 \times x^2 \\
&= 12x^2
\end{aligned}
$$

$x \times x = x^2$

Exercise 4D

Simplify these expressions:

1 $2 \times 5k$ **2** $3 \times 6b$ **3** $6 \times 2x$

4 $4a \times 5$ **5** $2h \times 7$ **6** $3m \times 4$

7 $3a \times 2b$ **8** $4c \times 3d$ **9** $6p \times 7q$

Use Example 7(b) to help.

10 $6g \times h$ **11** $x \times 5y$ **12** $7j \times 8k$

13 $3t \times 4t$ **14** $6x \times 7x$ **15** $5a \times 6a$

16 $4n \times n$ **17** $6c \times 4d$ **18** $x \times 7x$

$h = 1h.$

Expanding brackets

Key words:
expand
brackets

You can work out 6×34 by thinking of it as
'6 lots of 30' + '6 lots of 4'.

$$6 \times 34 = 6 \times (30 + 4)$$
$$= 6 \times 30 + 6 \times 4$$
$$= 180 + 24$$
$$= 204$$

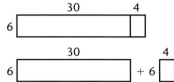

Multiply the 30 *and* the 4,
by 6.

Then add the two answers
together.

This is called **expanding** the **brackets** .

It is also sometimes called
'multiplying out the
brackets' or 'removing the
brackets'.

> **When you expand brackets you multiply each term
> inside the brackets by the term outside the bracket.**

Example 8

Expand the brackets to find the value of these expressions:

(a) $4(50 + 7)$ **(b)** $6(30 - 2)$

(a) $4(50 + 7) = 4 \times 50 + 4 \times 7$
$\qquad\qquad = 200 + 28$
$\qquad\qquad = 228$

(b) $6(30 - 2) = 6 \times 30 - 6 \times 2$
$\qquad\qquad = 180 - 12$
$\qquad\qquad = 168$

The minus sign means take
'6 lots of 2' away from
'6 lots of 30'.

We often use brackets in algebra.

$6(x + 4)$ means $6 \times (x + 4)$

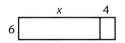

You usually write
expressions like this
without the $\times$.

You have to multiply each term in the brackets by 6.

$6(x + 4) = 6 \times x + 6 \times 4$
$\qquad\quad = 6x + 24$

The same as in the 6×34
example above.

Remember you write
$6 \times x$ as $6x$.

Example 9

Simplify the following by multiplying out the brackets:

(a) $5(a + 6)$ **(b)** $2(x - 8)$ **(c)** $3(2c - d)$

(a) $5(a + 6) = 5 \times a + 5 \times 6$
$$= 5a + 30$$

(b) $2(x - 8) = 2 \times x - 2 \times 8$
$$= 2x - 16$$

(c) $3(2c - d) = 3 \times 2c - 3 \times d$
$$= 6c - 3d$$

Multiply each term inside the bracket by the term outside the bracket.

A common mistake is to forget to multiply the second term in the bracket.

$3 \times 2c = 2c + 2c + 2c = 6c$,
or $3 \times 2c = 3 \times 2 \times c$
$$= 6 \times c$$
$$= 6c$$

Exercise 4E

1 Expand the brackets to find the value of these expressions:

(a) $2(50 + 7)$ **(b)** $5(40 + 6)$
(c) $4(60 + 8)$ **(d)** $6(70 + 3)$
(e) $3(40 - 2)$ **(f)** $7(50 - 4)$
(g) $6(50 - 1)$ **(h)** $8(40 - 3)$

Use Example 8 to help.

2 Write an expression for the area of each shape.

(a) 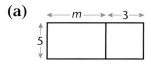   **(b)**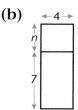

Simplify by removing the brackets.

3 Simplify the following by multiplying out the brackets:

(a) $5(p + 6)$ **(b)** $3(a + 5)$
(c) $7(k + 2)$ **(d)** $4(m + 9)$
(e) $5(7 + f)$ **(f)** $2(8 + q)$
(g) $2(a + b)$ **(h)** $5(x + y)$
(i) $8(g + h + i)$ **(j)** $4(u + v + w)$

Multiply each term inside the bracket by the term outside the bracket.

4 Simplify the following by removing the brackets:

(a) $2(y - 8)$ **(b)** $3(x - 5)$
(c) $6(b - 4)$ **(d)** $7(d - 8)$

(e) $2(7 - x)$ (f) $4(8 - n)$

(g) $5(a - b)$ (h) $2(x - y)$

(i) $7(4 + p - q)$ (j) $8(a - b + 6)$

5 Expand the brackets in these expressions:

(a) $3(2c + 6)$ (b) $4(3m + 2)$

(c) $5(4t + 3)$ (d) $6(4y + 9)$

(e) $4(3e + f)$ (f) $2(5p + q)$

(g) $3(2a - b)$ (h) $6(3c - 2d)$

(i) $2(m - 4n)$ (j) $7(2x + y - 3)$

(k) $6(3a - 4b + c)$ (l) $4(2u - 5v - 3w)$

> $3 \times 2c = 3 \times 2 \times c = 6c$

> Multiply each term inside the bracket by the term outside.

6 Simplify these expressions by removing the brackets:

(a) $2(x^2 + 3x + 2)$ (b) $3(x^2 + 5x - 6)$

(c) $2(a^2 - a + 2)$ (d) $4(y^2 - 3y - 10)$

7 Write down the pairs of cards that show equivalent expressions.

A	B	C	D
$4(x + 2y)$	$4x + 2y$	$2(4x + y)$	$4(2x - y)$

E	F	G	H
$8x - 8y$	$4x + 8y$	$8(x - y)$	$2x - 8y$

I	J	K	L
$8x + 2y$	$2(x - 4y)$	$2(2x + y)$	$8x - 4y$

> Equivalent expressions are the same when the brackets are expanded.

Letters outside brackets

Sometimes the term outside the bracket includes a letter.

You expand the brackets in the same way.

You multiply each term inside the brackets by the term outside.

Example 10

Expand the brackets in these expressions:

(a) $a(a + 4)$ (b) $x(2x - y)$ (c) $3k(2k + 5)$

(a) $a(a + 4) = a \times a + a \times 4$

$\qquad = a^2 + 4a$

(b) $x(2x - y) = x \times 2x - x \times y$

$\qquad = 2x^2 - xy$

(c) $3k(2k + 5) = 3k \times 2k + 3k \times 5$

$\qquad = 6k^2 + 15k$

$a \times a = a^2$

$x \times 2x = x \times 2 \times x$
$\qquad = 2 \times x \times x$
$\qquad = 2x^2$
$x \times y = xy$

$3k \times 2k = 3 \times 2 \times k \times k = 6k^2$
$3k \times 5 = 3 \times 5 \times k = 15k$

Exercise 4F

Expand the brackets in these expressions:

1 $b(b + 4)$ **2** $a(a + 5)$

3 $k(k - 6)$ **4** $m(m - 9)$

5 $a(2a + 3)$ **6** $g(4g + 1)$

7 $p(2p + q)$ **8** $t(t + 5w)$

9 $m(m + 3n)$ **10** $x(2x - y)$

11 $r(4r - t)$ **12** $a(a - 4b)$

13 $2t(t + 5)$ **14** $3x(x - 8)$

15 $5k(k + l)$ **16** $3a(2a + 4)$

17 $2g(4g + h)$ **18** $5p(3p - 2q)$

19 $3x(2y + 5z)$ **20** $4p(3p + 2q)$

Use Example 10(c) to help.

Negative numbers outside brackets

If you have an expression like $-3(2x - 5)$, multiply both terms in the brackets by -3.

$-3(2x - 5) = -3 \times 2x + (-3) \times (-5)$

$\qquad = -6x + 15$

$-3 \times 2x = -6x$
$-3 \times -5 = 15$

The term outside the bracket is negative.

$- \times + = -$
$+ \times - = -$
$- \times - = +$
$+ \times + = +$

Example 11

Expand these expressions:

(a) $-2(3t + 4)$ **(b)** $-3(4x - 1)$

(a) $-2(3t + 4) = -2 \times 3t + -2 \times +4$

$\qquad = -6t + -8$

$\qquad = -6t - 8$

(b) $-3(4x - 1) = -3 \times 4x + -3 \times -1$

$\qquad = -12x + 3$

Multiply both terms in the bracket by -2.

$-2 \times 3 = -6$
$-2 \times 4 = -8$

$-3 \times 4 = -12$
$(-3) \times (-1) = +3$

Exercise 4G

Expand these expressions:

1 $-2(3k + 4)$

2 $-3(2x + 6)$

3 $-5(3n + 1)$

4 $-4(3t + 5)$

5 $-3(4p - 1)$

6 $-2(3x - 7)$

7 $-6(x - 3)$

8 $-5(2x - 3)$

4.5 Adding and subtracting expressions with brackets

Adding

To add expressions with brackets you expand the brackets first, then collect like terms to simplify your answer.

Example 12

Expand then simplify these expressions:

(a) $3(a + 4) + 2a + 10$ **(b)** $3(2x + 5) + 2(x - 4)$

(a) $3(a + 4) + 2a + 10 = 3a + 12 + 2a + 10$

$\qquad = 3a + 2a + 12 + 10$

$\qquad = 5a + 22$

Expand the brackets first.
Then collect like terms.

(b) $3(2x + 5) + 2(x - 4) = 6x + 15 + 2x - 8$

$\qquad = 6x + 2x + 15 - 8$

$\qquad = 8x + 7$

Expand both sets of brackets first.

Subtracting

To subtract an expression with brackets you expand the brackets first, then collect like terms.

For example:

$3(2x + 3) - 2(x - 1) = 6x + 9 - 2x + 2$

$\qquad = 6x - 2x + 9 + 2$

$\qquad = 4x + 11$

$3(2x + 3) = 6x + 9$
$-2(x - 1) = -2x + 2$

Example 13

Expand then simplify these expressions:

(a) $3(2t + 1) - 2(2t + 4)$ **(b)** $8(x + 1) - 3(2x - 5)$

(a) $3(2t + 1) - 2(2t + 4) = 6t + 3 - 4t - 8$

$\qquad\qquad\qquad\qquad\quad = 6t - 4t + 3 - 8$

$\qquad\qquad\qquad\qquad\quad = 2t - 5$

(b) $8(x + 1) - 3(2x - 5) = 8x + 8 - 6x + 15$

$\qquad\qquad\qquad\qquad\quad = 8x - 6x + 8 + 15$

$\qquad\qquad\qquad\qquad\quad = 2x + 23$

Multiply both terms in the second bracket by -2.

Expand the brackets first. Then collect like terms.

Exercise 4H

Expand then simplify these expressions:

1 $3(y + 4) + 2y + 10$ **2** $2(k + 6) + 3k + 9$

3 $4(a + 3) - 2a + 6$ **4** $3(t - 2) + 4t - 10$

5 $3(2y + 3) + 2(y + 5)$ **6** $4(x + 7) + 3(x + 4)$

7 $3(2x + 5) + 2(x - 4)$ **8** $2(4n + 5) + 5(n - 3)$

9 $3(x - 5) + 2(x - 3)$ **10** $4(2x - 1) + 2(3x - 2)$

11 $3(2b + 1) - 2(2b + 4)$ **12** $4(2m + 3) - 2(2m + 5)$ Use Example 12(a) to help.

13 $5(2k + 2) - 4(2k + 6)$ **14** $2(4p + 1) - 4(p - 3)$ Use Example 12(b) to help.

15 $5(2g - 4) - 2(4g - 6)$ **16** $2(w - 4) - 3(2w - 1)$

4.6 Factorising algebraic expressions

Key words:
factor
common factor

Factors

A **factor** is a number or letter which divides exactly into another term.

2 is a factor of 4.
2 is a factor of $6x$.

Example 14

(a) If 2 is a factor of these terms, show it by rewriting the term.

     A   12   **B**   7   **C** $6t$ **D** $10x$ **E** $5y^2$

(b) If x is a factor of these terms, show it by rewriting the term.

     A $6t$ **B** $4x$ **C** x^2 **D**   10   **E** xy

(a) A $12 = 2 \times 6$

   B Not a factor

   C $6t = 2 \times 3t$

   D $10x = 2 \times 5x$

   E Not a factor

(b) A Not a factor

   B $4x = 4 \times x$

   C $x^2 = x \times x$

   D Not a factor

   E $xy = x \times y$

$5y^2 = 5 \times y \times y$

If two terms have the same factor, or a factor 'in common', you call this a **common factor** .

For example: 2 is a factor of $4x$

              2 is a factor of 6

              2 is a common factor of $4x$ and 6.

Example 15

(a) Is 2 a common factor of these pairs of terms?
   If yes, then show how you can rewrite the terms.
   A $8t$ and 12   **B** $3x$ and 6   **C** $20z$ and 4

(b) Is 3 a common factor of these pairs of terms?
   If yes, then show how you can rewrite the terms.
   A $5x$ and 9   **B** $6p$ and 12   **C** $3t$ and 15

(a) A Yes $8t = 2 \times 4t, 12 = 2 \times 6$

   B No

   C Yes $20z = 2 \times 10z, 4 = 2 \times 2$

(b) A No

   B Yes $6p = 3 \times 2p, 12 = 3 \times 4$

   C Yes $3t = 3 \times t, 15 = 3 \times 5$

Exercise 4I

1 If 2 is a factor of these terms, show by rewriting the term.
   A 10   **B** 9   **C** $4t$ **D** $8x$ **E** $3y^2$

2 If 3 is a factor of these terms, show by rewriting the term.
   A 10   **B** 9   **C** $4t$ **D** $8x$ **E** $3y^2$

3 If x is a factor of these terms, show by rewriting the term.
   A $6f$ **B** $5x$ **C** x^2 **D** 12   **E** wx

4 Is 3 a common factor of these pairs of terms?
   If yes, then show how you can rewrite the terms.
   A $12y$ and 6   **B** $13n$ and 9   **C** $6q$ and 21

5 Is x a common factor of these pairs of terms?
   If yes, then show how you can rewrite the terms.
   A $4x^2$ and $2x$ **B** xy and y **C** xy and tx

Factorising expressions

Factorising an algebraic expression is the opposite of expanding brackets.

Factorising an expression means you write it as one term multiplied by a simpler expression.

For example: $6x + 10$ can be written as $2(3x + 5)$
   because $6x = 2 \times 3x$
   and $10 = 2 \times 5$

2 is a factor of 6x.
2 is also a factor of 10.
2 is a common factor of 6x and 10.

Example 16

Copy and complete these factorised expressions.
(a) $3t + 15 = 3(\square + 5)$ **(b)** $4n + 12 = \square(n + 3)$

(a) $3t + 15 = 3(t + 5)$

(b) $4n + 12 = 4(n + 3)$

$3 \times t = 3t$ and $3 \times 5 = 15$

$4 \times n = 4n$ and $4 \times 3 = 12$

Example 17

Factorise these expressions:
(a) $5a + 20$ **(b)** $4x - 12$ **(c)** $x^2 + 7x$

(a) $5a + 20 = 5 \times a + 5 \times 4$
      $= 5(a + 4)$
      $= 5(a + 4)$
   Check: $5(a + 4) = 5 \times a + 5 \times 4$
      $= 5a + 20$ ✓

(b) $4x - 12 = 4 \times x - 4 \times 3$
      $= 4(x - 3)$
      $= 4(x - 3)$

(c) $x^2 + 7x = x \times x + x \times 7$
      $= x(x + 7)$
      $= x(x + 7)$

5 is a factor of 5a
   $5a = 5 \times a$
5 is a factor of 20
   $20 = 5 \times 4$
So 5 is a common factor of 5a and 20.

Check your answer by expanding the brackets.

2 is a common factor of 4x and 12,
4 is also a common factor of 4x and 12.
Use 4 because it is higher than 2.

x is a common factor of x^2 and 7x

Exercise 4J

Don't forget to check your answers by removing the brackets.

1 Copy and complete these factorised expressions:

    (a) $3x + 15 = 3(\square + 5)$ **(b)** $5a + 10 = 5(\square + 2)$

    (c) $2x - 12 = 2(x - \square)$ **(d)** $4m - 16 = 4(m - \square)$

    (e) $4t + 12 = \square(t + 3)$ **(f)** $3n + 18 = \square(n + 6)$

    (g) $2b - 14 = \square(b - 7)$ **(h)** $4t - 20 = \square(t - 5)$

> Use Example 16 to help you.

2 Factorise these expressions:

    (a) $5p + 20$ **(b)** $2a + 12$ **(c)** $3y + 15$

    (d) $7b + 21$ **(e)** $4q + 12$ **(f)** $6k + 24$

    (g) $5a + 5$ **(h)** $4g + 8$

> Use Example 17(a) to help you.

3 Factorise these expressions:

    (a) $4t - 12$ **(b)** $3x - 9$ **(c)** $5n - 20$

    (d) $2b - 8$ **(e)** $6a - 18$ **(f)** $7k - 7$

    (g) $4r - 16$ **(h)** $6g - 12$

> $5 = 5 \times 1$

> Use Example 17(b) to help you.

4 Factorise these expressions:

    (a) $y^2 + 7y$ **(b)** $x^2 + 5x$ **(c)** $t^2 + 2t$

    (d) $n^2 + n$ **(e)** $x^2 - 7x$ **(f)** $z^2 - 2z$

    (g) $p^2 - 8p$ **(h)** $a^2 - a$

> Use Example 17(c) to help you.

> $n = n \times 1$

> $6p = 2 \times 3p$

5 Factorise these expressions:

    (a) $6p + 4$ **(b)** $4a + 10$ **(c)** $4t - 6$

    (d) $8m - 12$ **(e)** $10x + 15$ **(f)** $6y - 9$

    (g) $4a + 8b$ **(h)** $10p + 5q$

6 Write down the pairs of cards that show equivalent expressions.

A	B	C	D	E	F
$4a - 12$	$2(2a - 3)$	$a(a - 4)$	$3a + 6$	$a^2 + 2a$	$6a + 9$

G	H	I	J	K	L
$4(a - 3)$	$a(a + 2)$	$3(a + 2)$	$a^2 - 4a$	$3(2a + 3)$	$4a - 6$

Expanding two brackets

You can use a grid method to multiply two numbers.

For example:

34×57

×	50	7
30	1500	210
4	200	28

$34 \times 57 = (30 + 4) \times (50 + 7)$

$= 30 \times 50 + 30 \times 7 + 4 \times 50 + 4 \times 7$

$= 1500 + 210 + 200 + 28$

$= 1938$

You can also use a grid method when you multiply two brackets together because you have to multiply each term in one bracket by each term in the other bracket.

To expand and simplify $(x + 2)(x + 5)$:

×	x	5
x	x^2	$5x$
2	$2x$	10

$(x + 2)(x + 5) = x \times x + x \times 5 + 2 \times x + 2 \times 5$

$= x^2 + 5x + 2x + 10$

$= x^2 + 7x + 10$

Example 18

Expand and simplify $(t + 6)(t - 2)$.

×	t	-2
t	t^2	$-2t$
6	$6t$	-12

$(t + 6)(t - 2) = t \times t + t \times (-2) + 6 \times t + 6 \times (-2)$

$= t^2 - 2t + 6t - 12$

$= t^2 + 4t - 12$

> You simplify the final expression by collecting the like terms: $5x + 2x = 7x$.

> Remember you are multiplying by -2. $+\text{ve} \times -\text{ve} = -\text{ve}$

Exercise 4K

Expand and simplify:

1 $(a + 2)(a + 7)$ **2** $(x + 3)(x + 1)$ **3** $(x + 5)(x + 5)$

4 $(t + 5)(t - 2)$ **5** $(x + 7)(x - 4)$ **6** $(n - 5)(n + 8)$

7 $(x - 4)(x + 5)$ **8** $(p - 4)(p + 4)$ **9** $(x - 9)(x - 4)$

> Be careful when there are negative signs – this is where a lot of mistakes are made.

Examination style questions

1 Jan is making small cakes for a Christmas fair.
Each cake uses 7 g of flour.
How much flour does she use to make x cakes? *(1 mark)*
AQA, Spec B, 5I, November 2003

2 Sam buys x packets of sweets.
Each packet of sweets costs 22 pence.
Sam pays with a £5 note.
Write down an expression for the change, in pence, Sam should receive.
(2 marks)
AQA, Spec A, I, June 2003

3 **(a)** Factorise $7x + 14$
(b) Expand and simplify $4(m + 3) + 3(2m - 5)$ *(3 marks)*
AQA, Spec A, I, June 2003

4 Expand and simplify $5(2x + 1) - 3(x - 4)$ *(2 marks)*
AQA, Spec B, 5I, June 2003

5 **(a)** Factorise $10a + 5$
(b) Factorise $c^2 - 4c$ *(3 marks)*
AQA, Spec B, 5I, November 2003

6 Show that $(x + 2)(x + 3)$ expands and simplifies to $x^2 + 5x + 6$. *(2 marks)*
AQA, Spec B, 5I, November 2003

7 Part of a number grid is shown below.

1	2	3	4	5	6	7	8	9	10
11	12	13	14	15	16	17	18	19	20
21	22	23	24	25	26	27	28	29	30
31	32	33	34	35	36	37	38	39	40
41	42	43	44	45	46	47	48	49	50

The shaded shape is called T_{16} because it has the number 16 on the left.
The sum of the numbers in T_{16} is 67.
(a) This is T_n:
Make a copy of T_n and fill in the
empty boxes.

(b) Find the sum of all the numbers in T_n in terms of n.
Give your answer in its simplest form. *(4 marks)*
AQA, Spec B, 5I, June 2003

Summary of key points

Using letters to write simple expressions (grade F)

$m + 4$ is called an expression in terms of m.

$3 \times x = x \times 3 = 3$ lots of $x = x + x + x = 3x$

y times $y = y \times y = y^2$ ('y squared')

$d \div 4 = \dfrac{d}{4}$

$g \times h = gh$, $\ x \times y = xy$ You leave out the '$\times$ sign'.

Collecting like terms (grade F/E)

$3a + 6b + 5a$ This expression has three terms.

Terms which use the same letter are called like terms, for example $3a + 6b + 5a$.

You can simplify algebraic expressions by collecting like terms together.

$3a + 6b + 5a = 3a + 5a + 6b$
$ = 8a + 6b$

Expanding brackets (grade D/C)

To multiply algebraic terms first multiply the numbers, then multiply the letters.

When you expand brackets you must multiply each term inside the brackets by the term outside the bracket.

$3(2a + 7) = 6a + 21$ $x(3x + 5) = 3x^2 + 5x$

Use a grid method to multiply two brackets together. Multiply each term in one bracket by each term in the other bracket.

$(x + 2)(x + 5) = x \times x + x \times 5 + 2 \times x + 2 \times 5 = x^2 + 7x + 10$

Factorising (grade D/C)

Factorising an algebraic expression is the opposite of expanding brackets.

$8t - 10 = 2(4t - 5)$

This chapter will show you how to:

✔ solve simple equations
✔ deal with equations that have negative, decimal or fractional answers
✔ solve equations combining two or more operations, or involving brackets
✔ write inequalities and represent them on a number line
✔ solve simple inequalities and find the solution set

5.1 Equations

Equations involving addition and subtraction

A **simple equation** is an equation involving an unknown, usually represented by a letter. A **linear equation** has no letters with powers.

In the equation the letter may be added to, subtracted from, multiplied by or divided by a whole number.

When you **solve** an equation you find the value of the unknown. Your aim is to get the unknown on its own on one side of the equation.

To solve a linear equation you must always do the same to both sides of the equation. This means do the same to the expression on either side of the equals sign.

Example 1

Solve the equation $b + 2 = 5$ to find the value of b.
2 has been added to b. To get b by itself you must 'undo' this operation. You perform the **inverse** or 'opposite' operation, so subtract 2.

$$b + 2 = 5$$
$$b + 2 - 2 = 5 - 2$$
$$b = 3$$

> **Key words:**
> simple equation
> linear equation
> solve
> inverse operation
> subject

$3x + 5 = 17$ is a linear equation.
$x^2 = 16$ is not.

$5x + 73 = 198$
$x = ?$

For example, if you add 6 to one side of an equation you must add 6 to the other side.

Subtract 2 from both sides of the equation.

Example 2

Solve the equation $p - 4 = 9$.

4 has been subtracted from p. The inverse operation is add 4.

Look for the inverse operation.

$$p - 4 = 9$$
$$p - 4 + 4 = 9 + 4$$
$$p = 13$$

Add 4 to both sides of the equation.

Write your equations with the equals signs underneath one another.

Sometimes the unknown letter appears on the right-hand side of the equation.

Example 3

Solve the equation $7 = c + 2$.

Subtract 2 from both sides of the equation.

$$7 = c + 2$$
$$7 - 2 = c + 2 - 2$$
$$5 = c$$
$$c = 5$$

You usually write the answer with the unknown letter (the **subject**) on the left-hand side of the equals sign. $5 = c$ is the same as $c = 5$.

Alternatively you could write the equation as $c + 2 = 7$ and solve it as in Example 1.

Exercise 5A

1 Solve each equation to find the value of the letter.

(a) $h + 4 = 6$ (b) $b + 1 = 12$ (c) $x + 9 = 10$

(d) $k - 5 = 2$ (e) $y - 9 = 11$ (f) $a - 20 = 30$

Make sure you treat both sides of the equation in the same way.

2 Find the value of the letter in each of these equations. Write the letter on the left-hand side of the equals sign in your answers.

(a) $4 = s + 3$ (b) $15 = d - 18$ (c) $6 = r + 4$

3 Find the value of the unknown letter in each of these equations.

(a) $5 + f = 32$ (b) $40 + w = 120$ (c) $14 = 8 + n$

4 Viewfone gave me an extra 25 free texts. I now have 43 free texts. How many did I have to start with? Write an equation and solve it.

Equations involving multiplication and division

Sometimes the unknown letter is multiplied or divided by a **whole number (integer)** .

Key words:
whole number
integer

Example 4

Solve the equation $4a = 12$.

$4a$ means $4 \times a$. The inverse operation is divide by 4.

In algebra you write $4a \div 4$ as $\dfrac{4a}{4}$.

$$4a = 12$$

$$\frac{4a}{4} = \frac{12}{4}$$

$$a = 3$$

Divide both sides of the equation by 4.

Example 5

Solve the equation $\dfrac{d}{5} = 8$.

$\dfrac{d}{5}$ means $d \div 5$. The inverse operation of $\div 5$ is $\times 5$.

$$\frac{d}{5} = 8$$

$$\frac{d}{5} \times 5 = 8 \times 5$$

$$d = 40$$

Multiply both sides of the equation by 5.

Exercise 5B

1 Solve these equations.
Write the letter on the left-hand side of the equals sign in your answers.

(a) $3b = 9$ (b) $5t = 25$ (c) $7p = 21$

(d) $4g = 28$ (e) $9x = 36$ (f) $8n = 24$

(g) $10m = 130$ (h) $6y = 360$ (i) $7k = 49$

(j) $20 = 5h$ (k) $64 = 8r$ (l) $32 = 4a$

$64 = 8r$ is the same as $8r = 64$

2 Find the value of the letter in each of these equations.

(a) $\dfrac{t}{2} = 12$ (b) $\dfrac{y}{6} = 3$ (c) $\dfrac{f}{4} = 9$

(d) $\dfrac{r}{3} = 11$ (e) $\dfrac{q}{6} = 6$ (f) $\dfrac{g}{9} = 8$

(g) $\dfrac{s}{2} = 56$ (h) $\dfrac{b}{5} = 30$ (i) $\dfrac{n}{8} = 1$

(j) $4 = \dfrac{m}{5}$ (k) $16 = \dfrac{v}{3}$ (l) $100 = \dfrac{u}{9}$

3 Find the value of the unknown letter in each equation. Write the letter on the left-hand side of the equals sign in your answers.

(a) $8t = 48$ (b) $4 + p = 7$ (c) $\dfrac{r}{7} = 8$

(d) $15 = \dfrac{m}{4}$ (e) $45 = x - 35$ (f) $g - 100 = 550$

(g) $\dfrac{s}{12} = 5$ (h) $9d = 99$ (i) $42 = 7j$

4 It took Joe 20 minutes to plant one quarter of his runner beans. How long will it take him to plant all of them? Write an equation and solve it.

Equations involving two operations

Some equations involve carrying out two **operations**.

Key words:
operation
inverse

In the equation $4x + 2 = 10$, the operations are

$x \rightarrow$ **multiply** by $4 \rightarrow 4x \rightarrow$ **add** $2 \rightarrow 4x + 2$

You solve the equation by 'undoing' the operations. The **inverse** operations are

$x \leftarrow$ **divide** by $4 \leftarrow 4x \leftarrow$ **subtract** $2 \leftarrow 4x + 2$

Example 6

Solve the equation $3b + 2 = 8$.

The expression $3b + 2$ has been formed like this:
$b \rightarrow$ **multiply** by $3 \rightarrow 3b \rightarrow$ **add** $2 \rightarrow 3b + 2$

So the inverse of this is:
$b \leftarrow$ **divide** by $3 \leftarrow 3b \leftarrow$ **subtract** $2 \leftarrow 3b + 2$

The order of this inverse process is very important.

$$3b + 2 = 8$$
$$3b + 2 - 2 = 8 - 2$$
$$3b = 6$$
$$\frac{3b}{3} = \frac{6}{3}$$
$$b = 2 \quad \text{Check: } 3b + 2 = 3 \times 2 + 2 = 6 + 2 = 8$$

First subtract 2 from both sides, then divide both sides by 3.

To check your answer try the value $b = 2$ in the expression.
You will cover substitution in more detail in Chapter 7.

Example 7

Solve the equation $5g - 4 = 36$

$$g \to \times 5 \to 5g \to -4 \to 5g - 4$$

Inverse

$$g \leftarrow \div 5 \leftarrow 5g \leftarrow +4 \leftarrow 5g - 4$$

g has been multiplied by 5 and then 4 has been subtracted. The inverse is add 4 and then divide by 5.

$$5g - 4 = 36$$
$$5g - 4 + 4 = 36 + 4$$
$$5g = 40$$
$$\frac{5g}{5} = \frac{40}{5}$$
$$g = 8$$
Check: $5 \times 8 - 4 = 40 - 4 = 36$

First add 4 to both sides and then divide both sides by 5.

Example 8

Solve the equation $\dfrac{k}{4} + 3 = 7$

The inverse of $\div 4$ is $\times 4$. The inverse of $+3$ is -3.

First do -3, then do $\times 4$.

Look for the opposite operations and carry them out in the correct order.

$$\frac{k}{4} + 3 = 7$$
$$\frac{k}{4} + 3 - 3 = 7 - 3$$
$$\frac{k}{4} = 4$$
$$\frac{k}{4} \times 4 = 4 \times 4$$
$$k = 16$$
Check: $\frac{16}{4} + 3 = 16 \div 4 + 3 = 4 + 3 = 7$

First subtract 3 from both sides, then multiply both sides by 4.

Sometimes the whole of one side of an equation is divided by a number.

For example $\dfrac{x + 2}{4} = 3$.

Operations: $x \to +2 \to x + 2 \to \div 4 \to \dfrac{x + 2}{4}$

Inverse: $x \leftarrow -2 \leftarrow x + 2 \leftarrow \times 4 \leftarrow \dfrac{x + 2}{4}$

> First multiply both sides of the equation by 4.

Example 9

Solve the equation $\dfrac{12 - 2n}{3} = 2$

$\dfrac{12 - 2n}{3} \times 3 = 2 \times 3$ —————— First multiply both sides of the equation by 3.

$12 - 2n = 6$

$12 - 2n + 2n = 6 + 2n$ —————— This is now an equation like the ones in Examples 6 and 7. Next add $2n$ to both sides. This means that the value of $2n$ remains positive.

$12 = 6 + 2n$

$12 - 6 = 6 - 6 + 2n$ —————— Subtract 6 from both sides.

$6 = 2n$

$3 = n$ —————— Divide both sides by 2.

$n = 3$ —————— Put n on the left-hand side of the equation.

Exercise 5C

1 Solve these equations:

(a) $3w + 2 = 14$ (b) $2u + 9 = 19$

(c) $5t - 4 = 36$ (d) $9m - 2 = 25$

(e) $8d + 3 = 19$ (f) $7g + 6 = 69$

(g) $4s + 5 = 17$ (h) $8b + 6 = 6$

(i) $20a - 30 = 70$

2 Find the value of the unknown letter in each equation.

(a) $\dfrac{r}{3} + 2 = 5$ (b) $\dfrac{h}{4} - 5 = 2$ (c) $\dfrac{f}{3} - 8 = 2$

(d) $\dfrac{x}{2} + 4 = 5$ (e) $\dfrac{d}{10} - 10 = 1$ (f) $\dfrac{a}{3} + 7 = 10$

3 Solve these equations:

(a) $\dfrac{3x-2}{11}=2$ (b) $\dfrac{6t+5}{7}=5$ (c) $\dfrac{4w+9}{7}=7$

(d) $\dfrac{14-3a}{2}=4$ (e) $\dfrac{30+10n}{5}=14$ (f) $\dfrac{22+8q}{9}=6$

Solutions involving fractions and decimals

You will need to know:
- **how to change between fractions and decimals**

So far you have solved equations with whole number (integer) answers.

> Some equations have answers that are **fractions** or **decimals** .

The question may ask for the answer as a fraction or as a decimal. If not, you can give your answer in either form.

Example 10
Solve the equation $9t+7=19$. Give your answer as a **mixed number** .

$$9t+7=19$$
$$9t+7-7=19-7$$
$$9t=12$$
$$\frac{9t}{9}=\frac{12}{9}$$
$$t=1\tfrac{3}{9}$$
$$t=1\tfrac{1}{3}$$

Example 11
Solve the equation $5x-6=3$.
Give your answer as a decimal.

$$5x-6=3$$
$$5x-6+6=3+6$$
$$5x=9$$
$$\frac{5x}{5}=\frac{9}{5}$$
$$x=1.8$$

Key words:
fractions
decimals
mixed number
lowest terms
simplest form

A mixed number has a whole number part and a fractional part. $1\tfrac{3}{4}$ is a mixed number.

First subtract 7 from both sides, then divide both sides by 9.

$\tfrac{3}{9}=\tfrac{1}{3}$ in its **lowest terms** or **simplest form** .

First add 6 to both sides, then divide both sides by 5.

$\tfrac{9}{5}=9\div5=1.8$
See Section 3.5 Fractions, decimals and percentages.

Exercise 5D

1 Solve these equations. Give your answers as mixed numbers.

(a) $2b - 4 = 5$ (b) $7c + 4 = 13$ (c) $3y + 6 = 16$

(d) $7j - 13 = 6$ (e) $3g - 2 = 3$ (f) $8p - 2 = 10$

2 Solve these equations. Give your answers as decimals.

(a) $8k - 3 = 7$ (b) $4e - 5 = 1$ (c) $10i - 10 = 12$

(d) $5s + 9 = 40$ (e) $4m - 2 = 5$ (f) $2a + 5 = 20$

Solutions involving negative values

> **Key words:**
> negative number

Some equations have **negative number** solutions.

Example 12

Solve the equation $2u + 7 = 1$.

$$2u + 7 = 1$$
$$2u + 7 - 7 = 1 - 7$$
$$2u = -6$$
$$\frac{2u}{2} = \frac{-6}{2}$$
$$u = -3$$

First subtract 7 from both sides, then divide both sides by 2.

Dividing a negative number by a positive number gives a negative number answer. See Chapter 1 on rules for positive numbers.

Example 13

Solve the equation $\dfrac{d}{2} - 4 = -5$.

$$\frac{d}{2} - 4 = -5$$
$$\frac{d}{2} - 4 + 4 = -5 + 4$$
$$\frac{d}{2} = -1$$
$$\frac{d}{2} \times 2 = -1 \times 2$$
$$d = -2$$

First add 4 to both sides, then multiply both sides by 2.

Multiplying a negative number by a positive number gives a negative number answer.

Exercise 5E

1 Solve these equations.

(a) $3k + 8 = 2$ (b) $5f + 12 = 7$ (c) $6m + 20 = 2$

(d) $10g - 5 = -55$ (e) $2w - 1 = -5$ (f) $4s - 4 = -12$

(g) $3u + 12 = 6$ (h) $5f + 13 = 3$ (i) $4v + 23 = 3$

(j) $19 + 2x = 7$ (k) $12 - 6a = 18$ (l) $3n + 7 = -8$

2 Find the value of the unknown letter in each equation.

(a) $\dfrac{m}{2} + 3 = 1$ (b) $\dfrac{q}{3} + 7 = 6$ (c) $\dfrac{x}{5} - 3 = -5$

(d) $\dfrac{b}{8} - 7 = -10$ (e) $20 + \dfrac{e}{4} = 17$ (f) $\dfrac{t}{6} + 10 = 4$

3 Sita gave $\frac{1}{20}$ of her wages, plus an extra £2, to charity last week. She gave the charity £18. Write an equation and solve it to find out Sita's wages last week.

Equations with unknowns on both sides

Some equations have the unknown letter on both sides of the equation.

You can still solve the equation by treating both sides of the equation in the same way.

For example: $5a - 2 = 2a + 4$.

It is useful to keep the letter on the side with the most, to keep it positive.

In this example, the left-hand side has $5a$ (the most) and the right-hand side has $2a$. So you collect the a terms on the left-hand side.

To 'remove' the a terms on the right-hand side you subtract them. Remember to do the same to both sides.

$$5a - 2 = 2a + 4$$
$$5a - 2a - 2 = 2a - 2a + 4$$
$$3a - 2 = 4$$

Subtract $2a$ from both sides of the equation.

See Example 6.

Now you have a linear equation to solve.

Example 14

Solve the equation

$$3x + 1 = x + 7$$

$3x - x + 1 = x - x + 7$

$2x + 1 = 7$

$2x + 1 - 1 = 7 - 1$

$2x = 6$

$\dfrac{2x}{2} = \dfrac{6}{2}$

$x = 3$

There are more xs on the left-hand side. Subtract x from both sides.

Subtract 1 from both sides, then divide by 2.

Exercise 5F

1 Solve these equations.

(a) $3x + 4 = 2x + 6$ (b) $5u + 9 = 3u + 17$

(c) $r + 2 = 5r - 10$ (d) $7p + 11 = 6p + 16$

(e) $4w - 6 = 3w + 1$ (f) $6b - 2 = 2b + 14$

2 Solve:

(a) $4a + 6 = 2a + 2$ (b) $7m + 10 = 4m + 1$

(c) $8d - 2 = 3d - 12$ (d) $y - 5 = 6y + 15$

Equations involving brackets

Some equations involve **brackets** .

> **To solve an equation with brackets:**
> **1 Expand the brackets.**
> **2 Solve by treating both sides of the equation in the same way.**

Key words:
brackets
expand

Example 15

Solve $4(c + 3) = 20$

$4(c + 3) = 20$

$4 \times c + 4 \times 3 = 20$

$4c + 12 = 20$

Expanding brackets was covered in Section 4.4.

First expand the brackets.

continued ▼

$$4c + 12 = 20$$
$$4c + 12 - 12 = 20 - 12$$
$$4c = 8$$
$$\frac{4c}{4} = \frac{8}{4}$$
$$c = 2$$

This equation is the same type as the one in Example 6.

First subtract 12 from both sides, then divide both sides by 4.

Example 16

Solve $5(3p - 4) = 10$.

$$5(3p - 4) = 10$$
$$5 \times 3p - 5 \times 4 = 10$$
$$15p - 20 = 10$$
$$15p - 20 + 20 = 10 + 20$$
$$15p = 30$$
$$\frac{15p}{15} = \frac{30}{15}$$
$$p = 2$$

Expanding the brackets.

First add 20 to both sides, then divide both sides by 15.

When there are two pairs of brackets expand both brackets first before solving.

Example 17

Solve $2(2m + 10) = 12(m - 1)$.

$$2(2m + 10) = 12(m - 1)$$
$$4m + 20 = 12m - 12$$
$$12 + 20 = 12m - 4m$$
$$32 = 8m$$
$$\frac{32}{8} = \frac{8m}{8}$$
$$4 = m$$
$$m = 4$$

Expand the brackets.

This equation has the unknown on both sides – see Example 14.

Subtract $4m$ from both sides.

Divide both sides by 8.

$4 = m$ is the same as $m = 4$.

Exercise 5G

1 Solve:

(a) $4(g + 6) = 32$ (b) $7(k + 1) = 21$

(c) $5(s + 10) = 65$ (d) $2(n - 4) = 6$

(e) $3(f - 2) = 24$ (f) $7(v - 3) = 42$

(g) $4(m + 6) = 20$ (h) $2(w + 7) = 8$

2 Find the value of the letters in these equations.

(a) $7(2b + 2) = 56$ (b) $2(5r + 4) = 18$

(c) $4(5t + 2) = 48$ (d) $8(2v - 9) = 24$

(e) $9(3k - 3) = 0$ (f) $2(5y - 12) = 6$

(g) $3(4x + 15) = 9$ (h) $8(12 + 3f) = 48$

3 Solve these equations.

(a) $2(b + 1) = 8(2b - 5)$ (b) $5(4a + 7) = 3(8a + 9)$

(c) $6(x - 2) = 3(3x - 8)$ (d) $5(2p + 2) = 6(p + 5)$

(e) $9(3s - 4) = 5(4s - 3)$ (f) $4(10t - 7) = 3(6t - 2)$

(g) $4(2w + 2) = 2(5w + 7)$ (h) $3(3t - 2) = 7(t - 2)$

5.2 Inequalities

Key words:
inequality
less than $<$
greater than $>$
less than or equal to $\leqslant$
greater than or equal to $\geqslant$
number line

An expression whose right- and left-hand sides are not equal is called an inequality .

You need to know these inequality symbols:

$<$ **means less than**

$>$ **means greater than**

$\leqslant$ **means less than or equal to**

$\geqslant$ **means greater than or equal to**

The pointed end points towards the smaller number. The open end points towards the larger number. You used the signs $<$ and $>$ in Chapter 1.

Example 18

Write the correct inequality sign between these pairs of numbers.

(a) 3, 4   (b) 21, 56   (c) 9, 4   (d) 3, −5

(a) 3 is less than 4, $3 < 4$

(b) 21 is less than 56, $21 < 56$

(c) 9 is greater than 4, $9 > 4$

(d) 3 is greater than -5, $3 > -5$

For negative numbers you can use a **number line**.

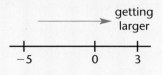

getting larger

Example 19

Write down the whole number values for each letter to make the inequality true.

(a) $n > 5$ and $n < 12$

(b) $b > 2$ and $b \leqslant 7$

(c) $x \geqslant -4$ and $x < 3$

The values of *n* are greater than 5 but less than 12.

(a) $6, 7, 8, 9, 10, 11$

(b) $3, 4, 5, 6, 7$

(c) $-4, -3, -2, -1, 0, 1, 2$

The values of *b* are greater than 2 but less than *or equal to* 7.

The values of *x* must be greater than *or equal to* -4 but less than 3.

You can show inequalities on a number line.

$n > 5$ and $n < 12$

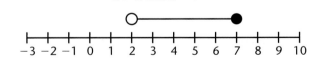

Use the symbol $\bigcirc$ when the end value *is not* included in the answer.

$b > 2$ and $b \leqslant 7$

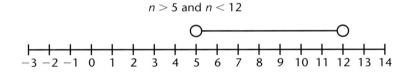

Use the symbol $\bullet$ when the end value *is* included in the answer.

$x \geqslant -4$ and $x < 3$

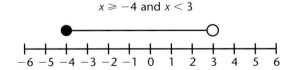

You can combine inequalities.
From the number lines you can see that:

$$n > 5 \text{ and } n < 12 \rightarrow 5 < n < 12$$

$$b > 2 \text{ and } b \leqslant 7 \rightarrow 2 < b \leqslant 7$$

$$x \geqslant 4 \text{ and } x < 3 \rightarrow -4 \leqslant x < 3$$

You say '*n* is greater than 5 but less than 12' or that '*n* lies between 5 and 12'.

$n > 5$ is the same as $5 < n$.

Exercise 5H

1 Write the correct inequality sign between these pairs of numbers in the order they are given.

(a) 5, 8 (b) 10, 5 (c) −7, 6
(d) 20, 80 (e) 2, −4 (f) 23, 21
(g) −5, −8 (h) 5.5, 6.4 (i) 12.7, 12.6
(j) 0, 0.01 (k) 0.01, 10.01 (l) 1112, 1121

2 Write down whether each statement is true (T) or false (F).

(a) $8 > 9$ (b) $12 < 17$ (c) $31 > 30$
(d) $5 \leqslant 4$ (e) $23 > 22.9$ (f) $0.01 > 0.001$
(g) $89 = 88$ (h) $45 \geqslant 44$ (i) $99.9 > 99.89$

3 Write down the whole number values for each letter to make the inequality true.

(a) $b > 4$ and $b < 9$ (b) $r \leqslant 12$ and $r > 7$
(c) $p > 0$ and $p \leqslant 5$ (d) $k \leqslant 100$ and $k > 94$
(e) $x > -2$ and $x \leqslant 3$ (f) $q \geqslant -9$ and $q < -4$

4 Write down the values for each letter, to 1 decimal place, to satisfy the inequalities.

(a) $h > 3.2$ and $h < 4.1$ (b) $t \leqslant 21.2$ and $t > 20.7$
(c) $y \leqslant 0.9$ and $y \geqslant 0.3$ (d) $m \geqslant 78.9$ and $m < 80.1$

5 Show each inequality on a separate copy of this number line.

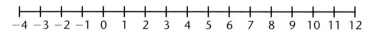

(a) $x > 4$ (b) $t > 2$ and $t < 11$
(c) $u \leqslant 12$ and $u > 6$ (d) $w \leqslant 9$ and $w > 0$
(e) $2 < j \leqslant 8$ (f) $8 \leqslant s < 12$
(g) $-3 \leqslant m < 6$ (h) $-4 \leqslant d \leqslant 0$

6 Write each pair of inequalities in questions 3 and 4 as a combined inequality. For example,
3(a) $b > 4$ and $b < 9 \rightarrow 4 < b < 9$.

Key words:
solution set
double inequality

Solving inequalities

Finding the solution to an inequality is very similar to finding the solution to an equation. You use the same rules of algebra and do the same to both sides.

Sometimes an inequality is more than one single value answer. You need to give all the values. The set of values is called the solution set .

In Example 19 $n > 5$ and $n < 12$ had the solution set 6, 7, 8, 9, 10, 11.

Example 20

Solve the inequality $4n \geqslant 20$, where n has integer values.

An integer is a whole number (negative or positive).

Divide both sides by 4.

$4n \geqslant 20$

$\dfrac{4n}{4} \geqslant \dfrac{20}{4}$

$n \geqslant 5$

In other words n can have the values 5, 6, 7, 8, 9, ... This is the solution set.

Example 21

Solve $3m - 6 < 9$, where m is an integer.

$3m - 6 < 9$

$3m - 6 + 6 < 9 + 6$

$3m < 15$

$\dfrac{3m}{3} < \dfrac{15}{3}$

$m < 5$

Add 6 to both sides of the inequality and then divide both sides by 3.

m can take any value in the list 4, 3, 2, 1, 0, −1, −2, ...

Example 22

Solve the inequality $5p \leqslant 2p + 12$, where p is an integer.

$5p \leqslant 2p + 12$

$5p - 2p \leqslant 2p - 2p + 12$

$3p \leqslant 12$

$p \leqslant 4$

Subtract $2p$ from both sides.

Divide both sides by 3.

p can take any value in the list 4, 3, 2, 1, 0, −1, −2, ...

Sometimes you need to solve a **double inequality** .

Example 23

Solve $4 < 2m \leqslant 16$, where m is an integer.

You can treat this as two separate inequalities and solve each one separately.

Or you can solve both sides at the same time.

(a) Solving the inequalities separately

$$4 < 2m \quad \text{and} \quad 2m \leqslant 16$$

$$\frac{4}{2} < \frac{2m}{2} \quad \text{and} \quad \frac{2m}{2} \leqslant \frac{16}{2}$$

$$2 < m \quad \text{and} \quad m \leqslant 8$$

The solution is $2 < m \leqslant 8$

There are two inequality signs in a double inequality.

Write the inequality in two parts.

Write these as a combined inequality.

m can take the values 3, 4, 5, 6, 7 and 8.

(b) Solving both sides at the same time

$$4 < 2m \leqslant 16$$

$$\frac{4}{2} < \frac{2m}{2} \leqslant \frac{16}{2}$$

$$2 < m \leqslant 8$$

Divide all terms by 2.

Exercise 5I

1 Solve these inequalities, where each letter has integer values.

    (a) $5k > 20$ **(b)** $3h < 27$ **(c)** $4g \geqslant 8$

    (d) $6m \leqslant 18$ **(e)** $2v > 6$ **(f)** $8x \leqslant 24$

    (g) $4w \leqslant -16$ **(h)** $9a > -81$ **(i)** $7t \leqslant -21$

2 Solve the following inequalities where each letter has integer values.

    (a) $2x + 5 > 9$ **(b)** $8s + 3 < 19$ **(c)** $5u - 7 \leqslant 8$

    (d) $4a - 22 \geqslant 18$ **(e)** $6q - 7 > 23$ **(f)** $12 + 4r \leqslant 16$

    (g) $3n + 7 \geqslant 1$ **(h)** $7p + 25 \leqslant 4$ **(i)** $10y + 50 > 10$

3 Solve the following inequalities. All the letters represent integer values.

    (a) $8q > 5q + 12$ **(b)** $3c \leqslant c + 20$

    (c) $12v \leqslant 7v + 100$ **(d)** $5n \geqslant 8n - 6$

    (e) $3x + 14 \leqslant 5x$ **(f)** $7g \leqslant 2 + 8g$

    (g) $4m + 10 \leqslant 2m$ **(h)** $7w + 12 > 4w$

4 Solve these double equalities.

    (a) $6 < 2z < 20$ **(b)** $21 \leqslant 7c < 28$

    (c) $6 < 3x \leqslant 18$ **(d)** $200 > 10b \geqslant 150$

    (e) $0 < 9h < 81$ **(f)** $3 < 4t \leqslant 10$

    (g) $-6 < 3p \leqslant 3$ **(h)** $-25 \leqslant 10q \leqslant -5$

Examination style questions

1 Solve the equations:

    (a) $6d = 18$ **(b)** $k - 5 = 7$

    (c) $y + 9 = 13$ **(d)** $\dfrac{n}{3} = 8$ *(5 marks)*

2 Solve the equations:

    (a) $3x = 12$ **(b)** $y + 7 = 13$

    (c) $8c - 5 = 11$ **(d)** $3(w - 2) = 9$ *(7 marks)*

AQA, Spec A, F, June 2003

3 Solve the equations:

    (a) $6r + 2 = 8$ **(b)** $7s + 2 = 5s + 3$

    (c) $\dfrac{12 - y}{3} = 5$ *(8 marks)*

AQA, Spec A, I, November 2003

4 (a) Solve the equation:

    $\dfrac{23 - 2x}{5} = 3$

    (b) Solve the inequality: $3x + 8 < 29$ *(5 marks)*

AQA, Spec A, I, June 2003

5 (a) Solve the inequality $2x + 3 \geqslant 1$.

    (b) Write down the inequality shown by the following diagram.

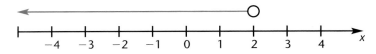

    (c) Write down three integers that satisfy both inequalities shown in parts (a) and (b).   *(4 marks)*

AQA, Spec A, I, June 2004

Summary of key points

Solving equations (grades F to C)

The solutions to equations may be positive or negative whole numbers (integers) or fractions or decimals.
To solve an equation:

1 Expand any brackets.
2 Solve by treating both sides of the equation in the same way.

Inequalities (grade C)

An expression whose left- and right-hand sides are not equal is called an inequality.

$<$ means 'less than'
$>$ means 'greater than'
$\leqslant$ means 'less than or equal to'
$\geqslant$ means 'greater than or equal to'
You can show inequalities on a number line

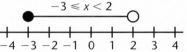

- means the end value *is* included in the answer
○ means the end value *is not* included in the answer

The complete set of answers for an inequality is called a solution set.

6 Angles, triangles and polygons

This chapter will show you how to:

✔ describe a turn, recognise and measure angles of different types
✔ discover angle properties
✔ use bearings to describe directions
✔ investigate angle properties of triangles, quadrilaterals and other polygons
✔ understand line and rotational symmetry

6.1 Angles

Turning

If you turn all the way round in one direction, back to your starting position, you make a **full turn** .

> **Key word:**
> full turn
> quarter turn
> half turn
> clockwise
> anti-clockwise

The minute hand of a clock moves a **quarter turn** from 3 to 6.

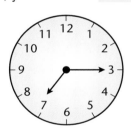 

> Minute hand makes $\frac{1}{4}$ turn.
> Time taken: $\frac{1}{4}$ hour.

The minute hand of a clock moves a **half turn** from 4 to 10.

> Minute hand makes $\frac{1}{2}$ turn.
> Time taken: $\frac{1}{2}$ hour.

The hands of a clock turn **clockwise** .
This fairground ride turns **anti-clockwise** .

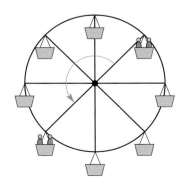

$\frac{1}{2}$ turn anti-clockwise

A ship sailing due South changes direction to sail due East.

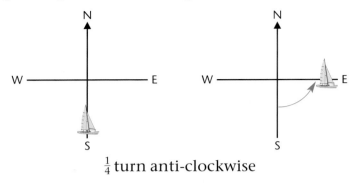

$\frac{1}{4}$ turn anti-clockwise

Exercise 6A

1 Describe the turn the minute hand of a clock makes between these times.

(a) 3 am and 3.30 am   **(b)** 6.45 pm and 7 pm

(c) 2215 and 2300   **(d)** 0540 and 0710

Look at the clock examples on page 133.

2 Here is a diagram of a compass.

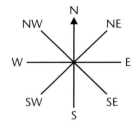

You are given a starting direction and a description of a turn.
What is the finishing direction in each case?

	Starting direction	Description of turn
(a)	N	$\frac{1}{4}$ turn clockwise
(b)	SE	$\frac{1}{4}$ turn anti-clockwise
(c)	SW	$\frac{1}{2}$ turn anti-clockwise
(d)	E	$\frac{3}{4}$ turn anti-clockwise
(e)	NE	$\frac{1}{4}$ turn clockwise
(f)	W	$\frac{1}{2}$ turn clockwise
(g)	NW	$\frac{3}{4}$ turn clockwise
(h)	S	$\frac{1}{4}$ turn anti-clockwise

Describing angles

Key word:
acute
right angle
obtuse
straight line
reflex
line segment

An angle is a measure of turn. Angles are usually measured in degrees. A complete circle (or full turn) is 360°.

The minute hand of a clock turns through 360° between 1400 (2 pm) and 1500 (3 pm).

You will need to recognise these types of angles:

acute		An angle between 0° and 90°, less than a $\frac{1}{4}$ turn.
right angle		An angle of 90°, a $\frac{1}{4}$ turn. A right angle is usually marked with this symbol.
obtuse		An angle between 90° and 180°, more than $\frac{1}{4}$ turn but less than $\frac{1}{2}$ turn.
straight line		An angle of 180°, a $\frac{1}{2}$ turn.
reflex		An angle between 180° and 360°, more than $\frac{1}{2}$ turn but less than a full turn.

You can describe angles in three different ways:

- 'Trace' the angle using capital letters. Write a 'hat' symbol over the middle letter: $A\hat{B}C$

  AB, BC, PQ and QR are called **line segments** .

- Use an angle sign or write the word 'angle': $\angle PQR$ or angle PQR

  The letter on the point of the angle always goes in the middle.

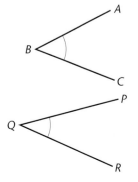

- Use a single letter:

Example 1

State whether these angles are acute, right angle, obtuse or reflex.

(a)
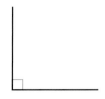

(b)
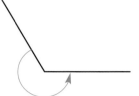

(c)
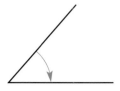

(d)
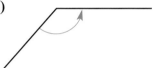

(e)
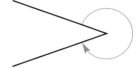

(a) right angle	The 'box' symbol shows the angle is 90°.
(b) reflex	More than a half turn.
(c) acute	Less than 90°.
(d) obtuse	More than 90° but less than 180°.
(e) reflex	Almost a full turn.

Exercise 6B

1 State whether these angles are acute, right angle, obtuse, reflex.

(a)
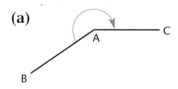

(b)
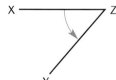

(c)
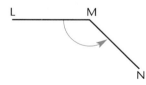

(d)
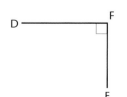

(e)
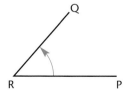

(f)
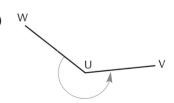

2 Describe each of the angles in question 1 using three letters.

For example
∠GHK   GĤK   angle GHK

Measuring angles

You use a **protractor** to measure angles accurately.

Follow these instructions carefully.

1   Estimate the angle first, so you don't mistake an angle
    of 30°, say, for an angle of 150°.

2   Put the centre point of the protractor exactly on top of
    the angle's point.

3   Place one of the 0° lines of the protractor directly on
    top of one of the angle 'arms'.
    If the line isn't long enough, draw it longer so that it
    reaches beyond the edge of the protractor.

4   Measure from the 0°, following the scale round the
    edge of the protractor.
    If you are measuring from the *left-hand* 0°, use the
    outside scale.
    If you are measuring from the *right-hand* 0°, use the
    inside scale.

5   On the correct scale, read the size of the angle in
    degrees, where the other 'arm' cuts the edge of the
    protractor.

> Use your estimate to help
> you choose the correct
> scale.

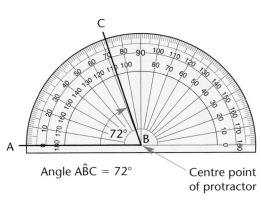

Angle AB̂C = 72°   Centre point
                         of protractor

Measuring from the left-hand 0°.

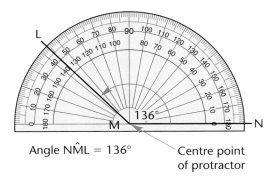

Angle NM̂L = 136°   Centre point
                          of protractor

Measuring from the right-hand 0°.

6   To measure a reflex angle (an angle that is bigger than
    180°), measure the acute or obtuse angle, and subtract
    this value from 360°.

Exercise 6C

1 Measure these angles using a protractor,

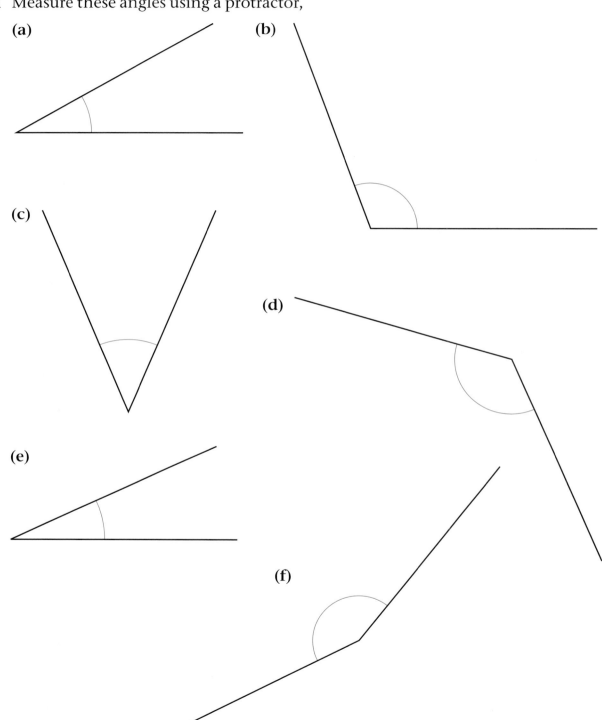

(a)

(b)

(c)

(d)

(e)

(f)

2 Draw these angles using a protractor. Label the angle in
  each case.

   (a) angle PQR = 54°

   (b) angle STU = 148°

   (c) angle MLN = 66°

   (d) angle ZXY = 157°

   (e) angle DFE = 42°

   (f) angle HIJ = 104°

For example

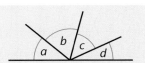

Angle properties

You need to know these angle facts:

Key words:
vertically opposite
perpendicular

Angles on a straight line add up to 180°.

These angles lie on a
straight line,
so $a + b + c + d = 180°$.

Angles around a point add up to 360°.

These angles make a full
turn,
so $p + q + r + s + t = 360°$.

Vertically opposite angles are equal.

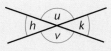

In this diagram $h = k$ and
$u = v$.

Equal angles are shown by
matching arcs:

Perpendicular lines intersect at
90° and are marked with the
right angle symbol.

Example 2

Calculate the size of the angles marked with letters.

(a)

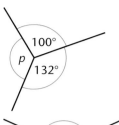

(b)

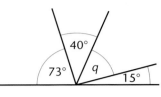

(c)

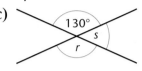

(d)

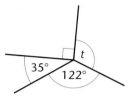

(a) $p + 100° + 132 = 360°$ (angles around a point)

$$p = 360° - 100° - 132°$$
$$p = 128°$$

Solve this equation to find *p*. Use Chapter 5 to help you.

(b) $q + 73° + 40° + 15° = 180°$ (angles on a straight line)

$$q = 180° - 73° - 40° - 15°$$
$$q = 52°$$

(c) $r = 130°$ (vertically opposite)

$s + 130° = 180°$ (angles on a straight line)

$$s = 180° - 130° = 50°$$

(d) $t + 122° + 35° + 90° = 360°$ (angles around a point)

$$t = 360° - 122° - 35° - 90°$$
$$t = 113°$$

⌐ means 90°.

Exercise 6D

Calculate the size of the angles marked with letters.

1

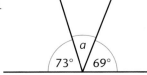

2

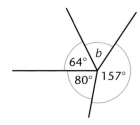

3

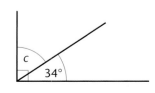

4

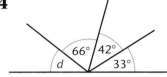

5

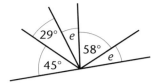

6

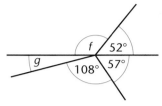

7

8

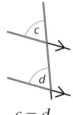

9

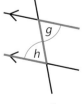

Angles in parallel lines

Key words:
parallel
transversal
corresponding
alternate
co-interior

Parallel lines are the same distance apart all along their length. You can use arrows to show lines are parallel.

A straight line that crosses a pair of parallel lines is called a **transversal.**

A transversal creates pairs of equal angles.

$$a = b \qquad c = d$$

a and b } are **corresponding** angles
c and d

The lines make an F shape.

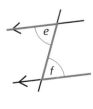

$$e = f \qquad g = h$$

e and f } are **alternate** angles.
g and h

The lines make a Z shape.

In this diagram the two angles are not equal, *j* is obtuse and *k* is acute.

The two angles lie on the *inside* of a pair of parallel lines. They are called **co-interior** angles or allied angles.

Co-interior angles add up to 180°.

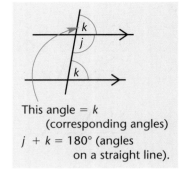

This angle = *k*
    (corresponding angles)
j + *k* = 180° (angles
        on a straight line).

$$j + k = 180°$$

Corresponding angles are equal.
Alternate angles are equal.
Co-interior angles add up to 180°.

Example 3

Calculate the size of the angles marked with letters.

(a)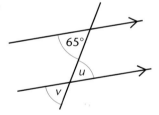

(b)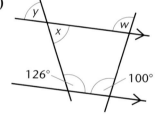

(a) Method A

$u = 65°$ (alternate)

$v = 65°$ (vertically opposite)

Method B

$v = 65°$ (corresponding)

$u = 65°$ (vertically opposite)

You could use method A or method B.

(b) $w = 100°$ (corresponding)

$x + 126° = 180°$ (co-interior angles)

$\qquad x = 180° - 126°$

$\qquad x = 54°$

$y = 54°$ (vertically opposite)

Exercise 6E

Calculate the size of the angles marked with letters.

1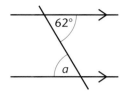

2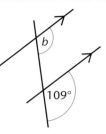

3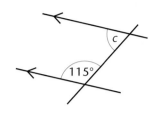

4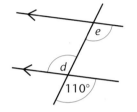

5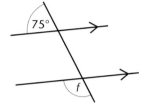

6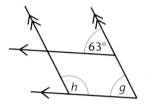

7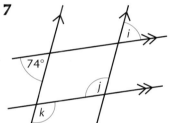

8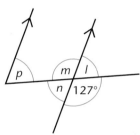

9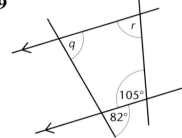

6.2 Three-figure bearings

Key words:
bearing

You can use compass points to describe a direction.
Another method is to use three-figure **bearings.**

> **A three-figure bearing gives a direction in degrees.
> It is an angle between 0° and 360°. It is always
> measured *from the North* in a *clockwise* direction.**

The diagram shows the cities of Manchester and
Sheffield.
The bearing of Sheffield from Manchester is 110°.
The bearing of Manchester from Sheffield is 290°.

A bearing must always be written with *three figures*.

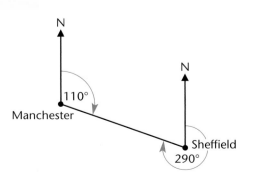

A bearing of 073°

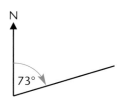

You write 073° because the bearing must have three figures.

A bearing of 254°

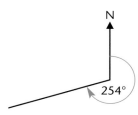

You need to be able to measure bearings accurately using a protractor.

Your answer needs to be within 2° of the correct value.

Example 4

(a) Write down the bearing of Q from P.
(b) Work out the bearing of P from Q.

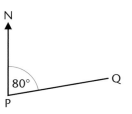

(a) *080°*

(b)

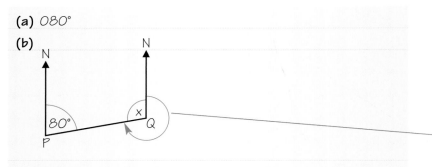

The two North lines are parallel, so
$80° + x = 180°$ (co-interior angles)
$80° - 80° + x = 180° - 80°$
$x = 100°$
Bearing of P from Q $+ x = 360°$ (angles around a point)
Bearing of P from Q $+ 100° = 360°$
Bearing of P from Q $= 260°$.

If you stand at P and face North, you need to turn through 80° to face Q.

Draw in the North line at Q.

This angle is the bearing of P from Q.

Exercise 6F

1 For each diagram:
- write down the bearing of Q from P
- work out the bearing of P from Q

(a)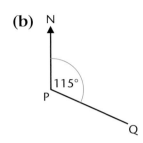

(b)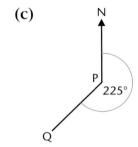

(c)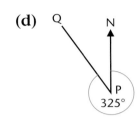

(d)

2 Draw accurate diagrams to show these three-figure bearings.

(a) 036°   **(b)** 145°

(c) 230°   **(d)** 308°

(e) 074°   **(f)** 256°

(g) 348°   **(h)** 115°

3 The diagram shows a triangle *LMP*.

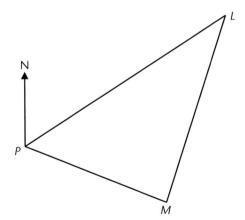

Use a protractor to measure these bearings:

(a) *L* from *P*   **(b)** *M* from *P*

(c) *L* from *M*   **(d)** *P* from *L*

(e) *P* from *M*   **(f)** *M* from *L*

6.3 Triangles

Types of triangle

There are four types of triangle. They can be described by their properties.

Triangle	Picture	Properties
scalene		The three sides are different lengths. The three angles are different sizes.
isosceles	A / B C	Two equal sides. $AB = AC$ Two equal angles ('base angles') angle ABC = angle ACB.
equilateral	60° / 60° / 60°	All three sides are equal in length. All angles 60°.
right-angled	X / Z Y	One of the angles is a right angle (90°). Angle $XZY = 90°$.

The marks ⋋ and ⋏ show that the sides are equal.

Angles in a triangle

1 Draw a triangle on a piece of paper.
2 Mark each angle with a different letter or shade them different colours.
3 Tear off each corner and place them next to each other on a straight line.

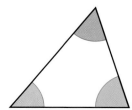
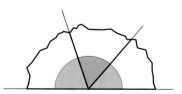

You will see that the three angles fit exactly onto the straight line.

So the three angles add up to 180°.

You can prove this for all triangles using facts about alternate and corresponding angles.

You are not *proving* this result, you are simply showing that it is true for the triangle you drew.

See page 141.

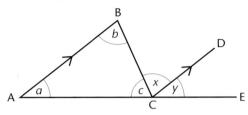

For the triangle ABC:
● Extend side *AC* to point *E*.
● From *C* draw a line CD parallel to *AB*.

Let ∠*BCD* = *x* and ∠*DCE* = *y*.

$$x = b \text{ (alternate angles)}$$
$$y = a \text{ (corresponding angles)}$$
$$c + x + y = 180° \text{ (angles on a straight line}$$
$$\text{at point } C)$$

which means that $c + b + a = 180°$.

> **The sum of the angles of a triangle is 180°.**

This *is* a proof.
Chapter 22 has more on proofs.

Example 5

Calculate the size of the angles marked with letters.

(a)
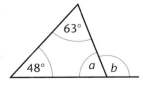

(b)
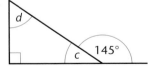

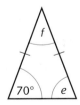

(c)
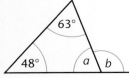

(a)

$$a = 180° - 63° - 48°$$
$$\text{(angle sum of } \triangle)$$
$$a = 69°$$
$$b = 180° - 69°$$
$$\text{(angles on a straight line)}$$
$$b = 111°$$

continued ▼

△ means 'triangle'.

(b)

$c = 180° - 145°$

       (angles on a straight line)

$c = 35°$

$d = 180° = 90° - 35°$

       (angle sum of $\triangle$)

$d = 55°$

(c)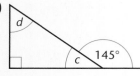

$e = 70°$ (base angles isosceles $\triangle$)

$f = 180° - 70° - 70°$

       (angle sum of $\triangle$)

$f = 40°$

The triangle has two equal sides. It is isosceles.

Exercise 6G

Calculate the size of the angles marked with letters.

1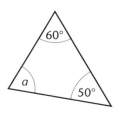

2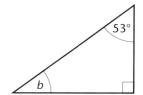

3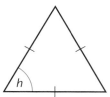

4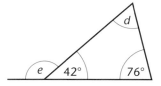

5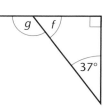

6

7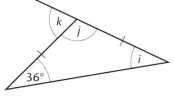

8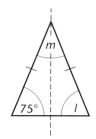

9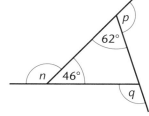

Interior and exterior angles in a triangle

The angles inside a triangle are called **interior** angles.

Key words:
interior
exterior

An **exterior** angle is formed by extending one of the sides of the triangle.

Angle *BCE* in this diagram is an exterior angle.

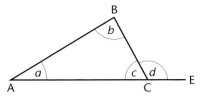

Let *BCE* = *d*

then $c + d = 180°$ (angles on a straight line)

but $c + b + a = 180°$ (angle sum of $\triangle ABC$)

which means that $d = a + b$

> **In a triangle, the exterior angle is equal to the sum of the two opposite interior angles.**

This is a *proof*.

Sometimes you need to calculate the size of 'missing' angles before you can calculate the ones you want. You can label the extra angles with letters.

This helps make your explanations clear.

Example 6

Calculate angles *g* and *h*.

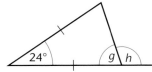

$g + x + 24° = 180°$ (angle sum of $\triangle$)

$g + x = 180° - 24°$

$g + x = 156°$

   $g = x$ (base angles isosceles $\triangle$)

so both *g* and $x = \frac{1}{2}$ of 156°   (or 156° ÷ 2)

$g = 78°$

$h = 180° - 78°$ (angles on a straight line)

$h = 102°$

Let the third angle of the triangle = *x*.

156 ÷ 2 = 78.

Example 7

Calculate angle *i*.

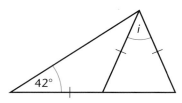

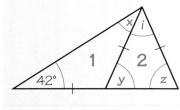

$x = 42°$

    (base angles isosceles △ 1)

$y = 42° + 42°$

    (exterior angle of △ 1)

$y = 84°$

$z = 84°$

    (base angles isosceles △ 2)

$i = 180° - 84° - 84°$

    (angle sum △ 2)

$i = 12°$

> Label the 'missing' angles *x*, *y* and *z*.

> Label the two triangles 1 and 2.

Example 8

Calculate angle *j*.

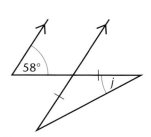

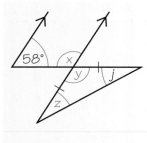

$x = 180° - 58°$

    (co-interior angles, parallel lines)

$x = 122°$

$y = 122°$ (vertically opposite)

$z + j = 180° - 122°$ (angle sum of △)

$z + j = 58°$

$z = j$ (base angles isosceles △)

so both *z* and $j = \frac{1}{2}$ of 58° (or 58° ÷ 2)

$j = 29°$

> Label the 'missing' angles *x*, *y* and *z*.

> 58 ÷ 2 = 29.

Exercise 6H

Calculate the size of the angles marked with letters.

Copy the diagram and label any 'missing' angles.

1

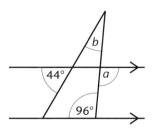

2

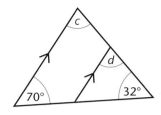

3

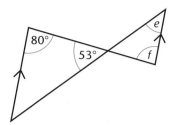

4

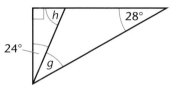

5

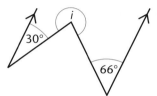

6

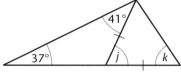

7

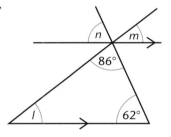

8

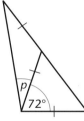

6.4 Quadrilaterals and other polygons

Quadrilaterals

Key words:
quadrilateral
bisect
diagonal
adjacent

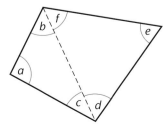

A **quadrilateral** is a 2-dimensional 4-sided shape.

This quadrilateral can be divided into 2 triangles:

You can divide *any* quadrilateral into 2 triangles.

In each triangle, the angles add up to 180°,
so $\quad a + b + c = 180°$
and $\quad d + e + f = 180°$

The 6 angles from the 2 triangles add up to 360°.
$a + b + c + d + e + f = 360°$

a, b + f, e, c + d are the angles of the quadrilateral.

In any quadrilateral the sum of the interior angles is 360°.

Once again, we have proved the result.

Some quadrilaterals have special names and properties.

Quadrilateral	Picture	Properties
square		Four equal sides. All angles 90°. Diagonals **bisect** each other at 90°.
rectangle		Two pairs of equal sides. All angles 90°. **Diagonals** bisect each other.
parallelogram		Two pairs of equal and parallel sides. Opposite angles equal. Diagonals bisect each other.
rhombus		Four equal sides. Two pairs of parallel sides. Opposite angles equal. Diagonals bisect each other at 90°.
trapezium		One pair of parallel sides.

Bisect means 'cut in half'.

Lines with equal arrows are parallel to each other.

Quadrilateral	Picture	Properties
kite		Two pairs of **adjacent** sides equal. One pair of opposite angles equal. One diagonal bisects the other at 90°.

Adjacent means 'next to'.

Exercise 6I

1 Write down the name of each quadrilateral:

(a)  (b) (c)

(d)   (e)  (f)

2 Write down the names of all the quadrilaterals with these properties:

(a) diagonals which cross at 90°

(b) all sides are equal in length

(c) only one pair of parallel sides

(d) two pairs of equal angles but not all angles equal

(e) all angles are equal

(f) two pairs of opposite sides are parallel

(g) only one diagonal bisected by the other diagonal

(h) two pairs of equal sides but not all sides equal

(i) the diagonals bisect each other

(j) at least one pair of opposite sides are parallel

(k) diagonals equal in length

(l) at least two pairs of adjacent sides equal.

Polygons

A **polygon** is a 2-dimensional shape with many sides and angles.

Here are some of the most common ones:

Key words:
polygon
regular
vertex
vertices

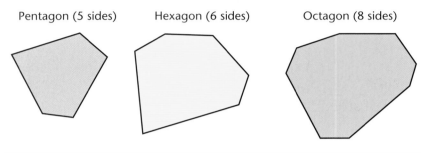

Pentagon (5 sides)   Hexagon (6 sides)   Octagon (8 sides)

Polygon means 'many angled'.

A polygon with all of its sides the same length and all of its angles equal is called regular .

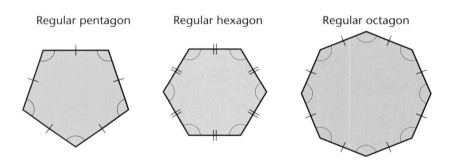

Regular pentagon   Regular hexagon   Regular octagon

The sum of the exterior angles of any polygon is 360°.

You can explain this result like this.

If you 'walk round' the sides of this hexagon until you get back to where you started, you complete a full turn or 360°.

The exterior angles represent your 'turn' at each corner, so they must add up to 360°.

On a regular hexagon, all 6 exterior angles are the same size.

For a regular hexagon:

$$\text{exterior angle} = \frac{360°}{6}$$

$$= 60°$$

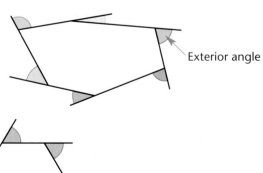

Exterior angle

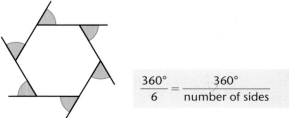

$$\frac{360°}{6} = \frac{360°}{\text{number of sides}}$$

For a regular octagon

$$\text{exterior angle} = \frac{360°}{8} = 45°$$

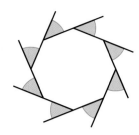

A regular octagon has 8 equal exterior angles.

For a regular polygon:

$$\text{exterior angle} = \frac{360°}{\text{number of sides}}$$

You can use this equation to work out the number of sides in a regular polygon.

For a regular polygon:

$$\text{number of sides} = \frac{360°}{\text{exterior angle}}$$

In a regular polygon with exterior angles of 18°:

$$\text{number of sides} = \frac{360°}{18°} = 20$$

Once you know the exterior angle, you can calculate the interior angle.

At each **vertex,** the exterior and interior angles lie next to each other (adjacent) on a straight line.

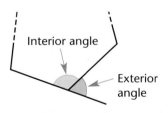

Interior angle

Exterior angle

In a polygon, each pair of interior and exterior angles adds up to 180°.

So, for any polygon

$$\text{interior angle} = 180° - \text{exterior angle}$$

In a regular polygon all the exterior angles are equal. So all the interior angles are equal.

For example:
- regular hexagon   interior angle = 180° − 60° = 120°
- regular octagon   interior angle = 180° − 45° = 135°

When a polygon is *not* regular the interior angles could all be different sizes. You can find the *sum* of the interior angles.

These diagrams show how you can divide any polygon into triangles.

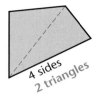
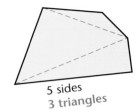
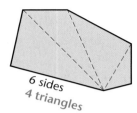

4 sides
2 triangles

5 sides
3 triangles

6 sides
4 triangles

Draw dotted lines from the **vertices** .

You can see that the number of triangles is always 2 less than the number of sides of the polygon.

The sum of the interior angles for each polygon is given by the formula

$$(n - 2) \times 180°$$ **where n is the number of sides of the polygon.**

You should learn this formula.

Name of polygon	Number of sides	Number of triangles	Sum of interior angles
Triangle	3	1	$1 \times 180° = 180°$
Quadrilateral	4	2	$2 \times 180° = 360°$
Pentagon	5	3	$3 \times 180° = 540°$
Hexagon	6	4	$4 \times 180° = 720°$
Heptagon	7	5	$5 \times 180° = 900°$
Octagon	8	6	$6 \times 180° = 1080°$
Nonagon	9	7	$7 \times 180° = 1260°$
Decagon	10	8	$8 \times 180° = 1440°$

Example 9

Calculate the size of the angles marked with letters in each of these diagrams.

(a)

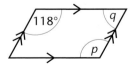

118°
q
p

$p = 118°$ (opposite angles of
      parallelogram are equal)
$q = 62°$ (p and q are co-interior
      angles, so $p + q = 180°$)

You could use the fact that the sum of the angles of a quadrilateral is 360° to calculate q.

continued ▼

(b)

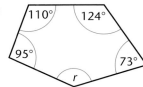

$r = 540° -$ (the sum of the other 4 angles)

$r = 540 - (95° + 110° + 124° + 73°)$

$r = 138°$

Sum of the interior angles of a pentagon (5 sides) = $3 × 180° = 540°$.

(c)

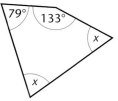

$x + x + 133° + 79° = 360°$

$2x = 360° - 133° - 79°$

$2x = 148°$

$x = 74°$

Sum of interior angles of quadrilateral = $360°$.

Example 10

(a) Calculate the size of the exterior and interior angles of a regular polygon with 20 sides.

(b) How many sides has a regular polygon with interior angle 168°?

(a)

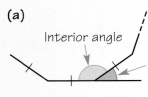

Interior angle

Exterior angle

Exterior angle $= \dfrac{360°}{20} = 18°$

Interior angle $= 180° -$ exterior angle

$= 180 - 18°$

$= 162°$

Exterior angle of regular polygon $= \dfrac{360°}{\text{Number of sides}}$

(b) Exterior angle $= 180° -$ interior angle

$= 180° - 168°$

$= 12°$

Number of sides $= \dfrac{360}{12°} = 30$

Exercise 6J

Calculate the size of the angles marked with letters.

1

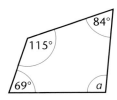

2

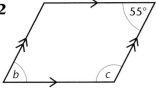

3

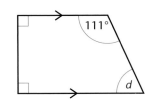

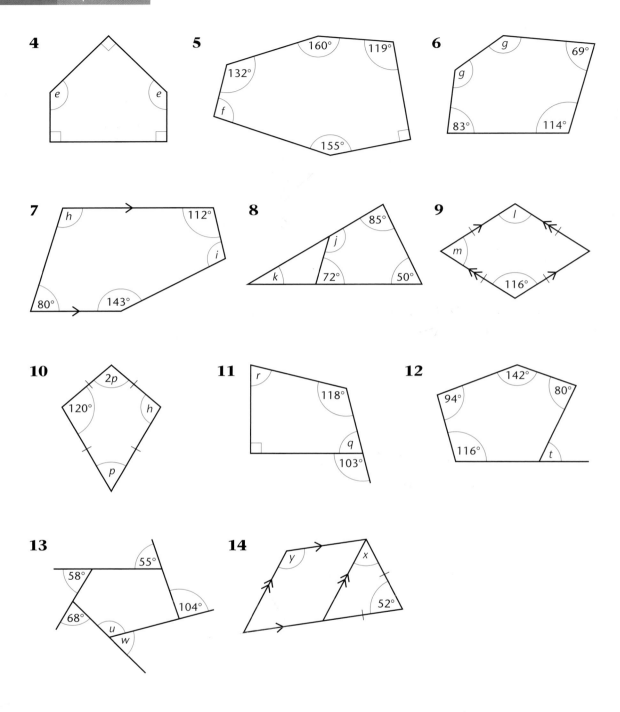

4

e   e

5

160° 119°
132°
f
155°

6

g
g   69°
83°   114°

7

h   112°
i
80°   143°

8

85°
j
k   72°   50°

9

l
m
116°

10

2p
120°   h
p

11

r
118°
q
103°

12

142°
94°   80°
116°   t

13

55°
58°
104°
68°
u
w

14

y   x
52°

15 Calculate the interior angle of a regular polygon with
18 sides.

16 Can a regular polygon have an interior angle of 130°?
Explain your answer.

6.5 Symmetry

Line symmetry

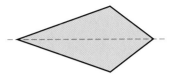

A kite is **symmetrical.**

If you fold it along the dashed line, one half fits exactly onto the other.

The dashed line is called a **line of symmetry**.

> **A line of symmetry divides a shape into two halves. One half is the mirror image of the other.**

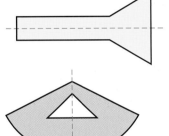

Some shapes have more than one line of symmetry.

Rectangle ... 2 lines of symmetry

You draw lines of symmetry with dashed lines.

Equilateral triangle ... 3 lines of symmetry

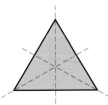

Regular hexagon ... 6 lines of symmetry

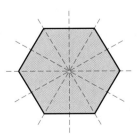

Some shapes have no lines of symmetry.

A parallelogram has no lines of symmetry.

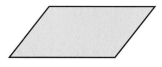 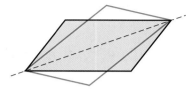

The dashed line is *not* a line of symmetry. If you reflect the parallelogram in it you get the red parallelogram.

These shapes have no lines of symmetry.

 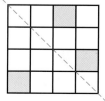 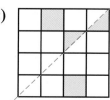

Exercise 6K

Copy these shapes and draw in all the lines of symmetry (if any).

Draw dashed lines for lines of symmetry.

1 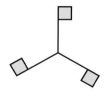   **2** 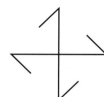   **3**

4   **5**   **6**

7 Copy and complete these grids so that they are symmetrical about the dashed line.

   (a) 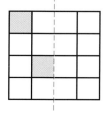   **(b)**   **(c)**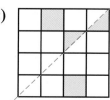

Rotational symmetry

Key words:
rotational symmetry
order

Look again at the shapes before Exercise 6K.
They have no lines of symmetry but they have
rotational symmetry.

You can turn them and they will fit exactly into their
original shape again.

The **order** of rotational symmetry is the number of
times a shape looks the same during one full turn.
This can be any number greater than or equal to 2.

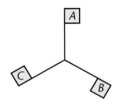

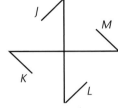

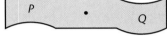

If *A* turns to
any of the
positions *A*, *B*
or *C* the
shape will
look exactly
the same.

Rotational
symmetry of
order 3.

If *J* turns to *J*, *K*,
L or *M* the
shape will look
exactly the
same.

Rotational
symmetry of
order 4.

If *P* turns to positions
P or *Q* the shape will
look exactly the same.

Rotational symmetry
of order 2.

Some shapes have line symmetry *and* rotational
symmetry.

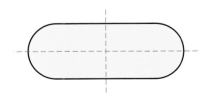

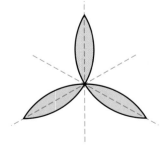

2 lines of symmetry.

Rotational symmetry of
order 2.

3 lines of symmetry.

Rotational symmetry of
order 3.

Exercise 6L

1 For each of the shapes in Exercise 6K, questions 1 to 6, state the order of rotational symmetry.

2 Draw shapes with these orders of rotational symmetry.
  (a) 2   **(b)** 3   **(c)** 4   **(d)** 5   **(e)** 6

3 Copy and complete these grids so that they have rotational symmetry of order **(i)** 2 and **(ii)** 4.

(a)   **(b)**

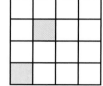

Examination style questions

1 (a) (i) Write down the order of rotational symmetry of this shape.

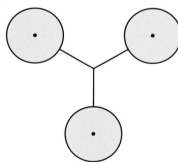

  (ii) Copy the shape and draw on all the lines of symmetry.
(b) Draw a quadrilateral with one line of symmetry but no rotational symmetry.

(5 marks)

2 *ABCD* is a quadrilateral.

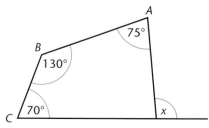

Work out the size of the exterior angle at *D*, marked *x* on the diagram.

(3 marks)

3 (a) The diagram shows four angles
meeting at a point.

Work out the value of *a*.

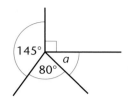

(b)

Work out the values of *b* and *c*.   *(4 marks)*

4 (a) The diagrams show the diagonals of two different
quadrilaterals.

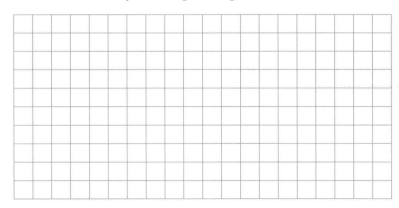

Write down the names of these quadrilaterals.

(b) (i) Use a grid like the one below to draw a
quadrilateral that has only one pair of parallel
lines and exactly two right angles.

(ii) Write down the name of this quadrilateral.   *(4 marks)*

AQA, Spec A, I, November 2004

5 *ABCDEF* is a hexagon.

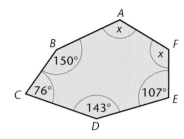

Work out the value of *x*. (3 marks)

6 The diagram shows part of a regular polygon.
Each interior angle is 150°.

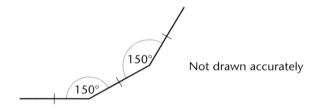

Not drawn accurately

Calculate the number of sides of the polygon. (3 marks)

7 The diagram shows a regular octagon.

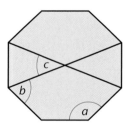

Calculate the size of angles *a*, *b* and *c*. (3 marks)

Summary of key points

Angle properties (grades F to D)

Angles on a straight line add up to 180°.

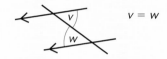

$$a + b + c = 180°$$

Alternate angles are equal.

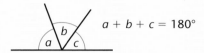

$$v = w$$

Vertically opposite angles are equal.

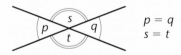

$$p = q$$
$$s = t$$

Angles around a point add up to 360°.

$$d + e + f + g = 360°$$

Corresponding angles are equal.

$$j = k$$

Co-interior angles add up to 180°.

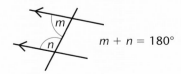

$$m + n = 180°$$

Three-figure bearings (grade E/D)

A three-figure bearing gives a direction in degrees.
It is an angle between 0° and 360°. It is always measured *from the North* in a *clockwise* direction.

This diagram shows that the bearing of the ship from the submarine is 248°.

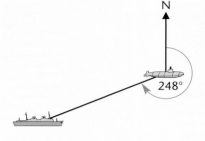

Triangle properties (grade E)

The sum of the angles of a triangle is 180°.

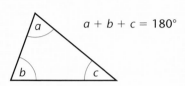

$a + b + c = 180°$

The exterior angle is equal to the sum of the two opposite interior angles.

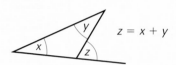

$z = x + y$

Quadrilateral and polygon properties (grades E to C)

The sum of the interior angles of a quadrilateral is 360°.

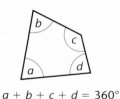

$a + b + c + d = 360°$

A polygon with all its sides the same length and all of its angles equal is called *regular*.

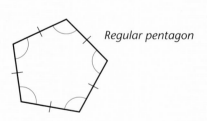

Regular pentagon

The sum of the exterior angles of any polygon is 360°.

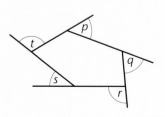

$p + q + r + s + t = 360°$

In a *regular* polygon

$$\text{exterior angle} = \frac{360°}{\text{number of sides}}$$

For example, for a regular polygon with 10 sides,

$$\text{exterior angle} = \frac{360°}{10} = 36°$$

In a *regular* polygon

$$\text{number of sides} = \frac{360°}{\text{exterior angle}}$$

For example, for a regular polygon with exterior angles of 18°

$$\text{number of sides} = \frac{360°}{18°} = 20$$

In any polygon, each pair of interior and exterior angles adds up to 180°.
The sum of the interior angles of a polygon with n sides is given by the formula $(n - 2) \times 180°$.

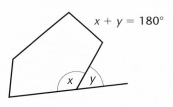

$x + y = 180°$

For example, this is a pentagon, so $n = 5$.
Sum of interior angles = $3 \times 180° = 540°$
So $a + b + c + d + e = 540°$

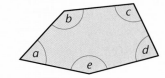

Symmetry (grades G to E)

A line of symmetry divides a shape into two halves.
One half is the mirror image of the other.

For example, this shape has 2 lines of symmetry.

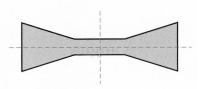

A shape has rotational symmetry if it can be rotated to fit exactly into its original shape.

This shape has rotational symmetry of order 3.

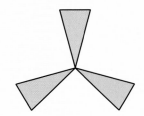

This chapter will show you how to:
- ✔ use index (power) notation and the index laws
- ✔ use formulae given in words or letters and symbols
- ✔ substitute numbers into formulae
- ✔ change the subject of a formula
- ✔ find approximate solutions to equations using trial and improvement

7.1 Indices

You will need to know:
- how to multiply positive and negative numbers
- how to use the power key on your calculator
- the correct order of operations

Using index notation to simplify expressions

6^2 (6 to the power 2) is called '6 **squared** '. $\quad 6^2 = 6 \times 6$

8^3 (8 to the power 3) is called '8 **cubed** '. $\quad 8^3 = 8 \times 8 \times 8$

You can write $5 \times 5 \times 5 \times 5$ as 5^4.

You say '5 raised to the **power** of 4', or '5 to the power 4'. The ^{4} is called the **index** .

You can also use powers (or **indices**) in algebraic expressions where there are unknown or **variable** values.

You can write:
- $d \times d \times d$ as d^3 ('d cubed', or 'd to the power 3').
- $y \times y \times y \times y \times y$ as y^5.

> **Key words:**
> squared cubed
> power index
> indices variable

A *variable* is a letter which stands for an unknown quantity.

d^3 and y^5 are written in index notation.

Example 1
Write these expressions using index notation.
(a) $d \times d \times d \times d \times d \times d$ **(b)** $2 \times x \times x \times x \times x$

(a) $d \times d \times d \times d \times d \times d = d^6$

(b) $2 \times x \times x \times x \times x = 2x^4$

It is all multiplied by 2 as well.
Write the 2 at the front of the expression.

Example 2

Simplify these expressions:

(a) $3 \times a \times a \times 5 \times a \times a \times a$

(b) $4t \times t \times 2t$

(c) $3x \times y \times x \times 2y \times y$

You can multiply the values in any order.

Multiply numbers first, then letters.

(a) $3 \times a \times a \times 5 \times a \times a \times a = 3 \times 5 \times a \times a \times a \times a \times a$

$$= 15 \times a^5$$

$$= 15a^5$$

$3 \times 5 = 15$ and
$a \times a \times a \times a \times a = a^5$

(b) $4t \times t \times 2t = 4 \times t \times t \times 2 \times t$

$$= 4 \times 2 \times t \times t \times t$$

$$= 8t^3$$

$4t$ means $4 \times t$.

Multiply numbers first, then letters.

(c) $3x \times y \times x \times 2y \times y = 3 \times x \times y \times x \times 2 \times y \times y$

$$= 3 \times 2 \times x \times x \times y \times y \times y$$

$$= 6 \times x^2 \times y^3$$

$$= 6x^2y^3$$

Multiply numbers, then the xs then the ys. You usually write letters in alphabetical order.

Exercise 7A

1 Write these expressions using index notation:

(a) $d \times d \times d \times d$

(b) $a \times a \times a \times a \times a \times a$

(c) $x \times x \times x \times x \times x$

(d) $y \times y \times y \times y \times y \times y \times y \times y$

(e) $b \times b$

(f) $m \times m \times m \times m$

(g) $p \times p \times p$

(h) $r \times r \times r \times r \times r \times r$

2 Simplify these expressions:

(a) $5 \times p \times p \times p \times p$

(b) $a \times 4 \times a \times a$

(c) $2 \times x \times 4 \times x \times x \times x$

(d) $f \times f \times 3 \times f \times 5 \times f \times f$

(e) $b \times 4 \times 2 \times b \times 1 \times b$

(f) $h \times h \times 6 \times h \times 5 \times 2 \times h \times h$

(g) $3 \times z \times 3 \times z \times 3 \times z$

(h) $5 \times r \times r \times 5 \times r \times r$

Use index notation.

3 Simplify these using index notation:
   (a) $b \times b \times c \times c \times c$
   (b) $f \times g \times f \times f \times g \times f$
   (c) $3 \times x \times x \times x \times 4 \times y$
   (d) $5 \times d \times d \times 2 \times e \times e \times d \times d$
   (e) $4 \times t \times 2 \times s \times s \times t \times s$
   (f) $k \times k \times 3 \times j \times 5 \times k \times 2 \times j$
   (g) $2 \times p \times p \times 2 \times q \times q \times q$
   (h) $4 \times a \times a \times 4 \times b \times b$

4 Simplify these algebraic expressions using index notation:
   (a) $2x \times 3x$ (b) $4y \times y \times 3y$
   (c) $a \times 3a \times 2a \times a$ (d) $3b \times 2b \times 4b$
   (e) $5n \times n \times n \times 4n \times 2n$ (f) $7k \times 3k \times 3k \times 2k$
   (g) $3x \times 3x \times 3x$ (h) $6z \times z \times z \times 6z \times z \times z$

5 Simplify these algebraic expressions:
   (a) $2a \times a \times 3b$ (b) $p \times 2q \times q \times 4p \times p$
   (c) $3x \times 2y \times 3y$ (d) $5r \times 2s \times 3s \times 2r$
   (e) $2d \times c \times 3d \times 7c$ (f) $5y \times 4 \times x \times 3x \times 2x$
   (g) $4p \times r \times 4p \times r$ (h) $7r \times r \times s \times 7r \times r \times s$

6 Match each question (1 to 10) with an answer (a to j).

Questions	Answers
(1) $y \times y^2$	(a) y^4
(2) $y \times 5y$	(b) y^6
(3) $2y \times 3y$	(c) $5y^2$
(4) $y^2 \times y^2$	(d) $2y^3$
(5) $2y \times 4y$	(e) $4y^4$
(6) $4y \times y^3$	(f) $8y^3$
(7) $y^3 \times y^2$	(g) y^3
(8) $2y^2 \times y$	(h) $8y^2$
(9) $y \times y^2 \times y^3$	(i) y^5
(10) $2y \times 2y \times 2y$	(j) $6y^2$

Multiplying and dividing expressions involving indices

You will need to know:
- **how to simplify (cancel down) fractions**

You can multiply different powers of the *same* variable.

Example 3

Simplify

(a) $r^3 \times r^2$ **(b)** $4h^2 \times 5h^4$

(a) $r^3 \times r^2 = r \times r \times r \times r \times r$ $r^3 = r \times r \times r$ and $r^2 = r \times r$

           $= r^5$

(b) $4h^2 \times 5h^4 = 4 \times h \times h \times 5 \times h \times h \times h \times h$ Numbers first, then letters.

              $= 4 \times 5 \times h \times h \times h \times h \times h \times h$

              $= 20 \times h^6$

              $= 20h^6$

You can divide different powers of the *same* variable.

Example 4

Simplify:

(a) $a^5 \div a^2$ **(b)** $c^3 \div c^4$ **(c)** $10b^2 \div 5b^4$

In algebra you write
$x \div y$ as $\dfrac{x}{y}$

(a) $a^5 \div a^2 = \dfrac{a \times a \times a \times a \times a}{a \times a}$

The top and bottom of the fraction are multiplications. You can cancel any terms which are on both the top and bottom.

          $= \dfrac{a \times a \times a \times {}^1\!\!\not{a} \times {}^1\!\!\not{a}}{{}^1\!\!\not{a} \times {}^1\!\!\not{a}}$

          $= a \times a \times a$

          $= a^3$

$a \div a = 1$

(b) $c^3 \div c^4 = \dfrac{c \times c \times c}{c \times c \times c \times c}$

          $= \dfrac{{}^1\!\!\not{c} \times {}^1\!\!\not{c} \times {}^1\!\!\not{c}}{c \times {}^1\!\!\not{c} \times {}^1\!\!\not{c} \times {}^1\!\!\not{c}}$

The numerator is now $1 \times 1 \times 1 = 1$.

          $= \dfrac{1}{c}$

continued ▼

(c) $10b^2 \div 5b^4 = \dfrac{10 \times b \times b}{5 \times b \times b \times b \times b}$

$= \dfrac{10 \times {}^1\!\!\not b \times {}^1\!\!\not b}{5 \times b \times b \times {}^1\!\!\not b \times {}^1\!\!\not b}$ Cancel the b terms.

$= \dfrac{10}{5 \times b \times b}$

$= \dfrac{10}{5b^2}$ Divide the 10 by 5.

$= \dfrac{2}{b^2}$

Exercise 7B

1 Simplify:
(a) $a^3 \times a^2$ **(b)** $b^4 \times b^3$ **(c)** $c^2 \times c^4$ **(d)** $d^5 \times d^5$

2 Simplify:
(a) $x^3 \times x$ **(b)** $y^2 \times y$ **(c)** $z \times z^4$ **(d)** $w \times w^3$

3 Simplify each of these: Use Example 3 to help.
(a) $2a^3 \times 3a^4$ **(b)** $5b^4 \times 3b^5$
(c) $4c^2 \times c^6$ **(d)** $3d^4 \times 3d^4$

4 Can you see a quicker way to answer questions 1 to 3? Describe it, using an example to help.

5 Simplify:
(a) $p^5 \div p^2$ **(b)** $q^6 \div q^2$ **(c)** $r^4 \div r^3$ **(d)** $s^6 \div s^3$

6 Simplify each of these: Use Example 4(b) to help.
(a) $e^2 \div e^5$ **(b)** $f^2 \div f^6$ **(c)** $g^3 \div g^4$ **(d)** $h \div h^6$

7 Simplify:
(a) $5x^5 \div x^2$ **(b)** $8y^6 \div 4y^2$
(c) $12r^4 \div 3r^3$ **(d)** $2s^2 \div 10s^5$

8 Can you see a quicker way to answer questions 5 to 7? Describe it, using an example to help.

Laws of indices

Any value raised to the power 1 is equal to the value itself.

For example: $3^1 = 3,$ $x^1 = x.$

You may have noticed some of the rules that make it easier to simplify expressions involving indices.

To multiply powers of the *same* number or variable, add the indices.

$$4^3 \times 4^2 = 4^{3+2}$$
$$= 4^5$$
$$x^3 \times x^2 = x^{3+2}$$
$$= x^5$$

In general: $x^n \times x^m = x^{n+m}$

$$4^3 \times 4^2 = 4 \times 4 \times 4 \times 4 \times 4$$
$$= 4^{3+2}$$
$$= 4^5$$

To divide powers of the *same* number or variable, subtract the indices.

$$6^5 \div 6^3 = 6^{5-3}$$
$$= 6^2$$
$$t^5 \div t^3 = t^{5-3}$$
$$= t^2$$

In general: $x^n \div x^m = x^{n-m}$

$$6^5 \div 6^3$$
$$= \frac{6 \times 6 \times \cancel{6} \times \cancel{6} \times \cancel{6}}{\cancel{6} \times \cancel{6} \times \cancel{6}}$$
$$= 6^{5-3}$$
$$= 6^2$$

Example 5

Simplify: **(a)** $d^2 \times d^3$ **(b)** $3a^4 \times 5a^3$ **(c)** $g^3 \times g \times g^2$

(a) $d^2 \times d^3 = d^{2+3}$
$$= d^5$$

(b) $3a^4 \times 5a^3 = 3 \times a^4 \times 5 \times a^3$
$$= 3 \times 5 \times a^4 \times a^3$$
$$= 15 \times a^{4+3}$$
$$= 15a^7$$

Multiply numbers first, then letters.

You can write the answer down without the lines of working.

(c) $g^3 \times g \times g^2 = g^3 \times g^1 \times g^2$
$$= g^{3+1+2}$$
$$= g^6$$

$g = g^1$
You do not usually write the index 1.

Example 6

Simplify:

(a) $y^6 \div y^2$ **(b)** $x^3 \div x^5$ **(c)** $12b^5 \div 4b^3$ **(d)** $\dfrac{t^4 \times t}{t^3}$

(a) $y^6 \div y^2 = y^{6-2}$

$\qquad\qquad = y^4$

(b) $x^3 \div x^5 = x^{3-5}$

$\qquad\qquad = x^{-2}$

(c) $12b^5 \div 4b^3 = \dfrac{12 \times b^5}{4 \times b^3}$

$\qquad\qquad\quad = \dfrac{12}{4} \times \dfrac{b^5}{b^3}$

$\qquad\qquad\quad = 3 \times b^{5-3}$

$\qquad\qquad\quad = 3 \times b^2$

$\qquad\qquad\quad = 3b^2$

(d) $\dfrac{t^4 \times t}{t^3} = \dfrac{t^4 \times t^1}{t^3}$

$\qquad\quad = \dfrac{t^5}{t^3}$

$\qquad\quad = t^{5-3}$

$\qquad\quad = t^2$

> You can have negative powers as well as positive ones.

> Numbers first, then letters.

> $12 \div 4 = 3$
> $b^5 \div b^3 = b^2$

> Simplify the top first,
> $t^4 \times t^1 = t^5$.

> Then the division,
> $t^5 \div t^3 = t^2$.

Exercise 7C

Simplify each of the following:

1 **(a)** $m^3 \times m^2$ **(b)** $a^4 \times a^3$ **(c)** $n^2 \times n^4$
   (d) $u^5 \times u^5$ **(e)** $t^3 \times t^6$

2 **(a)** $d^3 \times d$ **(b)** $a^2 \times a$ **(c)** $r \times r^4$
   (d) $e \times e^3$ **(e)** $t \times t^6$

3 **(a)** $2h^3 \times 3h^4$ **(b)** $5e^4 \times 3e^5$ **(c)** $4g^2 \times g^6$
   (d) $3r \times 3r^4$ **(e)** $6e^3 \times 4e$

4 **(a)** $a^5 \div a^2$ **(b)** $t^6 \div t^2$ **(c)** $e^4 \div e^3$
   (d) $s^6 \div s^3$ **(e)** $t^2 \div t^2$

5 **(a)** $e^2 \div e^5$ **(b)** $f^2 \div f^6$ **(c)** $g^3 \div g^4$
   (d) $h \div h^4$ **(e)** $w \div w^6$

6 (a) $5x^5 \div x^2$ **(b)** $6y^6 \div 2y^2$ **(c)** $12r^4 \div 4r^3$

   (d) $8t^4 \div 4t$ **(e)** $12s^2 \div 6s^5$

7 (a) $h^3 \times h^4 \times h^2$ **(b)** $e^4 \times e \times e^3$ **(c)** $4c^2 \times c^6 \times 6c$

8 (a) $\dfrac{d^4 \times d}{d^3}$ **(b)** $\dfrac{a^2 \times 3a^2}{a}$

   (c) $\dfrac{2r \times 3r^3}{r^2}$ **(d)** $\dfrac{4 \times e^3 \times 5e^2}{10e^4}$

   (e) $\dfrac{b^2 \times 4b}{b^5}$ **(f)** $\dfrac{2x^3 \times 6x^4}{x \times 4x^2}$

7.2 Using and writing formulae

You will need to know:

- how to use the correct order of operations

Using formulae given in words

Lisa has a job. It pays £6 an hour.
How much is Lisa's pay if she works for 20 hours?

You can write a **formula** to find her pay for any number of hours she works:

    pay = number of hours worked × rate of pay

You can use the formula to answer the question:

For 20 hours work   pay = 20 × £6
                   = £120

> **A formula is a general rule that shows the relationship between quantities.**

The quantities in a formula can vary in size.

These quantities are called **variables** .

You can use the formula to find Lisa's pay for any number of hours worked and any rate of pay.

Key words:
formula
formulae
variable

Formulae is the plural of formula.

Lisa's rate of pay is £6 an hour.

The number of hours Lisa works is a variable – it can change from day to day. Her total pay and her rate of pay are both variables as well.

Example 7

To work out his pay Reece uses the formula:

pay = hours worked × rate of pay + bonus

What is his pay for 10 hours' work at a rate of £5.50 an hour, with a bonus of £7?

Pay = hours worked × rate of pay + bonus

= 10 × £5.50 + £7

= £55 + £7

= £62

Remember the order of operations: you do the multiplication before the addition.

 ## Exercise 7D

1 To work out his pay Amit uses the formula:

pay = hours worked × rate of pay

(a) Work out his pay for 10 hours' work at £6 an hour.

(b) Work out his pay for 20 hours' work at £5.50 an hour.

2 Emma uses this formula to work out the cost of stamps:

cost = number of stamps × cost of one stamp

(a) Work out the cost of 20 stamps at 30p each.

(b) Work out the cost of 15 stamps at 35p each.

(c) Lee spends £6 on 25p stamps.
How many stamps does Lee buy?

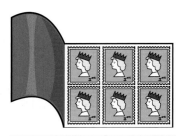

Use your equation solving skills (Chapter 5) to help.

3 To work out her pay April uses the formula:

pay = hours worked × rate of pay + bonus

(a) What is her pay when she works 10 hours at a rate of £6.50 an hour, with a bonus of £7.50.

Use Example 7 to help you.

(b) What is her pay when she works 15 hours at a rate of £6 an hour, with a bonus of £8?

4 Use this formula:

$$\text{Average speed} = \frac{\text{distance travelled}}{\text{time taken}}$$

to work out the average speed of these journeys:

(a) 100 miles from Leicester to Manchester in 2 hours.

(b) 180 miles by train in 3 hours.

(c) A sponsored walk of 12 miles that takes 4 hours.

(d) A marathon runner who takes 4 hours to run 26 miles.

Your answers will be in mph (miles per hour).

Writing formulae using letters and symbols

> **You can use letters for the variables in a formula.**

For example, for

pay = hours worked $\times$ rate of pay

you could write:

$p = hr$

$hr = h \times r$

where

p is the pay,
h is the number of hours worked,
r is the rate of pay

You must say what each letter stands for.

You can put in numbers for the hours worked and the rate of pay, to calculate the pay.

Example 8

The formula for the area of a rectangle is $A = lw$, where l is the length and w is the width.

Work out the value of A when $l = 8$ and $w = 4$.

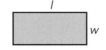

$A = lw$
$\quad = l \times w$
$\quad = 8 \times 4$
$\quad = 32$

Example 9

The perimeter of a rectangle is given by $P = 2l + 2w$, where l is the length and w is the width.

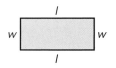

The perimeter is the distance around the outside.

Work out the perimeter when $l = 9$ and $w = 6$.

$P = 2l + 2w$

$\quad = 2 \times 9 + 2 \times 6$

$\quad = 18 + 12$

$\quad = 30$

Order of operations: you do Multiplication before Addition.

Exercise 7E

1 The formula for the area of a rectangle is $A = lw$, where l is the length and w is the width.
Work out the value of A when:

(a) $l = 8$ and $w = 5$ (b) $l = 7$ and $w = 6$

(c) $l = 10$ and $w = 6$ (d) $l = 9$ and $w = 7$

2 The formula for the voltage, V, in an electrical circuit is $V = IR$, where I is the current and R is the resistance.
Work out the voltage when:

(a) $I = 2$ and $R = 6$ (b) $I = 3$ and $R = 8$

(c) $I = 1.5$ and $R = 9$ (d) $I = 4$ and $R = 60$

3 The perimeter of a square is given by $P = 4l$, where l is the length of the square.
Work out the value of P when:

(a) $l = 12$ (b) $l = 25$ (c) $l = 3.5$ (d) $l = 5.2$

4 The formula for the area of a triangle is $A = \frac{1}{2}bh$, where b is the length of the base and h is the height. Work out the value of A when:

(a) $b = 8$ and $h = 6$ (b) $b = 10$ and $h = 5$

(c) $b = 9$ and $h = 8$ (d) $b = 7$ and $h = 5$

5 Ben uses the formula $P = 2l + 2w$ to work out the perimeter of a rectangle, where l is the length and w is the width. Work out the value of P when:

(a) $l = 8$ and $w = 5$ (b) $l = 12$ and $w = 5$

(c) $l = 7$ and $w = 4.5$ (d) $l = 7.5$ and $w = 3.5$

Use Example 9 to help.

6 Jenna uses the formula $P = 2(l + w)$ to work out the perimeter of rectangles, where l is the length and w is the width.
Use Jenna's formula to work out the perimeter of each rectangle in question 5 to show that she gets the same answers as Ben.
You must show your working.

> Order of operations: work out the bracket before doing the multiplication.

Writing your own formulae

Example 10

Alex buys x packets of sweets.
Each packet of sweets costs 45 pence.
Alex pays with a £5 note.
Write a formula for the change, C, in pence, Alex should receive.

> £5 = 500p

$$C = 500 - 45x$$

> The sweets cost 45p per packet so the cost, in pence, for x packets is $45x$.

Exercise 7F

1 Nilesh buys y packets of sweets.
Each packet of sweets costs 48 pence.
Nilesh pays with a £5 note.
Write a formula for the change, C, in pence, Nilesh should receive.

2 Apples cost r pence each and bananas cost s pence each.
Sam buys 7 apples and 5 bananas.
Write a formula for the total cost, t, in pence, of these fruit.

3 To roast a chicken you allow 45 minutes per kg and then a further 20 minutes.
Write a formula for the time, t, in minutes, to roast a chicken that weighs w kg.

4 To roast lamb you allow 30 minutes plus a further 65 minutes per kg.
Write a formula for the time, t, in minutes, to roast a joint of lamb that weighs w kg.

5 A rectangle has a length of $3x + 1$ and a width of $x + 2$.
Write down a formula for the perimeter, p, of this rectangle.

7.3 Substitution

Substitution into simple expressions

You can **substitute** numbers for the variables in an expression.

Key words:
substitute

Substitute means 'replace'. You replace each letter with a number.

> Use the correct order of operations to do the calculations.

Example 11

If $a = 4$, $b = 2$ and $c = 3$,
work out the value of these expressions:

(a) $3a$

(b) $5b + 4c$

(c) $ab - c$

(a) $3a = 3 \times 4$
    $= 12$

(b) $5b + 4c = 5 \times 2 + 4 \times 3$
    $= 10 + 12$
    $= 22$

(c) $ab - c = 4 \times 2 - 3$
    $= 8 - 3$
    $= 5$

Substitute the value 4 for a in the expression
$3a = 3 \times a$

$ab = a \times b$

Exercise 7G

If $a = 4$, $b = 2$ and $c = 3$,
work out the value of these expressions:

1 $2a$ 　　　　**2** $5b$ 　　　　**3** $6c$ 　　　　**4** $a + c$

5 $a + b + c$ 　**6** $a - 3$ 　　**7** $3a - 5$ 　　**8** $5b + 6$

9 $4c - 7$ 　　**10** $5a + 4b$ 　**11** $3b + 7c$ 　**12** $8a + 5c$

13 $6a - c$ 　　**14** $5b - 2c$ 　**15** $9c - 4a$ 　**16** $ab + c$

17 $ac - b$ 　　**18** $ab + ac$ 　**19** $bc - 2b$ 　**20** abc

You can substitute positive and negative integers, fractions and decimals into expressions.

Use the rules for adding, subtracting, multiplying and dividing negative numbers.

Example 12

If $x = 5$, $y = \frac{1}{2}$ and $z = -2$,
work out the value of these expressions:

(a) $3x + 4y$ **(b)** $6x + 2z$ **(c)** $4x - 6y + 3z$

(a) $3x + 4y = 3 \times 5 + 4 \times \frac{1}{2}$
$\qquad = 15 + 2$
$\qquad = 17$

(b) $6x + 2z = 6 \times 5 + 2 \times (-2)$
$\qquad = 30 + (-4)$
$\qquad = 30 - 4$
$\qquad = 26$

(c) $4x - 6y + 3z = 4 \times 5 - 6 \times \frac{1}{2} + 3 \times (-2)$
$\qquad = 20 - 3 + (-6)$
$\qquad = 20 - 3 - 6$
$\qquad = 11$

$4 \times \dfrac{1}{2} = \dfrac{4 \times 1}{2} = \dfrac{4}{2} = 2$

Adding -4 is the same as subtracting 4.

Exercise 7H

If $x = 5$, $y = \frac{1}{2}$ and $z = -2$,
work out the value of these expressions:

1 $6x$ **2** $2y$

3 $5z$ **4** $3x + 4$

5 $4y + 7$ **6** $3z + 14$

7 $3x + 10y$ **8** $3x + 5y$

9 $4x + 3z$ **10** $12y + 2z$

11 $7x - 8y + 4z$ **12** $4x - 5y + 2z$

If $f = 6$, $g = 1.5$ and $h = -3$,
work out the value of these expressions:

13 $7f$ **14** $4g$

15 $6h$ **16** $f + g$

17 $4g + h$ **18** $3f + 5h$

19 $2f - 5g$ **20** $fg + 8g + 3h$

Substitution into more complicated expressions

You can substitute values into expressions involving brackets and powers (indices).

Example 13

If $a = 5$, $b = 4$ and $c = 3$,
work out the value of these expressions:

(a) $\dfrac{a + 3}{2}$ (b) $3b^2 - 1$ (c) $\dfrac{5c + 1}{b}$

(a) $\dfrac{a + 3}{2} = \dfrac{5 + 3}{2}$

$= 8 \div 2$

$= 4$

(b) $3b^2 - 1 = 3 \times 4^2 - 1$

$= 3 \times 16 - 1$

$= 48 - 1$

$= 47$

(c) $\dfrac{5c + 1}{b} = (5 \times 3 + 1) \div 4$

$= (15 + 1) \div 4$

$= 16 \div 4$

$= 4$

$\frac{8}{2} = 8 \div 2$

Order of operations: indices ($4^2 = 16$), then the multiplication (3×16), then the subtraction ($48 - 1$).

Example 14

Evaluate: (a) $f^3 + f$ when $f = 2.3$ (b) $2x^3$ when $x = -4$

(a) $f^3 + f = 2.3^3 + 2.3$

$= 14.467$

b) $2x^3 = 2 \times x^3$

$= 2 \times (-4)^3$

$= 2 \times -64$

$= -128$

Evaluate means 'work out the value of'.

Use the power key on your calculator.

You need to know how to input a negative number in your calculator.
You may have to enter $(-4)^3$ like this:

(+/- 4) x^y 3 =

Exercise 7I

If $r = 5$, $s = 4$ and $t = 3$,
work out the value of these expressions:

1 $\dfrac{r + 3}{2}$ **2** $\dfrac{s + 5}{3}$ **3** $\dfrac{t + 7}{2}$

4 $3r^2 + 1$ **5** $4t^2 - 6$ **6** $2s^2 + r$

7 $4(5s + 1)$ **8** $t(r + s)$ **9** $5(2s - 3t)$

10 $\dfrac{5t + 1}{s}$ **11** $\dfrac{4r - 2}{t}$ **12** $\dfrac{3s + t}{r}$

$3r^2 = 3 \times r^2 = 3 \times r \times r$

Order of operations: brackets first.

If $a = 5, b = 1.5, c = -2,$
work out the value of these expressions:

13 $3a^2 + b$ **14** $c^2 + a$ **15** $2c^3 + b$

16 $10b + 2a^2$ **17** $4c^2 - a + b$ **18** $b^3 - c$

19 Copy and complete this table:

x	1	2	3	4	5
$x^2 + 2x$			15		

$3^2 + 2 \times 3 = 9 + 6 = 15$

20 Copy and complete this table:

x	3	4	3.5	3.7	3.8
$x^3 - x$		60			

$x^3 = x \times x \times x$
$4^3 - 4 = 64 - 4 = 60$

If $A = 6, B = -4, C = 3$ and $D = 30,$
work out the value of these expressions:

21 $D(B + 7)$ **22** $A(B + 1)$ **23** $A^2 + 2B + C$

24 $\dfrac{2A + 3}{C}$ **25** $\dfrac{4B + D}{2}$ **26** $\dfrac{A^2 + 3B}{C}$

Substitution into formulae

You can substitute values into a formula to work out the
value of a variable.

Example 15

A formula for working out acceleration is:

$$a = \frac{v - u}{t}$$

where v is the final velocity, u is the initial velocity and t is
the time taken.

Work out the value of a when $v = 50, u = 10, t = 8.$

$$a = \frac{v - u}{t}$$

$$= \frac{(50 - 10)}{8}$$

$$= 40 \div 8$$

$$= 5$$

Work out the numerator
first, then divide.

Example 16

A formula for working out distance travelled is:

$$s = ut + \tfrac{1}{2}at^2$$

where u is the initial velocity, a is the acceleration and t is the time taken.

Initial velocity = starting velocity.

Work out the value of s when $u = 3$, $a = 8$, $t = 5$.

$$s = ut + \tfrac{1}{2}at^2$$
$$= 3 \times 5 + \tfrac{1}{2} \times 8 \times 5^2$$
$$= 15 + 4 \times 25$$
$$= 15 + 100$$
$$= 115$$

$ut = u \times t$

Order of operations: indices ($5^2 = 25$) then multiplication ($3 \times 5 = 15$, $\tfrac{1}{2} \times 8 = 4$) then addition ($15 + 100 = 115$).

Exercise 7J

1 Use the formula $a = \dfrac{v - u}{t}$ to work out the value of a when:

Use Example 15 to help you.

   (a) $v = 15$, $u = 3$, $t = 2$ **(b)** $v = 29$, $u = 5$, $t = 6$
   (c) $v = 25$, $u = 7$, $t = 3$ **(d)** $v = 60$, $u = 10$, $t = 4$

2 The formula for the area of a trapezium is:

$$A = \tfrac{1}{2}(a + b)h$$

Work out the value of A when:

   (a) $a = 10$, $b = 6$, $h = 4$ **(b)** $a = 13$, $b = 9$, $h = 8$
   (c) $a = 9$, $b = 6$, $h = 4$ **(d)** $a = 15$, $b = 10$, $h = 6$

3 Use the formula $s = ut + \tfrac{1}{2}at^2$ to work out the value of s when:

Use Example 16 to help you.

   (a) $u = 3$, $a = 10$, $t = 2$ **(b)** $u = 7$, $a = 6$, $t = 5$
   (c) $u = 2.5$, $a = 5$, $t = 4$ **(d)** $u = -4$, $a = 8$, $t = 3$

 4 Body Mass Index, b, is calculated using the formula

$$b = \dfrac{m}{h^2}$$

where m is mass in kilograms and h is height in metres.
Work out the value of b when:

   (a) $m = 70$, $h = 1.8$ **(b)** $m = 38$, $h = 1.4$
   (c) $m = 85$, $h = 1.9$ **(d)** $m = 59$, $h = 1.7$

Round your answers to the nearest whole number.

5 A formula for working out the velocity of a car is:

$$v = \sqrt{(u^2 + 2as)}$$

$\sqrt{}$ is the symbol for square root.

where u is the initial velocity, a is the acceleration and s is the distance travelled.
Work out the value of v when:

(a) $u = 3, a = 4, s = 5$ (b) $u = 6, a = 8, s = 4$

(c) $u = 9, a = 10, s = 2$ (d) $u = 7, a = 4, s = 15$

6 *Talkalot* calculates telephone bills using this formula:

$$C = \frac{7.5n + 995}{100}$$

where C is the total cost (in £) and n is the number of calls.
Work out the telephone bill for

(a) 84 calls   (b) 156 calls   (c) 328 calls

Give your answers correct to the nearest penny.

7 My water company calculates water bills each quarter using this formula:

$$A = 2.87V + 4.94$$

where A is the amount to pay (in £) and V is the volume of water used (in m³).
In January I used 10 m³, in February I used 11 m³ and in March I used 8 m³.

(a) Work out my bill for the first quarter of the year.

(b) How many m³ of water would I need to use to make my bill for the quarter more than £100?

The first quarter is January, February and March.

7.4 Changing the subject of a formula

Key words:
subject
rearrange

In the formula $v = u + at$ the variable v is called the **subject** of the formula.

A variable is a letter that can take different values.

The subject of a formula is always the letter on its own on one side of the equation. This letter only appears once in the formula.

P is the *subject* of the formula $P = 2l + 2w$. You can use the formula to find the value of *P*.

You can rearrange a formula to make a different variable the subject. This is called 'changing the subject' of a formula.

This uses equation solving skills (Chapter 5).

If you are given some values, you can substitute these before you rearrange.

Example 17

A formula for working out the perimeter of a regular hexagon is: $P = 6x$, where x is the length of each side.

Work out the value of x when $P = 48$.

$P = 6x$
$48 = 6x$
$48 \div 6 = x$
$8 = x$

Substitute the value you know into the formula.

Solve the equation to find x.

Example 18

The perimeter of a rectangle is given by $P = 2y + 2w$, where y is the length and w is the width.

Work out the value of y when $P = 24$ and $w = 5$.

$P = 2y + 2w$
$24 = 2y + 2 \times 5$
$24 = 2y + 10$
$24 - 10 = 2y$
$14 = 2y$
$14 \div 2 = y$
$7 = y$

Substitute the values you know into the formula.

Solve the equation to find y.

You can use the same techniques to rearrange a formula when you don't know any of the values.

Example 19

Rearrange $d = a + 8$ to make a the subject.

You need to end up with a on its own on one side.

$d = a + 8$
$d - 8 = a + 8 - 8$
$d - 8 = a$
$a = d - 8$

Subtract 8 from both sides as when solving an equation.

Example 20

Rearrange $A = lw$ to make l the subject.

$A = lw$

$\dfrac{A}{w} = \dfrac{lw}{w}$

$\dfrac{A}{w} = l$ or $l = \dfrac{A}{w}$

lw means $l \times w$, l is multiplied by w so divide both sides by w to leave l on its own.

Example 21

Make x the subject of the formula $y = 5x - 2$.

$y = 5x - 2$

$y + 2 = 5x - 2 + 2$

$y + 2 = 5x$

$\dfrac{y + 2}{5} = \dfrac{5x}{5}$

$\dfrac{y + 2}{5} = x$ or $x = \dfrac{y + 2}{5}$

Add 2 to both sides to get $5x$ on its own.

Now divide both sides by 5 to leave x on its own.

Exercise 7K

1 A formula for working out the perimeter of a regular hexagon is: $P = 6l$, where l is the length of each side. Work out the value of l when:
   (a) $P = 60$ **(b)** $P = 30$ **(c)** $P = 120$

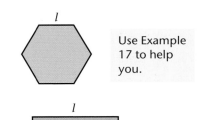

Use Example 17 to help you.

2 The formula for the area of a rectangle is $A = lw$, where l is the length and w is the width. Work out the value of w when:
   (a) $A = 12$ and $l = 4$ **(b)** $A = 36$ and $l = 9$
   (c) $A = 42$ and $l = 7$ **(d)** $A = 60$ and $l = 15$

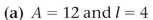

3 The perimeter of a rectangle is given by
   $$P = 2l + 2w$$
   where l is the length and w is the width. Use the formula to:
   (a) Find l when $P = 18$ and $w = 4$
   (b) Find l when $P = 32$ and $w = 7$
   (c) Find w when $P = 60$ and $l = 17$
   (d) Find w when $P = 50$ and $l = 13.5$

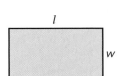

Use Example 18 to help.

4 Use the formula $v = u + at$

    (a) to find u when $v = 30$, $a = 8$ and $t = 3$

    (b) to find u when $v = 47$, $a = 4$ and $t = 9$

    (c) to find a when $v = 54$, $u = 19$ and $t = 7$

    (d) to find t when $v = 60$, $u = 15$ and $a = 5$

    (e) to find u when $v = 20$, $a = 7$ and $t = 4$

5 Rearrange each of these formulae to make a the subject:

    (a) $d = a + 8$ **(b)** $t = a + 12$

    (c) $k = a - 6$ **(d)** $w = a - 7$

6 Rearrange each of these formulae to make w the subject:

    (a) $P = 4w$ **(b)** $a = 3w$

    (c) $A = lw$ **(d)** $h = kw$

7 Make x the subject of each of these formulae:

    (a) $y = 5x - 6$ **(b)** $y = 4x - 7$

    (c) $y = 2x + 1$ **(d)** $y = 6x + 5$

Use Example 21 to help you.

8 The formula $F = 1.8C + 32$ can be used to convert degrees Celsius, °C, to degrees Fahrenheit, °F.

    (a) Convert 15°C to degrees Fahrenheit.

    (b) Rearrange the formula to make C the subject.

    (c) Use your new formula to convert 68°F to °C.

    (d) What is 82°F in degrees Celsius to the nearest degree? You must show your working.

9 Make r the subject of each of these formulae:

    (a) $p = 4r + 2t$ **(b)** $v = 7r + 4h$

    (c) $w = 3r - 2s$ **(d)** $y = 6r - 5p$

10 A formula used to calculate velocity is:

    $v = u - at$

    where u is the initial velocity, a is the acceleration and t is the time taken.

    (a) Rearrange the formula to make a the subject.

    (c) Rearrange the formula to make t the subject.

11 Rearrange these formulae to make a the subject:

(a) $b = \frac{1}{2}a + 6$ (b) $b = \frac{1}{2}a + 7$ (c) $b = \frac{1}{3}a - 1$

(d) $b = \frac{1}{4}a - 3$ (e) $b = 2(a + 1)$ (f) $b = 3(a - 5)$

$\frac{1}{2}a = \frac{a}{2}$
So $b = \frac{1}{2}a + 6$
$b - 6 = \frac{1}{2}a$
$a = \ldots\ldots$

Key words:
trial and improvement

7.5 Trial and improvement

You will need to know:
- **how to substitute values into algebraic expressions**
- **how to use your calculator effectively**

You cannot always solve equations using algebraic methods (like those you learnt in Chapter 5).

You can use **trial and improvement** to find an approximate solution.

Guess/estimate what the answer might be.

↓

Work out the value of the expression using your estimate.

Try a value. *Improve* your estimate and try again.

↓

Is this answer too big or too small?

↓

Use your answer to improve your estimate.

↓

Work out the value of the expression using your estimate.

↓

Is this answer too big or too small?

↓

Improve your estimate again.

Continue the process until your estimate gives an answer to the required degree of accuracy.

Example 22

Use trial and improvement to find the solution to the equation $x^3 + x = 80$.

Give your answer to 1 decimal place

x	$x^3 + x$	Comment
4	$4^3 + 4 = 68$	too small
5	$5^3 + 5 = 130$	too big
4.5	$4.5^3 + 4.5 = 95.625$	too big
4.2	$4.2^3 + 4.2 = 78.288$	too small
4.3	$4.3^3 + 4.3 = 83.807$	too big
4.25	$4.25^3 + 4.25 = 81.015...$	too big
$x = 4.2$ to 1 decimal place		

Try a value for x which you think is close. $4^3 = 64$ so try $x = 4$.

The answer for $x = 4$ is too small.

Try $x = 5$. As 5 is too big the solution must be between 4 and 5.

Try $x = 4.5$.

Still too big. Try between 4 and 4.5.

$x = 4.2$ is too small, but close. Try $x = 4.3$.

The solution must be between 4.2 and 4.3. You need to know which of these is closest so you *must* try half way between. Try $x = 4.25$.

4.25 is too big so the solution must be closer to 4.2 than to 4.3.

Exercise 7L

1 Use trial and improvement to find the solution to the equation

$$x^3 + x = 100$$

Give your answer to 1 decimal place.
Copy and continue this table to help you

x	$x^3 + x$	Comment
4	68	too small
5		

2 Use trial and improvement to find the solution to the equation

$$x^3 + x = 25$$

Give your answer to 1 decimal place.

Use a table like the one in Example 22 or question 1.

3 Use trial and improvement to find the solution to the equation

$$x^3 - x = 30$$

Give your answer to 1 decimal place.

4 Use trial and improvement to find the solution to the equation

$$x^3 + 2x = 240$$

Give your answer to 1 decimal place.

5 Use trial and improvement to find the solution to the equation

$$x^3 + x^2 = 1000$$

Give your answer to 1 decimal place.

Examination style questions

1 A company uses this formula to find the cost, in pounds, to hire out a car.

Cost = 25 × number of days hire + 20

(a) Calculate the cost of hiring a car for

(i) 2 days

(ii) 1 week

(b) Linda hires a car for her holiday.
She pays the company £270.
For how many days does she hire the car? *(6 marks)*
AQA, Spec A, F, June 2003

2 The cost, in pounds, to hire a conference centre is calculated by using this formula.

Cost = 4 × number of people attending + 250

(a) Find the cost of hiring the conference centre when 200 people attend.

(b) A company pays £650 to hire the conference centre.
How many people attend the conference? *(4 marks)*
AQA, Spec A, F, June 2005

3 Find the value of $3x + 4y$ when $x = 6$ and $y = -3$. *(2 marks)*
AQA, Spec A, I, June 2003

4 Ali is x cm tall.
   (a) Suki is 5 cm taller than Ali.
   Write down an expression in x for Suki's height.
   (b) Ali's sister is 2 cm shorter than Ali.
   Write down an expression in x for the height of Ali's sister.
   (b) Ali's father is twice as tall as Ali.
   Write down an expression in x for the height of Ali's father. *(3 marks)*
AQA, Spec A, F, June 2005

5 **(a)** Find the value of $3x + 4y$ when $x = 2$ and $y = 5$.
   (b) Find the value of $a^3 + b^2$ when $a = 2$ and $b = 5$. *(4 marks)*
AQA, Spec B, 5F, June 2003

6 An approximate rule for converting degrees Fahrenheit into degrees Centigrade is

$$C = \frac{F - 30}{2}$$

Use this rule to convert 22°F into °C. *(2 marks)*
AQA, Spec A, I, June 2003

7 **(a)** Use the formula $a = 5b + 2c$ to work out a when $b = 3$ and $c = -4$.
   (b) Use the formula $a = 5b + 3c$ to work out c when $a = 16$ and $b = 2$. *(5 marks)*
AQA, Spec A, F, June 2005

8 Use the formula $\qquad v = u + at$
   To find the value of v when $u = -10$, $a = 1.8$ and $t = 3.7$. *(2 marks)*
AQA, Spec A, F, June 2005

9 Here are 4 expressions: $\qquad n^2 \qquad \dfrac{n}{3} \qquad n + 3 \qquad \dfrac{3}{n}$
   (a) If $n = 3$, which expression has the greatest value?
   Show your working.
   (b) If $n = 0.3$, which expression has the greatest value?
   Show your working. *(4 marks)*
AQA, Spec B, 5I, November 2003

10 You are given that $m = \frac{3}{4}$, $p = \frac{1}{2}$, and $t = 2$
Find the value of

(a) $mp + t$

(b) $\dfrac{m + p}{t}$

(4 marks)
AQA, Spec A, I, November 2003

11 (a) Find the value of a^3 when $a = 4$.

(b) Find the value of $5x + 3y$ when $x = -2$ and $y = 4$.

(c) There are p seats in a standard class coach and q seats in a first class coach.
A train has 5 standard class coaches and 2 first class coaches.
Write down an expression in terms of p and q for the number of seats in a train.

(5 marks)
AQA, Spec A, I, November 2003

12 Make r the subject of the formula $p = 3 + 2r$ *(2 marks)*
AQA, Spec A, I, June 2005

Summary of key points

Indices (grade E)

6^2 is called '6 squared', 8^3 is called '8 cubed'.
$5 \times 5 \times 5 \times 5 = 5^4$. You say '5 to the power 4'.
The ^{4} is called the index or power.

Laws of indices (grade C)

$x^1 = x$
$x^n \times x^m = x^{n+m}$
To *multiply* powers of the *same* number or variable *add* the indices.
$x^n \div x^m = x^{n-m}$
To *divide* powers of the *same* number or variable *subtract* the indices.

Formulae given in words (grade G)

A formula is a general rule that shows the relationship between quantities that can vary:
For example, pay = number of hours worked × rate of pay

Formulae written using letters and symbols (grade F to D)

You can use letters for the variables in a formula.
For example $p = hr + b$ where p is the pay,
h is the number of hours worked,
r is the rate of pay,
b is the bonus.

Substitution (grades F to C)

Use the correct order of operations to help you do the calculations when you substitute values into an algebraic expression.

Changing the subject of a formula (grade C)

The subject of a formula only appears once, and only on its own on one side of the formula.
In the formula

$$v = u + at$$

the variable v is the subject.

You can rearrange a formula to make a different variable the subject:

You can rearrange $v = u + at$ as $a = \dfrac{(v - u)}{t}$

Trial and improvement (grade C)

You can use a trial and improvement method to find an approximate solution to an equation.

8 Measurement

This chapter will show you how to:

✔ estimate length, volume (capacity) and mass
✔ calculate with time and read timetables
✔ use metric units of length, volume and mass and convert between them
✔ convert between metric and imperial units
✔ understand the accuracy of measurement
✔ handle the compound measures of speed and density

8.1 Estimating length, volume and mass

Key words:
metric
imperial

You need to be able to estimate length, volume (or capacity) and mass in everyday life.

Will this book fit on the bookshelf?

Is there enough space in my glass to pour in the rest of my can of lemonade?

Will my suitcase be too heavy to take on the plane?

These quantities can be measured in **metric** or **imperial** units and you need to be familiar with both.

Metric measure is the most common nowadays. You may use imperial measure in height (feet and inches), in weight (stones and pounds). Petrol consumption for cars is sometimes still given in miles per gallon.

To estimate a measurement, compare with a measurement you already know.

For example, compare with:
● the length of a ruler (30 cm)
● the mass of a bag of sugar (1 kg)
● the capacity of a carton of juice (1 ℓ)

Estimating length

You can measure lengths and distances in these units:

Metric: **kilometres (km), metres (m), centimetres (cm), millimetres (mm)**

Imperial: **miles, yards, feet, inches**

Imperial measures:
12 inches = 1 foot
3 feet = 1 yard
1760 yards = 1 mile
5280 feet = 1 mile
63 360 inches = 1 mile

	metric	imperial
	15 cm	6 inches
	2 m	6 feet 6 inches
	4.5 m	15 feet

Exercise 8A

1 Estimate these lengths. Give an answer in metric units or imperial units for each.
  (a) The width of a classroom door.
  (b) The length of your middle finger.
  (c) The thickness of this book.
  (d) The height of your teacher.
  (e) The diameter of a football.
  (f) The height of a house.
  (g) The length of a bus.
  (h) The height of a street light.
  (i) The distance from London to Edinburgh.
  (j) The width of a pencil point.

2 Check your estimates for **(a)**, **(b)**, **(c)** and **(d)** in question 1 by measuring.

Estimating volume (capacity)

Volume or capacity is a measure of the amount of space inside a container, or the amount it can hold. You can measure it in these units.

The capacity of a bath is the volume of water it can hold.

Metric: **litres (ℓ), centilitres (cℓ), millilitres (mℓ)**

Imperial: **gallons, pints, fluid ounces (fl. oz)**

Imperial measures:
20 fluid ounces = 1 pint
8 pints = 1 gallon

	metric	imperial
(teaspoon)	5 mℓ	5 fl. oz
SODA	300 mℓ	½ pint
(petrol can)	4.5 ℓ	1 gallon

Exercise 8B

1 Estimate the volume of these liquids. Give an answer in metric units and imperial units for each.

Estimating mass

You can measure mass in these units:

Metric: tonnes (t), kilograms (kg), grams (g), milligrams (mg)

Imperial: tons, hundredweight, stones (st), pounds (lb), ounces (oz)

Imperial measures:
16 ounces = 1 pound
14 pounds = 1 stone
8 stone = 1 hundredweight
20 hundredweight = 1 ton
2240 pounds = 1 ton

	metric	imperial		metric	imperial
CHUNKY CHOC	62.5 g	2.5 ounces		80 kg	12½ stone
SUGAR 1kg	1 kg	2.2 pounds			
(vitamin)	500 mg	$\frac{1}{50}$ oz			

Exercise 8C

1 Estimate the mass of the following. Use metric units.

  (a) this book   **(b)** a large dog
  (c) a pair of trainers   **(d)** a bag of potatoes
  (e) a small car   **(f)** a 2p coin
  (g) an average adult female **(h)** a jar of jam
  (i) a birthday cake   **(j)** a bicycle

2 Write down an estimate for **(a)**–**(e)** in question 1 in imperial units.

Sensible estimates

You should always give estimates in sensible units.

For the length of a journey, sensible units are miles or kilometres. For the volume of a teaspoon, sensible units are millilitres. For mass of a sugar cube, sensible units are grams.

> You could say that a distance is 100 000 m, but this is not sensible.

Example 1

For each statement, say whether the estimate is sensible. Give a better estimate where necessary.

(a) A man is 2.8 m tall.

(b) My cat weighs 5.3 kg.

(c) A mug holds 2 ℓ of liquid.

(a) A man is not as tall as a classroom door. A classroom door is about 2 m (6 feet 6 inches). 2.8 m is not a sensible estimate. 1.8 m (about 6 feet) is reasonable.

> 1.65 m (about 5 ft 5 in) is reasonable for a woman.

(b) A bag of sugar weighs 1 kg, so 5.3 kg is equivalent to just over 5 bags of sugar. This seems reasonable.

(c) A carton of juice holds 1 litre. So 2ℓ = 2 cartons. This is more than a mug. A drinks can holds 300 mℓ. 200 mℓ would be a reasonable estimate for a mug.

> Compare the mug with a capacity you know.

Example 2

Write down the most appropriate metric and imperial units of measurement for these.

(a) The distance from Newcastle to Leeds.

(b) The mass of a bag of potatoes.

(c) The amount of liquid in a flask of tea.

'appropriate' means 'reasonable'.

	metric	imperial
(a)	kilometres	miles
(b)	kilograms	pounds
(c)	litres	pints

Exercise 8D

1 For each statement, say whether the estimate is sensible. Give a better estimate where necessary.

  (a) Mrs Chavda is about 110 cm tall.

  (b) A dog weighs 100 kg.

  (c) The mass of a raisin is 3 grams.

  (d) A tablespoon holds 20 mℓ of milk.

  (e) The distance from John O'Groats to Land's End is 200 km.

2 Peter has estimated the measurements in question 1, using imperial units. Which ones are sensible? Give a better estimate where necessary.

  (a) Mrs Chavda is about 5 feet 10 inches tall.

  (b) A dog weighs 4 stone.

  (c) The mass of a raisin is 1 ounce.

  (d) A tablespoon holds $\frac{1}{2}$ pint of milk.

  (e) The distance from John O'Groats to Land's End is 2000 miles.

3 For each statement, give the most appropriate metric and imperial units of measurement.

  (a) The mass of your chair.

  (b) The distance from the door to the window in your classroom.

  (c) The capacity of a tablespoon.

Use Example 2 to help you.

(d) The mass of this textbook.

(e) The length of this textbook.

(f) The mass of a Jumbo Jet.

(g) The distance from Paris to London.

(h) The amount of water in a swimming bath.

4 These statements are all true. Are they in appropriate *units*? Give more appropriate units where necessary.

(a) The length of our drive is 15 000 millimetres.

(b) Wes weighs 0.08 tonnes.

(c) A bag of flour weighs 2 pounds.

(d) A tablespoon holds 0.025 litres.

(e) A man is 1.8 metres tall.

(f) The height of the generators on a wind farm is 350 feet.

(g) My teapot holds 0.3 gallons of tea.

(h) I walk 120 000 cm to school each day.

8.2 Reading scales

You need to be able to read different types of scales. Some of the most common are:

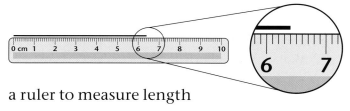

There are 10 spaces between the 6 and the 7.
10 spaces = 1 cm.
1 space = 1 ÷ 10 = 0.1 cm.
The line shown is 6.4 cm long.

a ruler to measure length

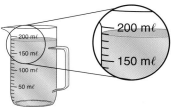

a measuring jug to measure capacity

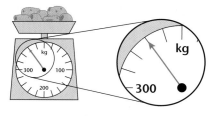

kitchen scales to weigh ingredients.

There are 5 spaces between the 150 and the 200.
5 spaces = 50 mℓ.
1 space = 50 ÷ 5 = 10 mℓ.
The scale shows a reading of 190 mℓ.

There are 5 spaces between the 300 and the 400.
5 spaces = 100 g.
1 space = 100 ÷ 5 = 20 g.
The scale shows a reading of 360 g.

To read a scale, you need to look at how it is marked.

How many spaces are there between the numbers?
Work out what one space represents.

Does it go up in 1s, 10s, 50s, 100s or something else?

Exercise 8E

1 What is the reading on the following scales?

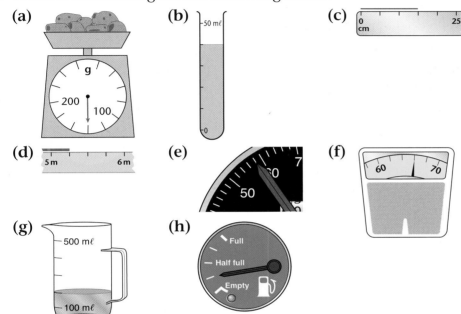

(a)

(b) 50 mℓ / 0

(c) 0 cm / 25

(d) 5 m / 6 m

(e)

(f)

(g) 500 mℓ / 100 mℓ

(h) Full / Half full / Empty

2 Copy the scales used in question 1 and mark on them
   (a) 180 grams   (b) 20 mℓ   (c) 20 cm
   (d) 5.75 m   (e) 54 mph   (f) 62 kg
   (g) 375 mℓ   (h) $\frac{3}{4}$ full

Some scales have very few markings on them and you need to estimate a reading.

This speedometer scale is marked in 10s. There are 2 spaces between each marked value, so each space represents 10 ÷ 2 = 5 mph.

The pointer lies between 35 and 40. It is slightly nearer to 35 than to 40. A sensible estimate is 37 mph.

Exercise 8F

1 Make a sensible estimate for each measurement shown.

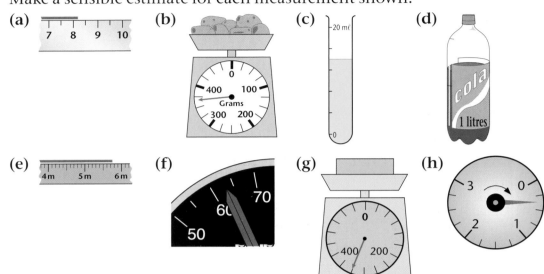

(a)   (b)   (c)   (d)

(e)   (f)   (g)   (h)

8.3 Time, dates and timetables

You will need to know:

- **how to write times in the 24-hour clock system**
- **how to write times in am or pm**

12-hour and 24-hour clock

There are two ways of measuring time:
- using **am** for morning and **pm** for afternoon
- using the **24-hour clock**

am and pm	24-hour clock
12 midnight	0000 or 2400
7.15 am	0715
10 am	1000
12 noon (midday)	1200
3.40 pm	1540
9.55 pm	2155

Key words:
am   pm
24-hour clock

The '12-hour clock' means time in am and pm.

Most digital watches use the 24-hour clock.

You say:
'quarter past seven in the morning'

'twenty to four in the afternoon'

'five to ten in the evening'

Exercise 8G

1 Write these times using **(i)** am and pm (the 12-hour clock)
(ii) the 24-hour clock.
  (a) twenty past six in the morning
  (b) quarter to eleven in the morning
  (c) five past three in the afternoon
  (d) ten minutes to seven in the evening
  (e) half past ten in the evening

2 Write these times as you would say them.
  (a) 9.15 pm   **(b)** 7.40 am   **(c)** 1200
  (d) 1430   **(e)** 0000

3 Change these times into 24-hour clock times.
  (a) 7 pm   **(b)** 6 am   **(c)** 8.30 am   **(d)** 12.45 pm
  (e) 2.15 pm   **(f)** 2.15 am   **(g)** 11.23 pm   **(h)** 4.55 pm
  (i) 11.47 am   **(j)** 12.15 am   **(k)** 9.45 pm   **(l)** 2.20 am

4 Write these 24-hour clock times as am or pm times.
  (a) 1100   **(b)** 1300   **(c)** 0600   **(d)** 1630
  (e) 0920   **(f)** 0030   **(g)** 2315   **(h)** 1005
  (i) 1545   **(j)** 0420   **(k)** 1822   **(l)** 2400

Calculating time

Take care when using a calculator to calculate times.
You may fall into the trap of thinking that there are
100 minutes in *1 hour* instead of *60 minutes*.

Example 3

My journey was due to take 2 hours and
50 minutes but I was delayed for 45 minutes.
How long did my journey take?

Writing the times as 2.50 and 0.45 does not give the correct answer.

Correct:

Journey time
= 2 hours and 50 minutes
  + 45 minutes
= 2 hours and 50 minutes
  + 10 minutes + 35 minutes
= 3 hours and 35 minutes

Incorrect:

Journey time
= 2.50 + 0.45
= 2.95 hours

This is 2½ hours, *not* 2 hours 50 minutes.

This is not the same as 45 minutes.

2 hours and 50 minutes + 10 minutes is exactly 3 hours.

Split the 45 minutes into (10 + 35) to make the calculation easier.

Example 4

My plane leaves Manchester airport at 0723.
I have to be there $1\frac{1}{2}$ hours beforehand.
What time is this?

$\frac{1}{2}$ hour = 60 ÷ 2 = 30 mins. | 30 − 23 = 7

Using a calculator to work out
7.23 − 1.5 gives a ridiculous answer!

Correct:

1 hour before 0723 is 0623.

I need to be 30 minutes earlier than this.

23 minutes earlier is 0600.

So I must be there 7 minutes before 0600 which is 0553.

Incorrect:

Arrival time = 7.23 − 1.5

= 5.73

I need to be there by 0573.

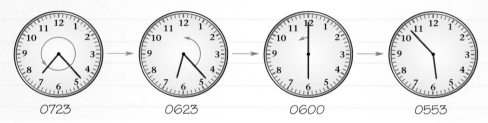

0723   0623   0600   0553

To calculate times you
need to remember:

60 seconds	= 1 minute
60 minutes	= 1 hour
24 hours	= 1 day
7 days	= 1 week
365 days	= 1 year
366 days	= 1 leap year

In some calculations, you
might use
52 weeks = 1 year.
This is not quite true
because
52 weeks = 52 × 7 days
= 364 days!

Example 5

(a) Write 3 hours 49 minutes in minutes.

(b) Write 438 minutes in hours and minutes.

(a) 3 hours = 3 × 60 minutes = 180 minutes

So 3 hours 49 minutes = 180 + 49 minutes

= 229 minutes

(b) Try some numbers of hours:

5 × 60 = 300   6 × 60 = 360

7 × 60 = 420   8 × 60 = 480

438 comes between 420 and 480 so it is more than 7 hours.

438 − 420 = 18, so 438 minutes = 7 hours 18 minutes

If you use your calculator to
work out 438 ÷ 60 you will
get an answer of 7.3.
What will you write as your
answer?
7 hours 3 minutes?
7 hours 30 minutes?
Neither of these is correct.

If you do use your calculator to get 7.3 you can get the correct answer if you then do this:

The '7' is the whole number of hours.

Work out 0.3×60 to change 0.3 hours into minutes.

This gives 18 minutes.

So 438 minutes = 7 hours 18 minutes.

Multiply by 60 to change hours into minutes.

Exercise 8H

1 The bus normally takes 1 hour and 52 minutes to travel from Stockham to Worthington. It was held up for 13 minutes in a traffic jam. How long did the journey take?

2 I usually work for 4 hours 30 minutes each day. On Fridays I work 1 hour 40 minutes less. How long do I work on a Friday?

3 Copy and complete the table.

55 minutes before	Time	45 minutes after
	0920	
	1135	
	1317	
	1648	
	2121	
	2352	

4 My journey to work takes 35 minutes by bus plus 8 minutes' walk. I want to be at work by 8.25 am. What time should I catch the bus?

5 Write the following in the units given.
   (a) 3 weeks in days
   (b) 2 years in days (*not* leap years)
   (c) 4 hours in seconds
   (d) 1 week in hours
   (e) 3 hours and 20 minutes in minutes
   (f) $2\frac{1}{2}$ minutes in seconds
   (g) 250 minutes in hours and minutes
   (h) 343 minutes in hours and minutes
   (i) 91 days in weeks
   (j) 1 leap year in minutes

Handling dates

To work out dates, you need to know how many days there are in each month. This rhyme will help you to remember.

> *30 days have September,*
> *April, June and November.*
> *All the rest have 31*
> *except in February alone*
> *which has but 28 days clear*
> *and 29 in each leap year.*

Example 6

Jim is going on holiday on 5th August.
He is counting the days until he goes.
Today is 12th April. He counts this as 'day 1'.
How many days are there until he goes away?

April	M	T	W	T	F	S	S
						1	2
	3	4	5	6	7	8	9
	10	11	12	13	14	15	16
	17	18	19	20	21	22	23
	24	25	26	27	28	29	30

There are 19 days left in April

May 31 days   June 30 days   July 31 days

August 5 days to the 5th

Total = 19 + 31 + 30 + 31 + 5 = 116 days

Exercise 8I

1 Use the calendar for April in Example 6 to answer the following.

(a) Today is Thursday the 6th April. What will be the date two weeks from today?

(b) If it is Wednesday the 5th today, what will be the date a week on Friday?

(c) My last day at work is the 20th April. I have 10 days' holiday. What is the date of the last day of my holidays?

(d) Today is Sunday 2nd April. My driving lesson is three weeks on Saturday. What date is this?

2 How many days are there between:
  (a) 25th November and the 3rd December?
  (b) 19th March and the 8th April?
  (c) 8th June and the 17th August?
  (d) 20th July and the 16th October?
  (e) 25th September and Christmas Day?

Include both start and end dates.

A calendar will help.

3 I finish my exams on 26th June. I do not start college until 11th September. How many days holiday do I have?

4 I was born on 7th December 1995. My brother is 1 year 3 months and 4 days younger than me. When is my brother's birthday?

Timetables

Bus and train timetables help you plan a journey.
They list the places where the bus or train stops and the departure time for each of these places.
Times are usually shown as 24-hour clock times.

This train timetable shows some of the train times from Newcastle to London.

This train leaves Newcastle at 1130.

Newcastle	1103	1130	1157	1233	1259
Durham		1143		1246	
Darlington	1131	1202		1305	
York	1202	1231	1250	1335	1354
Doncaster		1255		1401	
Peterborough	1310	1343	1359	1450	1501
London (King's Cross)	1412	1443	1455	1546	1559

This is the 1401 from Doncaster.

The 1401 from Doncaster arrives in London at 1546.

The times for any one train are written in a *column*. You read the time the train leaves each place in the *row* opposite the place name.

The 1130 train from Newcastle stops at all the stations listed. It leaves Darlington at 1202 and Peterborough at 1343. It arrives in London at 1443.

The journey starts at 1130 and ends at 1443, this is a total time of 3 hours and 13 minutes.

1130 to 1430 is 3 hours.

If you live in Doncaster and want to go to London you can either catch the 1255 train or the 1401 train. These are the only two trains that stop at Doncaster.

If you catch the 1401 train and arrive in London at 1546 the journey time is 1 hour and 45 minutes.

Example 7

Use the train timetable for Newcastle to London to answer these questions.

(a) Mike arrives at the train station at York at 1 pm.
   (i) What time is the next train he could catch to London?
   (ii) What is the earliest time he could arrive in London?

(b) Which train from Newcastle has the shortest journey time to London?

(a) (i) 1 pm = 1300

      The next train to London after 1300 is the 1335 train.

  (ii) The earliest time he could be in London is 1546.

(b) The trains that stop the least number of times will take the shortest time.

These are the 1157 and 1259 trains.

The 1157 train gets in at 1455. The journey time is 2 minutes less than 3 hours, which is 2 hours and 58 minutes.

The 1259 train gets in at 1559. The journey time is exactly 3 hours.

The 1157 train has the shortest journey time.

> Convert to 24-hour clock time.

> This is if he catches the first possible train (the 1335).

Exercise 8J

Use the train timetable for Newcastle to London for questions 1 and 2.

1 (a) How many trains on this timetable stop at Durham?
   (b) What time does the 1103 from Newcastle leave Peterborough?
   (c) What time does the 1305 from Darlington leave Newcastle?
   (d) What time will the 1250 from York arrive in London?
   (e) How long does it take the 1131 from Darlington to get to London?

2 Sally has a job interview in Peterborough at 3 pm.

(a) What is the last train from Newcastle she can catch to get to Peterborough before 3 pm?

(b) How long does this train take from Newcastle to Peterborough?

(c) It takes 15 minutes to get from Peterborough station to the interview. Which train should Sally catch? Give a reason for your answer.

3 This is a bus timetable from Chester to Ramsholt.

Chester	0845	0914	0934	0954		1614	1634	1654	1714	1739
Barton	0900	0920	0940	1000		1620	1640	1700	1720	1745
Holm	0911	0931	0951	1011	at	1631	1651	1711	1731	1756
Whitby	0920	0940	1000	1020	every	1640	1700	1720	1740	1805
Ellesmere (arr)	0926	0946	1006	1026	20	1646	1706	1726	1746	1811
Ellesmere (dep)	0929	0949	1009	1029	mins.	1649	1709	1729	1749	1814
Pooltown	0934	0954	1014	1034	until	1654	1714	1734	1754	1819
Overpool	0938	0958	1018	1038		1658	1718	1738	1758	1823
Ramsholt	0940	1000	1020	1040		1700	1720	1740	1800	1825

(a) What time does the 0914 from Chester arrive at Ramsholt?

(b) What time does the 1700 from Barton arrive in Overpool?

(c) How long do these journeys take:
  (i) from Chester to Ramsholt (if you leave after 0900)
  (ii) from Whitby to Pooltown?

(d) I arrive in Ellesmere at 1646. What time did I catch the bus in Holm?

(e) In Chester, what time is the next bus after the 0954?

(f) How long does the bus stop in Ellesmere?

(g) Is there a bus at 1230? Explain your answer.

4 Use the bus timetable in question 3 for this question. I live in Chester and I am meeting my friend in Barton at 4.45 pm.

(a) What is the latest bus I could catch?

We are going to the cinema in Ramsholt. The film starts at 6.15 pm.

(b) How long will the journey take?

(c) What is the time of the last bus we can catch?

(d) Another friend is getting on the bus at Ellesmere to travel with us. What time would he catch the bus?

(e) How long do we have to get from the bus to the cinema if the bus is on time?

5 Here is a train timetable.

Paddington	1400	1500	1600	1635	1700	1800	1830
Slough	–	–	–	1649	–	–	1854
Reading	1434	1534	1632	1701	–	1825	1908
Swindon	1506	1606	–	1735	–	–	1939
Bristol	1530	1630	1722	1759	1806	1911	2004
Newport	1553	1653	1745	–	1829	1935	2027
Cardiff	1610	1709	1801	–	1845	1952	2044

(a) How many trains stop at Slough?

(b) Which is the fastest train from Paddington to Reading?

(c) Which trains from Paddington will get me into Bristol after 7 pm?

(d) I live in Swindon and have to be in Cardiff by 6.30 pm. Which train should I catch?

(e) I am meeting a friend in Swindon at 5 pm. What time train will I catch from Paddington?

(f) Which are the quickest trains from Bristol to Cardiff?

6 Use the train timetable for question 5 to answer these questions.
I want to go to Cardiff but I catch the 1635 from Paddington by mistake.

(a) How far can I go on the 1635 from Paddington?

(b) Where will I have to change trains?

(c) How long will I have to wait for the next train?

(d) What time will I get into Cardiff?

8.4 Converting metric units

You will need to know:

- how to multiply and divide whole numbers and decimals by 10, 100 and 1000.

Converting length, mass and capacity

You need to learn these conversions.

Metric units of length:

10 mm = 1 cm	100 cm = 1 m
1000 mm = 1 m	1000 m = 1 km

1 cm³ is shown by this cube whose sides are all 1 cm.

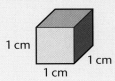

1 cm   1 cm   1 cm

Metric units of capacity:

100 cℓ = 1 ℓ	1000 mℓ = 1 ℓ

1 mℓ = 1 cm³ = 1 cc.

cm³ and cc both stand for cubic centimetres.

Metric units of mass:

1000 mg = 1 g   1000 g = 1 kg   1000 kg = 1 tonne

For the units of length:

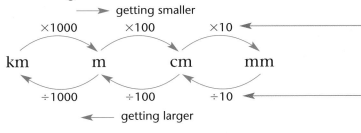

To change from larger units to smaller units you *multiply*.

To change from smaller units to larger units you *divide*.

Example 8

(a) How many metres are there in 4 kilometres?

(b) How many millimetres are there in 3 metres?

(c) Change 160 millimetres into centimetres.

(d) Change 8 kilometres into centimetres.

(a) 4 km = 4 × 1000 m = 4000 m

(b) 3 m = 3 × 100 cm = 300 cm

     300 cm = 300 × 10 mm = 3000 mm

(c) 160 mm = 160 ÷ 10 cm = 16 cm

(d) 8 km = 8 × 1000 m = 8000 m

     8000 m = 8000 × 100 cm = 800 000 cm

Move digits to the left to multiply.

Move digits to the right to divide.

See Section 1.5 for a reminder of the rules.

You can use the same ideas to convert metric units of capacity and mass:

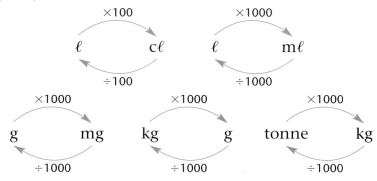

Example 9

Change these volumes to litres:

(a) 4000 mℓ **(b)** 72 000 cℓ

(a) 4000 mℓ = 4000 ÷ 1000 ℓ = 4 ℓ

(b) 72 000 cℓ = 72 000 ÷ 100 ℓ = 720 ℓ

Example 10

Change these masses to kilograms:

(a) 3 tonnes **(b)** 83 000 g

(a) 3 tonnes = 3 × 1000 kg = 3000 kg

(b) 83 000 g = 83 000 ÷ 1000 kg = 83 kg

A tonne is **more** than a kilogram. To change tonnes to kg you **multiply**.

 Exercise 8K

1 (a) How many centimetres are there in 6 metres?
(b) How many metres are there in 8 kilometres?

2 Change these lengths into the units given.
(a) 3 metres → millimetres
(b) 5 metres → centimetres
(c) 12 cm → mm **(d)** 4000 m → km
(e) 6000 mm → m **(f)** 300 mm → cm

3 Change these masses into the units given.
(a) 5 tonnes → kilograms **(b)** 8000 grams → kg
(c) 60 kg → grams **(d)** 16 000 kg → tonnes

4 Change these capacities into the units given.
   (a) 7000 mℓ → ℓ   **(b)** 800 cℓ → ℓ
   (c) 30 cℓ → mℓ   **(d)** 5 ℓ → cℓ

5 Find the missing values.
   (a) 7000 mm = ___ m   **(b)** 50 mℓ = ___ cℓ
   (c) 4 kg = ___ g   **(d)** 15 km = ___ m
   (e) 5000 kg = ___ t   **(f)** 2000 mℓ = ___ ℓ

6 A small bag of chocolates weighs 0.4 kg. I have 3 bags.
   How many grams of chocolates do I have altogether?

7 How many 150 mℓ glasses can be filled from a 3 litre
   bottle?

Converting and comparing measurements

> To compare measurements, they must be in the
> same units.

Example 11

Change these lengths to centimetres:

(a) 43 mm   **(b)** 7.8 m   **(c)** 0.75 km
(d) 0.645 m   **(e)** 136 mm   **(f)** 9 mm

Which is the longest length? Which is the shortest?

(a) 43 mm = 43 ÷ 10 cm = 4.3 cm

(b) 7.8 m = 7.8 × 100 cm = 780 cm

(c) 0.75 km = 0.75 × 1000 m = 750 m

   and 750 m = 750 × 100 cm = 75 000 cm

(d) 0.645 m = 0.645 × 100 cm = 64.5 cm

(e) 136 mm = 136 ÷ 10 cm = 13.6 cm

(f) 9 mm = 9 ÷ 10 cm = 0.9 cm

The lengths are   4.3 cm   780 cm   75 000 cm   64.5 cm

                  13.6 cm   0.9 cm

The longest is 75 000 cm or 0.75 km.

The shortest is 0.9 cm or 9 mm.

Look at Section 1.5 if you
need help with × or ÷ by
10, 100, 1000.

For your answer, write the
lengths with the units from
the question.

Example 12

Put these masses in order, smallest first:

(a) 950 g   **(b)** 0.003 tonne   **(c)** 2250 g   **(d)** 2.16 kg

(a) 950 g

(b) 0.003 tonne = 0.003 × 1000 kg = 3 kg
    and 3 kg = 3 × 1000 g = 3000 g

(c) 2250 g

(d) 2.16 kg = 2.16 × 1000 g = 2160 g

So 950 g < 2.16 kg < 2250 g < 0.003 tonne

You can only compare if they are all written in the same units.
Always change them to the *smallest* unit.

950 g < 2160 g < 2250 g < 3000 g
Write the measurements in the units from the question.

Exercise 8L

1 Change these lengths to metres:
   (a) 320 cm   **(b)** 4500 mm   **(c)** 4.2 km
   (d) 0.485 km   **(e)** 87 cm   **(f)** 750 mm
   Which is the longest length? Which is the shortest?

2 Change these masses to kilograms:
   (a) 5640 g   **(b)** 2.6 t   **(c)** 800 g
   Which is the smallest?

3 Change these capacities to centilitres:
   (a) 163 mℓ   **(b)** 9.3 ℓ   **(c)** 0.7 ℓ
   Which is the largest?

4 Put these lengths in order from smallest to largest.
     450 cm   0.05 km   4620 mm   5.1 m

UAM **5** Two lengths of timber are the same price. One has a label saying 2450 mm. The other says 2.4 m. Which one is the best buy?

UAM **6** Place these bottles in order from the smallest amount to the largest amount of foam bath.

8.5 Converting between metric and imperial units

You need to be able to convert between metric and imperial units.

To help you to do this, you need to remember these approximate metric and imperial equivalents:

Metric	Imperial
* 8 km	5 miles
* 30 cm	1 foot
* 1 litre	$1\frac{3}{4}$ pints
* 4.5 litres	1 gallon
* 1 kg	2.2 pounds
2.5 cm	1 inch
1 m	39 inches
25 g	1 ounce

You need to remember the ones marked * for your exam.

The others are useful.

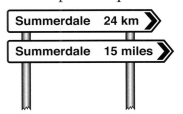

Example 13

A marathon is 26 miles. How many kilometres is this?

```
        5 miles = 8 km
So    25 miles = 40 km   (multiplying by 5)
        5 miles = 8 km  so  1 mile = 8/5 km = 1.6 km
So    26 miles = 40 + 1.6 = 41.6 km
```

A marathon is actually 26 miles and 385 yards but 26 miles is accurate enough for this calculation.

Example 14

Ahmed and Susie each guess the mass of a cake.
Ahmed says it is 9 pounds. Susie says it is 3.5 kg.
The correct answer is 8 pounds. Who is nearer?

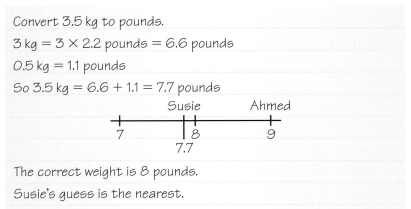

Convert 3.5 kg to pounds.

3 kg = 3 × 2.2 pounds = 6.6 pounds

0.5 kg = 1.1 pounds

So 3.5 kg = 6.6 + 1.1 = 7.7 pounds

The correct weight is 8 pounds.

Susie's guess is the nearest.

The correct mass was given in pounds so convert all the mass to pounds.

Exercise 8M

1 It is 23 miles from Dover to Calais (by sea).
How many kilometres is this?

2 Which is bigger?
  (a) 1 inch or 1 centimetre   **(b)** 1 pound or 1 kilogram
  (c) 1 mile or 1 kilometre   **(d)** 1 litre or 1 pint
  (e) 1 yard or 1 metre   **(f)** 1 litre or 1 gallon

Use the conversion table to help you.

3 Change these into the units given.
  (a) 2 inches → cm   **(b)** 7.5 cm → inches
  (c) 2 feet → cm   **(d)** 35 cm → inches
  (e) 5 miles → km   **(f)** 24 km → miles
  (g) 2.5 miles → km   **(h)** 2 km → miles

4 Change these into the units given.
  (a) $3\frac{1}{2}$ pints → litres   **(b)** 5 litres → pints
  (c) 5 gallons → litres   **(d)** 13.5 litres → gallons
  (e) 2.25 litres → gallons   **(f)** 4.2 gallons → litres

5 Change these into the units given.
  (a) 3 kilograms → pounds   **(b)** $6\frac{1}{2}$ kg → pounds
  (c) 11 pounds → kg   **(d)** 44 pounds → kg
  (e) 8 ounces → grams   **(f)** 375 grams → ounces

6 I have a picture 8 inches by 6 inches. Frames are
measured in metric units. Change my measurements
into centimetres to find the size of frame I need.

7 A caravan water tank holds 12 gallons. Will it hold 50
litres of water? Explain your answer.

8 Change this recipe into
metric measurements.

8 ounces flour
5 ounces butter
2 ounces cocoa
4 ounces sugar
1 egg

9 Sam weighs 43 kg. Lynn weighs 95 pounds.
Who is the heaviest?

10 A rainwater butt contains 200 litres of water. Another contains 360 pints. Which one contains the most?

UAM 11 A bicycle wheel has a diameter of 18 inches. A new tyre is marked 'diameter 45 cm'. Will it fit?

UAM 12 Sara's car does 32 miles on one gallon of petrol. John's car does 8 miles on 1 litre of petrol. Which car has the better petrol consumption?

How far does each car travel on 1 gallon of petrol?

8.6 Converting areas and volumes

Key words:
square
cube
cubic

These two squares are the same size:

Area is measured in **square** units: cm^2, m^2 or km^2.

Area = 1 m × 1 m
= 1 m²

Area = 100 cm × 100 cm
= 10 000 cm²

1 m² = 10 000 cm²

The most common mistake is to think that because there are 100 cm in 1 m there must be 100 cm² in 1 m². You can see that the correct answer is 10 000 cm² = 1 m².

In the same way: 1 km = 1000 m
so   1 km² = 1000 m × 1000 m

1 km² = 1 000 000 m²

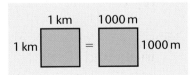

You can use the same method for imperial measures.
For example

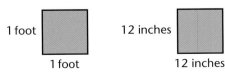

For imperial units you write 'square feet' 'square miles'. You do not use ².

1 foot = 12 inches
So   1 square foot = 12 inches × 12 inches
So   **1 square foot = 144 square inches**

Example 15

(a) Convert 20 m² to cm².   **(b)** Convert 15 000 cm² to m².

> **(a)** $20 \, m^2 = 20 \times 10\,000 \, cm^2 = 200\,000 \, cm^2$
>
> **(b)** $15\,000 \, cm^2 = 15\,000 \div 10\,000 \, m^2 = 1.5 \, m^2$

> To change from larger to smaller units you *multiply*.

You can use the same ideas for volumes.

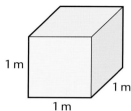 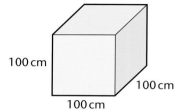

> Smaller unit → larger unit
> → *divide*.

> 1 m = 100 cm.

Volume
= 1 m × 1 m × 1 m
= 1 m³

Volume
= 100 cm × 100 cm × 100 cm
= 1 000 000 cm³

> The **cubes** are the same size. They have the same volume.

> Volume is measured in **cubic** units: cm³ or m³.

$$1 \, m^3 = 1\,000\,000 \, cm^3$$

> A common mistake is to think that 1 m³ = 100 cm³. The correct answer is much larger than this.

Example 16

(a) Convert 3 m³ to cm³.   **(b)** Convert 4000 cm³ to m³.

> **(a)** $3 \, m^3 = 3 \times 1\,000\,000 \, cm^3 = 3\,000\,000 \, cm^3$
>
> **(b)** $4000 \, cm^3 = 4000 \div 1\,00\,000 \, m^2 = 0.004 \, m^3$

Exercise 8N

1 (a) Convert 4 m² to cm².

(b) Convert 1900 cm² to m².

2 (a) Convert 2.5 m³ to cm³.

(b) Convert 300 000 cm³ to m³.

3 Copy and complete these conversions:

   (a) 3 m² = ___ cm²   **(b)** 5 m³ = ___ cm³

   (c) 70 000 cm² = ___ m² **(d)** $\frac{1}{2}$ m³ = ___ cm³

   (e) $\frac{1}{2}$ m² = ___ cm²   **(f)** 3 000 000 cm³ = ___ m³

4 How many:
  (a) mm^2 in 1 cm^2 **(b)** cm^2 in 1 km^2
  (c) square feet in a square yard?

Draw squares to help you.

1 yard = 3 feet

5 How many:
  (a) m^3 in 1 km^3 **(b)** mm^3 in 1 m^3
  (c) cubic inches in 1 cubic foot?

6 A lake covers an area of 3.25 million square metres.
What is its area in square kilometres?

7 A rectangular room measures 420 cm by 380 cm.
  (a) What is the area of the floor in cm^2?

A shop sells carpet by the square metre.
  (b) What does the room measure in square metres?
  (c) How many whole square metres of carpet should I buy?
  (d) If the carpet costs £10.50 per square metre, how much do I pay?

8 A rectangular tank for a poisonous spider measures 85 cm by 65 cm by 45 cm.
  (a) What is the volume of the container?
  (b) This type of spider needs at least $\frac{1}{4}$ m^3 of space. Is this container suitable?

8.7 Accuracy of measurement

There are two kinds of measurement: discrete and continuous.

 Discrete measure is for quantities that can be counted or only have certain values. For example, the number of spectators at a tennis final was 12 416.

 Continuous measure can take any values in a range. The accuracy of the measure depends on the measuring instrument. For example, if you measure your height as 174 cm, it may not be *exactly* 174 cm.
It is likely to be 174 cm **to the nearest** centimetre.

Key words:
discrete
continuous
to the nearest

You will meet these again in Chapter 10.

Notice that the ≤ sign is used for the 'smaller' value but the < sign is used for the 'larger' value.

Think what this really means.

The true height lies between 173.5 cm and 174.5 cm.

This is written as 173.5 cm ≤ height < 174.5 cm.

If the value was 174.5 cm, it would round up to 175 cm to the nearest cm.

In the same way, if your mass is 68 kg to the nearest kilogram, it must be nearer to 68 kg than to either 67 kg or 69 kg.

Your true mass is
67.5 kg ≤ mass < 68.5 kg.

Any measurement given to the nearest whole unit may be inaccurate by up to one half in either direction.

Example 17

A rectangle has sides of 8 cm and 5 cm, each measured to the nearest centimetre.

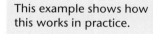

This example shows how this works in practice.

(a) Write the range within which each measurement must lie.

(b) What is its smallest possible area?

5 cm

8 cm

> 8 cm means a length in the range 7.5 cm ≤ length < 8.5 cm
>
> 5 cm means a length in the range 4.5 cm ≤ width < 5.5 cm
>
> The rectangle could be as small as 7.5 cm by 4.5 cm
>
> The smallest possible area = 7.5 × 4.5 cm^2
>
>   = 33.75 cm^2

Area of rectangle
= length × width

Exercise 80

1 These have been measured to the nearest whole unit. Write the range within which each measurement must lie.

For example,
4.5 cm ≤ 5 cm < 5.5.

    (a) 9 cm   **(b)** 12 kg   **(c)** 65 cℓ   **(d)** 10 seconds

2 These have been measured to the nearest millimetre. Write the range within which each measurement must lie.

Change them all to mm.

(a) 14 mm   (b) 1.4 cm   (c) 3.8 cm   (d) 7.5 cm

3 These have been measured to the nearest centimetre. Write the range within which each measurement must lie.

What must you remember to do first?

(a) 125 cm   (b) 1.25 m   (c) 4.07 m   (d) 6.20 m

4 For each question state whether the measurement is discrete or continuous.

(a) The height of a door

(b) The number of sweets in a bag

(c) The mass of a cake

(d) The cost of pairs of shoes

(e) The capacity of a glass

5 A rectangular box is 17 cm long by 11 cm wide by 5 cm high to the nearest centimetre.

(a) Write down the range within which each length must lie.

(b) What is the smallest possible volume of the box?

6 A concrete paving stone measures 1.4 m by 1.2 m to the nearest 10 cm.

(a) What is its smallest possible area?

(b) What is its largest possible area?

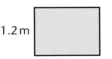

Remember, the largest area will have the longest possible sides.

8.8 Compound measures

Speed

You will need to know:

- **how to convert times between 'hours' and 'hours and minutes'**

Key words:
speed   distance   time

Speed is a measurement of how fast something is travelling.

It involves two other measures, **distance** and **time** . If you travel from Newcastle to Leeds, a distance of 100 miles, and it takes 2 hours, you have averaged 50 miles per hour.

Your speed would not be constant for the whole journey so the speed calculated is an *average* speed.

To calculate a speed you divide a distance by a time.

There are three formulae connecting speed, distance and time:

$$\text{speed} = \frac{\text{distance}}{\text{time}} \qquad \text{time} = \frac{\text{distance}}{\text{speed}}$$

$$\text{distance} = \text{speed} \times \text{time}$$

100 miles in 2 hours:
$$\frac{100}{2} = 50 \text{ miles per hour.}$$

This triangle will help you to remember the formulae.

If you cover up the one you want to find you are left with the formula you need.

For example, cover up 'T', you are left with $\dfrac{D}{S}$.

which means that $T = \dfrac{D}{S}$ or $\text{time} = \dfrac{\text{distance}}{\text{speed}}$.

Speed can be measured in many different units.

The most common are:

metres per second (m/s), kilometres per hour (km/h) centimetres per second (cm/s), miles per hour (mph)

When you do calculations involving speed, distance and time you must make sure that you are consistent with the units that you use.

The word 'per' means divide. Kilometres per hour (km/h) means you divide a distance in km by a time in hours.

For example:

- to calculate a speed in **metres per second** you need the distance in **metres** and the time in **seconds**.
- if you divide a distance in **miles** by a speed in **miles per hour** your answer for the time taken will be in **hours**.

Example 18

Jo went on her holidays and travelled a distance of 180 miles. The journey took $4\frac{1}{2}$ hours.

What was her average speed for the journey?

$$\text{Average speed} = \frac{\text{total distance travelled}}{\text{total time taken}}$$

$$= \frac{180 \text{ miles}}{4.5 \text{ hours}}$$

$$= 40 \text{ mph}$$

Distance in miles divided by time in hours gives speed in miles per hour.

Example 19

Simon averaged a speed of 80 km/h for a journey of 208 km. How long did the journey take?

$$\text{Time} = \frac{distance}{speed} = \frac{208}{80} = 2.6$$

0.6 hours = 0.6 × 60 minutes = 36 minutes

Time taken for journey = 2 hours 36 minutes

> Distance in km divided by speed in km/h gives time in hours.

> You must *not* read this as 2 hours 6 minutes.
> It is 2.6 hours.

> To convert the decimal part to minutes you multiply by 60.
> Look back at Example 5 to help you.

Example 20

Ian drove for 2 hours 45 minutes at an average speed of 48 mph. How far did he travel?

2 hours 45 minutes = $2\frac{3}{4}$ hours = 2.75 hours

Distance = speed × time

        = 48 × 2.75

        = 132

Distance = 132 miles

> 2 hours 45 needs to be written in hours only.
> 45 minutes is $\frac{3}{4}$ or 0.75 of an hour, so the time is 2.75 hours.

> To change minutes to hours, divide by 60:
> $\frac{45}{60} = \frac{3}{4} = 0.75$.

Exercise 8P

1 I cycled 60 miles in 4 hours. What was my average speed?

> Remember to state the units in your answers.

2 John ran the last 400 metres to school in 50 seconds. What was his average speed?

3 It took me 4 hours to travel to Coventry at an average speed of 65 mph. How far did I travel?

4 It is 14 miles along the Castle Wood Walkway. I walk at an average speed of 4 mph. How long will it take me to complete the walk?

5 I left home at 9 am to drive to Bristol 150 miles away. I arrived there at 1 pm. What was my average speed?

6 A long-distance lorry drove 320 miles in 5 hours 45 minutes. What was its average speed? Give your answer correct to 1 d.p.

7 How long would it take to travel 270 miles at a speed of 50 mph?

UAM **8** I cycled for 3 hours at a speed of 12 mph. I then walked for 2 hours at a speed of 3 mph. How far did I travel altogether?

UAM **9** During a race Ali ran the first 300 metres in 2 minutes 10 seconds and the last 100 metres in 30 seconds. What was his average speed in m/s?

> Look at Example 18 to help you.

10 The journey to a conference was in three parts.
- 50 mile drive to the airport took 1 hour 20 minutes.
- 1200 mile flight took 2 hours.
- 42 mile bus trip from the airport to the conference venue took 55 minutes.

(a) What was the total travelling time?

(b) What was the total distance travelled?

(c) What was the average speed for the whole journey?

Density

Density is defined as '**mass** per unit **volume**'.

It is the mass (usually given in grams) of one unit of volume of material (usually given in cm³).

This means that density is calculated by dividing the mass of an object by its volume.

There are three formulae connecting density, mass and volume.

$$\text{density} = \frac{\text{mass}}{\text{molume}} \qquad \text{volume} = \frac{\text{mass}}{\text{density}}$$

$$\text{mass} = \text{density} \times \text{volume}$$

> **Key words:**
> density
> mass
> volume

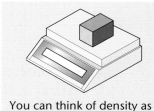

You can think of density as the mass of 1 cm³.

This triangle will help you to remember the formulae. You use it like the one for speed, distance and time.

You need to be consistent with the units that you use.

- Density is usually in g/cm³
- Volume is usually in cm³
- Mass will then be calculated in g.

Cover the one you want to find. Read the formula you need. For example, to find density, cover D.

Read $\dfrac{M}{V} = \dfrac{mass}{volume}$.

If mass is in kg and volume is in m³, density will be in kg/m³.

Example 21

A piece of wood weighs 124 g and has a volume of 140 cm³.

What is the density of the wood?

$$\text{Density} = \frac{mass}{volume} = \frac{124}{140} = 0.886 \ (3 \ sf)$$

$$\text{Density} = 0.886 \ g/cm^3 \ (3 \ sf)$$

The terms 'mass' and 'weight' are not exactly the same but in this section we shall take them to be the same.

Example 22

A measuring cylinder contains a liquid whose density is 1.18 g/cm³. The volume of the liquid is 0.35 ℓ.

What is the mass of the liquid?

$$0.35 \ \ell = 0.35 \times 1000 \ cm^3 = 350 \ cm^3$$

$$\text{Mass} = \text{density} \times \text{volume} = 1.18 \times 350 = 413$$

$$\text{Mass} = 413 \ g$$

Change the units of volume from litres to cm³.
1 ℓ = 1000 cm³.

Exercise 8Q

1 What is the density of a piece of material with a volume of 50 cm³ and a mass of 225 g?

Remember to give your answer using the units g/cm³.

2 Gold has a density of approximately 19 g/cm³. What is the mass of a 250 cm³ gold bar?

3 Flour has a density of 1.5 g/cm³. What is the volume of a bag weighing 825 g?

4 Calculate the missing measurements in this table,

	Mass	Density	Volume
(a)	24 g	1.6 g/cm³	
(b)		2.05 g/cm³	40 cm³
(c)	28 g		13.3 cm³
(d)		3.2 g/cm³	1.5 litres
(e)	5.7 kg	1.24 g/cm³	

Change the units first in parts **(d)** and **(e)**.

5 (a) Which has the greatest mass, a cubic metre of feathers or a cubic metre of steel?

(b) Which has the greatest density, feathers or steel?

6 What is the volume of a concrete block whose mass is 50 kg and density is 1600 kg/m³? Give your answer in m³, correct to 2 decimal places.

7 A sculpture in a museum has a volume of about 3 m³. The material from which it is made has a density of 11.4 g/cm³.
What is the approximate mass of the sculpture? Give your answer in kg.

Examination style questions

1 A speed limit in France is 100 kilometres per hour. The speedometer shows the speed of a lorry.

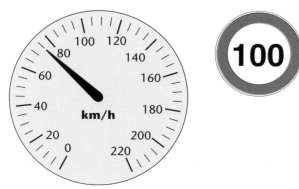

How much slower than the speed limit is the lorry travelling? *(2 marks)*

AQA, Spec B, 5F, June 2003

2 A printing machine prints 8 pages per minute.
How long will it take to print 1000 pages?
Give your answer in hours and minutes. *(3 marks)*
AQA, Spec B, 5F, June 2003

3 Dave drives 15 miles to work.
The journey takes 20 minutes.
What is Dave's average speed in miles per hour? *(3 marks)*
AQA, Spec B, 3F, March 2003

4 The diagram shows some kitchen scales.

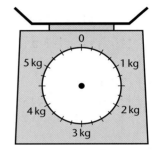

(a) Mrs Hall weighs a chicken on the scales.
The chicken weighs $3\frac{1}{2}$ kilograms.

(i) Draw an arrow on the diagram to show
$3\frac{1}{2}$ kilograms.

(ii) Change $3\frac{1}{2}$ kilograms into pounds.
Give your answer to the nearest pound.

(b) Mrs Hall's recipe book states:
To cook a chicken: allow 20 minutes per pound and add 30 minutes.
How long does it take to cook a chicken that weighs five pounds?
Give your answer in hours and minutes. *(4 marks)*
AQA, Spec A, F, June 2003

5 A sports pitch has a length of 75 metres, correct to the nearest metre.
Write down the least and the greatest possible length of this pitch. *(2 marks)*
AQA, Spec B, 3I, June 2002

6 This scale shows pints and litres.

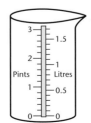

(a) Sketch the scale, then draw an arrow on the scale to show 2.5 pints.
(b) Use the scale to estimate how many pints are in 1 litre.
(c) Estimate the number of litres in 8 pints. *(4 marks)*
AQA, Spec A, F, June 2004

7 Apples are sold in a farm shop at £1.76 per kilogram.
Calculate the price of 1 pound of apples.
Use the conversion   1 kg = 2.2 pounds

(3 marks)
AQA, Spec B, 3F, June 2004

8 Charles drove 132 miles at an average speed of 55 mph.
Calculate the time taken for this journey.
Give your answer in hours and minutes.

(4 marks)
AQA, Spec B, 3I, June 2004

9 Henry arrives at Fareham station at 9.30 am.
He catches the first train to Newport.
This train is on time throughout the whole of its journey.
How long is the train journey to Newport?

Portsmouth Harbour	dep	0600	0708		0824	0924		1024	1124		1224		1324
Portsmouth & Southsea	dep	0604	0712		0828	0928		1028	1128		1228		1328
Cosham	dep	0614	0722		0839	0939		1039	1139		1239		1339
Fareham	dep	0624	0730		0847	0947	1006	1047	1147		1247		1347
London Waterloo ⊖	dep									1217			
Southampton Central	dep	0652	0754	0809	0909	1009	1033	1109	1209		1309	1319	1409
Romsey	dep	0703	0805	0834	0920	1020	1044	1120	1220		1320	1344	1420
Dunbridge	dep		0810	0840			1049					1350	
Dean	dep		0815	0846								1356	
Salisbury	arr	0721	0828	0859	0940	1040	1104	1140	1240		1340	1409	1440
Salisbury	dep	0723	0830	0908	0940	1040	1107	1140	1240		1340	1410	1440
Warminster	dep	0743	0850	0928	1000	1100	1127	1200	1300	1354	1400	1430	1500
Westbury	arr	0751	0856	0936			1133	1206	1306		1406	1438	
Westbury	dep	0756	0858	0937			1136	1207	1307		1407	1439	
Trowbridge	dep	0802	0904	0943	1011	1111	1142		1313	1405		1445	1511
Bradford-on-Avon	dep	0808	0910	0949			1148			1411		1451	
Bath Spa	dep	0824	0922	1002	1027	1127	1204	1228	1328	1423	1428	1512	1528
Bristol Temple Meads	arr	0842	0940	1020	1045	1145	1222	1245	1345	1441	1445	1530	1545
Newport	arr	1020	1119		1219	1319		1444	1519	1513	1624		1722
Cardiff Central	arr	1037	1135		1235	1338		1502	1537	1529	1639		1739

(6 marks)
AQA, Spec B, 3F, November 2002

Summary of key points

Estimating (grade G/F)

Lengths are measured in

| Metric: kilometres (km), metres (m), centimetres (cm), millimetres (mm) |
| Imperial: miles, yards, feet, inches |

Capacity is measured in

| Metric: litres(ℓ), centilitres(cℓ), millilitres (mℓ) |
| Imperial: gallons, pints, fluid ounces |

Mass is measured in

| Metric: tonnes (t), kilograms (kg), grams (g) |
| Imperial: tons, hundredweight, stones, pounds, ounces |

To estimate length, capacity and mass, compare with familiar objects (e.g. bag of sugar = 1 kg or 2.2 pounds). Give estimates in sensible units (e.g. measure long distances in miles or km, not in inches or cm).

Reading scales (grade G)

To read a scale, you need to look at how it is marked.
How many spaces are there between the numbers?
Work out what one space represents.
Sometimes you have to estimate a reading from a scale.
On this scale, each space represents 2 units.

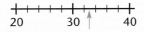

A sensible estimate is 33.

Times, dates and timetables (grade E)

There are two ways of measuring time:

- using am for morning and pm for afternoon

- using the 24-hour clock.

am and pm	24-hour clock
12.00 midnight	0000 or 2400
7.15 am	0715
12.00 noon (midday)	1200
3.40 pm	1540
9.55 pm	2155

You need to remember these facts:

60 seconds = 1 minute 60 minutes = 1 hour 24 hours = 1 day 7 days = 1 week 365 days = 1 year 366 days = 1 leap year	30 days have September, April, June and November, All the rest have 31, except in February alone which has but 28 days clear and 29 in each leap year.

In a timetable, times for any one train (or bus) are usually written in *columns*.

Newcastle	1103	1130	1157
Durham		1143	
Darlington	1131	1202	
York	1202	1231	1250

You read the time the train leaves each place in the *row* opposite the place name. For example, the 1130 train from Newcastle leaves Darlington at 1202.

Conversions between metric units of length, capacity and weight (grade G/F)

You need to know:

Metric units of length: 10 mm = 1 cm, 100 cm = 1 m 1000 mm = 1 m, 1000 m = 1 km	Metric units of capacity: 100 cℓ = 1 ℓ 1000 mℓ = 1 ℓ	Metric units of mass 1000 mg = 1 g 1000 g = 1 kg 1000 kg = 1 tonne

To compare measurements, they must be in the same units.

Conversions between metric and imperial units (grade F/E)

You need to remember these for your exam:

Metric	8 km	30 cm	1 litre	4.5 litres	1 kg
Imperial	5 miles	1 foot	$1\frac{3}{4}$ pints	1 gallon	2.2 pounds

Conversions of areas and volumes (grade D/C)

$1\,m = 100\,cm$

so $1\,m^2 = 100\,cm \times 100\,cm = 10\,000\,cm^2$.

$1\,m^3 = 100\,cm \times 100\,cm \times 100\,cm$
$\quad = 1\,000\,000\,cm^3$

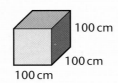

Accuracy of measurement (grade C)

Any measurement given to the nearest whole unit
may be inaccurate by up to one half in either direction.
For example a time of 21 seconds (to the nearest second)
means $20.5\,sec \leqslant time < 21.5\,sec$.

Compound measures (grade D/C)

$$\text{speed} = \frac{\text{distance}}{\text{time}} \qquad \text{time} = \frac{\text{distance}}{\text{speed}}$$

$$\text{distance} = \text{speed} \times \text{time}$$

$$\text{density} = \frac{\text{mass}}{\text{volume}} \qquad \text{volume} = \frac{\text{mass}}{\text{density}}$$

$$\text{mass} = \text{density} \times \text{volume}$$

This chapter will show you how to:
- ✔ find perimeters and areas of simple and compound shapes
- ✔ learn the vocabulary of the circle
- ✔ calculate the circumference and area of a circle
- ✔ draw and interpret 3-D shapes
- ✔ calculate the volumes of prisms and cylinders

9.1 Perimeters and areas

Rectangles, parallelograms and triangles

> Key words:
> perimeter
> area
> parallelogram
> perpendicular height

The **perimeter** of a shape is the sum of the lengths of all its sides.

For a rectangle of length *l* and width *w*

perimeter $= l + w + l + w$
$\quad\quad\quad = 2l + 2w$

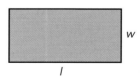

Perimeter of a rectangle = $2(l + w)$

The **area** of a shape is the amount of space inside it.

This rectangle has length 4 cm and width 3 cm.

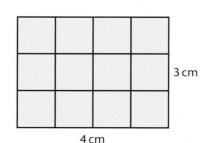

4 cm

3 cm

The rectangle is drawn on a centimetre square grid.

The area is 12 squares. Each square is 1 cm².

The area is 12 cm².

You can calculate area using the formula

Area of a rectangle = length × width
= l × w

The base of a **parallelogram** is b and its
perpendicular height is h.

You could cut triangle *ABE* from one end of the
parallelogram and place it on the other end, to make a
rectangle.

The perpendicular height is
perpendicular (at 90°) to
the base.

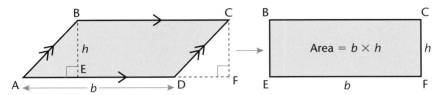

The parallelogram and the rectangle have the *same area*
so

Area of a parallelogram = base × perpendicular
height
= b × h

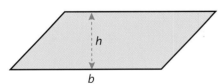

You must remember to use
the *perpendicular* height
not the slant height.

The diagonal in this parallelogram splits it into two
identical triangles.

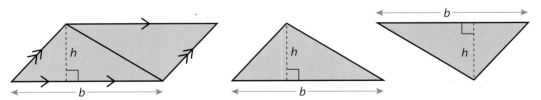

The two triangles have equal areas.
Each one is half the area of the parallelogram.

Area of a triangle = $\frac{1}{2}$ × base × perpendicular
height
= $\frac{1}{2}$ × b × h

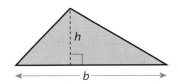

Example 1

Find the perimeter and area of these shapes:

(a)

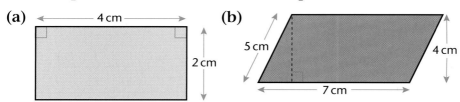

← 4 cm →

2 cm

(b)

5 cm

4 cm

← 7 cm →

(c)

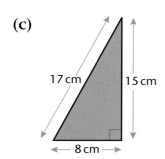

17 cm

15 cm

← 8 cm →

(a) Perimeter $= 2(l + w)$

$= 2(4 + 2)$

$= 2 \times 6$

$= 12 \ cm$

Area $= l \times w$

$= 2 \times 4$

$= 8 \ cm^2$

> Order of operations: brackets first.

> Square units for area.

(b) Perimeter $= 5 + 7 + 5 + 7$

$= 24 \ cm$

Area $= b \times h$

$= 7 \times 4$

$= 28 \ cm^2$

> Start at the bottom left vertex and go round the shape.

(c) Perimeter $= 17 + 15 + 8$

$= 40 \ cm$

Area $= \frac{1}{2} \times b \times h$

$= \frac{1}{2} \times 8 \times 15$

$= 60 \ cm^2$

Exercise 9A

1 Find the perimeter and area of these shapes.

All lengths are in centimetres.

(a)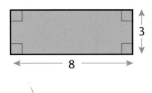

(b)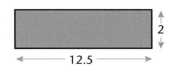

(c)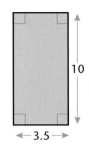

(d)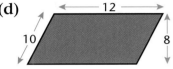

(e)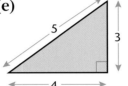

(f)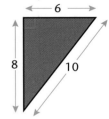

(g)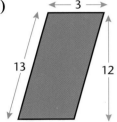

(h)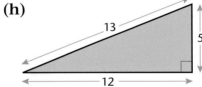

Substitute the values into the formula for area. Rearrange to solve the equation.

2 Each triangle has area 20 cm². Calculate the lengths marked x in each case.

(a)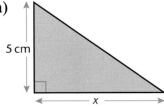

(b)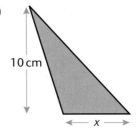

(c)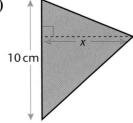

Trapezia

A **trapezium** is a quadrilateral with *one* pair of parallel sides.

This trapezium has:
- parallel sides of length a and b
- perpendicular height h.

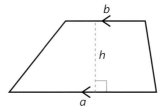

If you put an identical trapezium upside down next to it, you get a parallelogram:

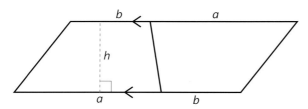

The parallelogram is made of two identical **trapezia**.

The base of this parallelogram is $(a + b)$ and its perpendicular height is h.

So the area of this parallelogram is
base × perpendicular height $= (a + b) \times h$.

The area of the trapezium is half the area of this parallelogram.

> **Area of a trapezium** $= \frac{1}{2} \times$ **(sum of parallel sides)**
> $\times$ **perpendicular height**
> $= \frac{1}{2} \times (a + b) \times h$

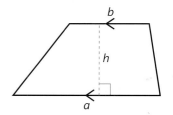

Another way of saying this is:
$\frac{1}{2} \times$ sum of parallel sides × distance between them

Example 2

Find the area of these shapes.

(a)

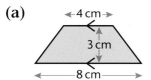

(b)

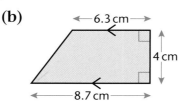

(a) Area $= \frac{1}{2}(a + b) \times h$

$\qquad = \frac{1}{2}(4 + 8) \times 3$

$\qquad = \frac{1}{2} \times 12 \times 3$

$\qquad = 6 \times 3$

$\qquad = 18 \ cm^2$

(b) Area $= \frac{1}{2}(a + b) \times h$

$\qquad = \frac{1}{2}(8.7 + 6.3) \times 4$

$\qquad = \frac{1}{2} \times 15 \times 4$

$\qquad = \frac{1}{2} \times 4 \times 15$

$\qquad = 2 \times 15$

$\qquad = 30 \ cm^2$

Order of operations: brackets first.

Remember the units.

15×4 is the same as 4×15. It is easier to multiply 4 by $\frac{1}{2}$.

Example 3

For each of these shapes you are given the area.
Calculate the value of the unknown length x in each case.

(a) Area $= 35 \ cm^2$

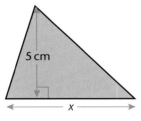

(b) Area $= 27 \ cm^2$

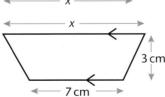

(a) Area of triangle $= \frac{1}{2} \times b \times h$

$\qquad 35 = \frac{1}{2} \times x \times 5$

$\qquad 70 = 5x$

$\qquad 14 = x$

(b) Area of trapezium $= \frac{1}{2} \times (a + b) \times h$

$\qquad 27 = \frac{1}{2} \times (x + 7) \times 3$

$\qquad 54 = 3(x + 7)$

$\qquad 18 = x + 7$

$\qquad 11 = x$

Multiply both sides by 2.

Divide both sides by 5.

Multiply both sides by 2.

Divide both sides by 3.

Subtract 7 from both sides.

Exercise 9B

1 Find the area of these shapes:

(a)
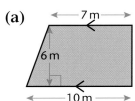

(b)
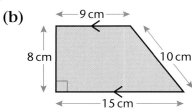

2 (a) This trapezium has area 15 cm².

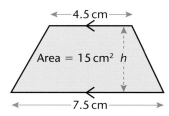

Calculate the height, *h*, of the trapezium.

(b) This trapezium has area 30 cm².

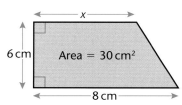

Calculate the length marked *x*.

3 Work out the area of these shapes.

All lengths are in cm.

(a)
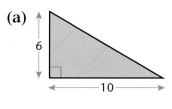

(b)
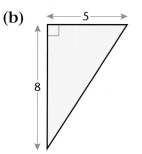

(c)
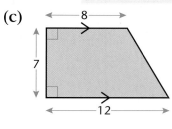

(d)
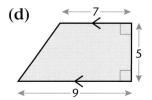

(e)
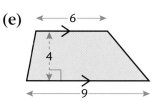

(f)
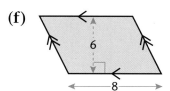

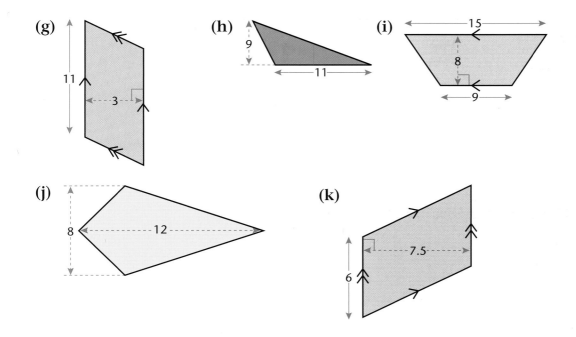

(g) **(h)** **(i)**

(j) **(k)**

Areas of compound shapes

A **compound shape** is made from simple shapes.

Key words:
compound shape

To find the area of a compound shape, you split it into simple shapes.

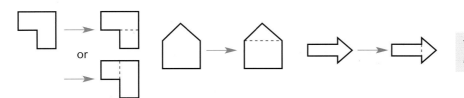

The simple shapes here are *rectangles* and *triangles*.

Use the formulae for areas of the simple shapes.

Example 4

Find the perimeter and area of this compound shape.

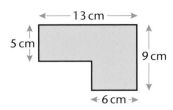

Split the shape into two rectangles *A* and *B*.

Total width of shape = 13 cm

So   x + 6 = 13

      x = 7 cm

Total height of shape = 9 cm

So   y + 5 = 9

      y = 4 cm

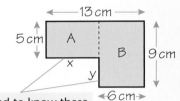

You need to know these lengths to work out the perimeter. Label them x and y.

You could split it like this instead:

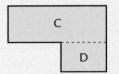

You would get the same answers.

Work out the missing lengths.

Perimeter = 5 + 13 + 9 + 6 + 4 + 7

         = 44 cm

      y = 4   x = 7

Starting at the bottom left corner and working clockwise.

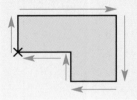

Area = area of A + area of B

    = (5 × 7) + (9 × 6)

    = 35 + 54

    = 89 cm²

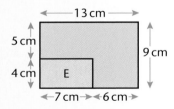

Area of large rectangle = 13 × 9 = 117 cm²

Area of E = 4 × 7 = 28 cm²

Area of shape = area of large rectangle − area of E

         = 117 − 28 = 89 cm²

Order of operations: brackets first.

Another way to work this out is:

add a small rectangle *E* to 'fill in' the missing part.

Example 5

Find the area of this shape:

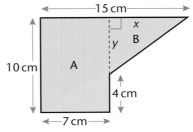

> Split the shape into rectangle *A* and triangle *B*.

> You need to find the lengths marked *x* and *y* before you can find the area of the triangle.

Total width of shape = 15

So x + 7 = 15

     x = 8

Total height of shape = 10

So y + 4 = 10

     y = 6

Area = rectangle A + triangle B

     = (10 × 7) + ($\frac{1}{2}$ × 6 × 8) = 70 + 24 = 94 cm²

> Area of triangle = $\frac{1}{2} \times h \times b$

Exercise 9C

> All lengths are in cm.

1 Find the perimeter and area of each shape.

(a)

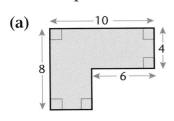

(b)

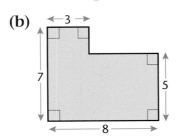

(c)

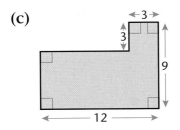

(d)

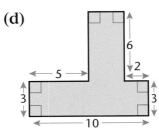

(e)

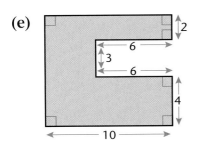

(f)

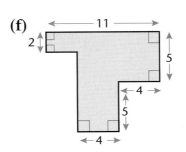

2 Work out the area of these shapes.

All lengths are in cm.

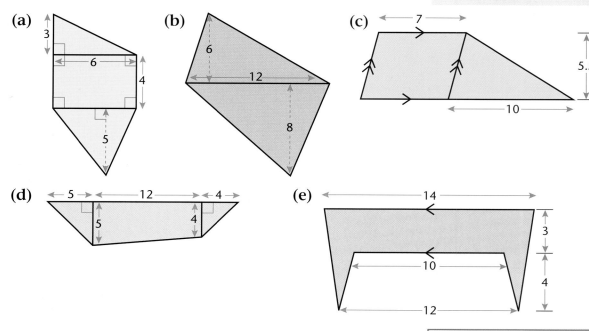

9.2 The circle

The **diameter** is twice the length of the **radius** .

The **circumference** is the distance around the outside.

$d = 2 \times r = 2r$

You need to know these parts of a circle:

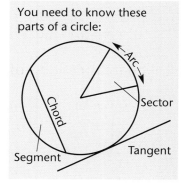

Circumference of a circle

1 Draw three circles, each with a different radius, using a pair of compasses.
2 Draw a diameter on each circle and measure its length.
3 Use a piece of string to measure the length of the circumference of each circle.

4 Work out $\dfrac{\text{circumference of circle}}{\text{diameter of circle}}$ for each circle.

The answer will be approximately 3 each time.

In fact the answer is just over 3 and is given a special symbol π.

The diameter passes through the centre of the circle.

π (pi), a letter of the Greek alphabet.

$$\frac{\text{circumference}}{\text{diameter}} = \pi$$

You can rearrange this formula as
Circumference $= \pi \times$ diameter

$C = \pi d$, where C is circumference and d is diameter.

Diameter $d = 2r$, where r is the radius, so $C = \pi \times 2r$ or

$C = 2\pi r$

> The value of π is 3.141592654... π is a non-recurring, non-terminating decimal.
> Your calculator has a $\boxed{\pi}$ key.
> In calculations, you often use 3.14 or 3.142 as an approximation for π.

> You can use these formulae to calculate the circumference of a circle when you know either the radius or the diameter.

Example 6

A circle has diameter 36 cm.
Calculate its circumference.
Leave your answer as a multiple of π.

$C = \pi d$

$\quad = \pi \times 36$

$\quad = 36\pi$ cm

> On a calculator paper, you would have to calculate the answer using the $\boxed{\pi}$ key on your calculator.
> $C = \pi d$
> $\quad = \pi \times 36$
> $\quad = 113.097$
> $\quad = 113$ cm (to the nearest cm).

Example 7

The distance around the edge of a circular pond is 10.5 metres.
Calculate the radius of the pond.

Using the formula $\quad C = \pi d$

$\qquad\qquad\qquad 10.5 = \pi \times d$

$\qquad\qquad\qquad \dfrac{10.5}{\pi} = d$

which gives $\qquad\quad d = 3.342...$

and so $\qquad\qquad r = \dfrac{3.342...}{2} = 1.67$ m (3 sf)

> Dividing by π rearranges the formula to make d the subject.

> Remember that $d = 2 \times r$, so divide d by 2 to get r.

Example 8

This protractor is a semicircle with radius 5 cm.
Calculate the perimeter of the protractor.

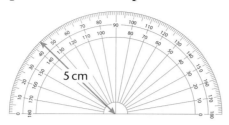

The circumference of a whole circle of radius 5 cm is $2 \times \pi \times 5$

So the distance around a semicircle of radius 5 cm is

$\pi \times 5 = 15.7$ cm

| Leaving out the '2' means you calculate half of the circumference. |

Perimeter = semicircular distance + diameter

$\qquad = 15.7 + 10$

$\qquad = 25.7$ cm

Don't forget to add on the straight edge.

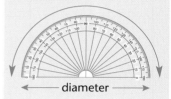

Exercise 9D

1 Calculate the circumference of these circles, leaving your answer as a multiple of π.

  (a) Diameter = 12 cm   **(b)** Radius = 4 cm
  (c) Radius = 15 m

diameter = 2 × radius
$\qquad$ = 2 × 5 cm
$\qquad$ = 10 cm

2 Calculate the circumference of these circles:
  (a) Diameter = 23.5 cm
  (b) Radius = 6.7 cm
  (c) Radius = 0.84 m

3 Calculate the diameter of these circles:
  (a) Circumference = 48 cm
  (b) Circumference = 18.2 cm
  (c) Circumference = 57 mm

Use Example 7 to help you.

4 Calculate the radius of these circles:
  (a) Circumference = 21 cm
  (b) Circumference = 70.8 mm
  (c) Circumference = 6.25 m

5 Calculate the perimeter of these:

(a) A 'quarter-light' window in the shape of a quarter of a circle of radius 52 cm.

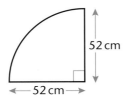

52 cm

52 cm

Use Example 8 to help you.

(b) A church window in the shape of a semicircle on top of a rectangle.

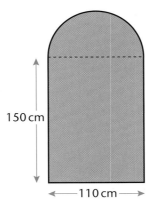

150 cm

110 cm

You need the diameter of the semicircle. Look at measurements on other parts of the diagram to find this.

(c) A running track in the shape of two semicircular ends joined by two 'straights'.

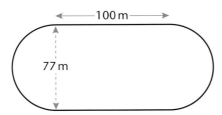

100 m

77 m

6 A simple dartboard has two circles of wire. The larger one has diameter 40 cm. The smaller one has diameter 20 cm. They are held together by four straight pieces, as shown in the diagram.
Calculate the total length of wire needed to make the dartboard.

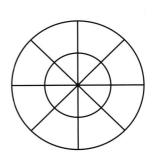

UAM **7** A bicycle wheel has a diameter of 65 cm. How many complete revolutions does the wheel make when the bicycle travels 800 metres?

You will need to change the units.

8 This Penny-Farthing bicycle has a front wheel of radius 90 cm and a rear wheel of radius 21 cm.

In a journey, the front wheel turns 170 times.

(a) How far was the journey?

(b) How many complete turns did the rear wheel make?

> Give your answer in metres.

Area of a circle

1 Draw a circle of radius 8 cm.
2 Mark off points on the circumference at 30° intervals.
3 Draw the 12 sections.

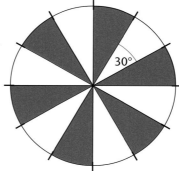

> Circumference
> $= 2\pi r$
> $= 2 \times \pi \times 8$
> $= 50.265...$
> $= 50$ cm (to the nearest cm)

4 Cut out the sections and arrange them like this:

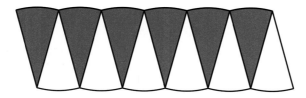

This is approximately the shape of a rectangle.
The width of the 'rectangle' is 8 cm.
The length of the 'rectangle' is approximately 25 cm.

So the area of the 'rectangle' $\approx 25 \times 8 = 200$ cm^2

> The radius of the original circle.

> $\frac{1}{2}$ the circumference of the original circle.

> $\approx$ means 'is approximately equal to'.

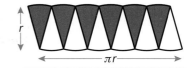

The formula for the circumference of the circle is $2\pi r$, so the length of the 'rectangle' $= \pi r$.
The width of the 'rectangle' is r.

So the area of the 'rectangle' = length $\times$ width
$$= \pi r \times r = \pi r^2$$

This is the same as the area of the original circle.

Area of a circle $A = \pi r^2$

For a circle of radius 8 cm: $A = \pi \times 64 = 201.06... \text{ cm}^2$

This is very close to the approximate answer of 200 cm^2.

Example 9

Calculate the area of these circles:

(a) Radius $= 10$ cm

(b) Diameter $= 18.6$ cm

(a) $A = \pi r^2$ or $A = \pi r^2$

$\quad = \pi \times 10^2$ $= \pi \times 10^2 = 314.159...$

$\quad = \pi \times 100$ $= 314 \text{ cm}^2$

$\quad = 100\pi \text{ cm}^2$ (to the nearest cm)

(b) $r = d \div 2 = 18.6 \div 2 = 9.3$ cm

$\quad A = \pi r^2$

$\quad\quad = \pi \times 9.3^2$

$\quad\quad = 272 \text{ cm}^2 \text{ (3 sf)}$

Answer left as a multiple of π on the non-calculator paper.

Using the π button on the calculator.

$A = \pi r^2$ uses the radius.

First find the radius.

Using the π button on the calculator.

Example 10

A circle has area 215 cm^2. Calculate its diameter.

Using the formula $A = \pi r^2$

$\quad\quad\quad 215 = \pi \times r^2$

$\quad\quad\quad \dfrac{215}{\pi} = r^2$

$\quad r^2 = 68.436...$

so $r = \sqrt{68.436...}$

$\quad\quad = 8.272$

$d = 2r = 2 \times 8.272... = 16.5 \text{ cm} \text{ (3 sf)}$

Rearranging to make r^2 the subject.

$\sqrt{}$ means square root.

Keep the full calculator display at this stage.

Only round your answer at the end.

You can calculate the area of compound shapes involving circles or parts of circles.

Example 11

A garden is in the shape of a rectangle with two semi-circles, one on the length of the rectangle and one on the width, as shown in the diagram.

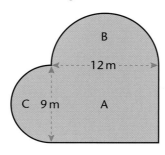

Calculate the area of the garden, giving your answer to the nearest square metre.

Area of rectangle $A = 12 \times 9 = 108 \text{ m}^2$

Area of semicircle $B = \frac{1}{2} \times \pi \times 6^2 = 56.548\ldots$

Area of semicircle $C = \frac{1}{2} \times \pi \times 4.5^2 = 31.808\ldots$

Total area $= 108 + 56.548\ldots + 31.808\ldots$

$\qquad = 196.35\ldots$

$\qquad = 196 \text{ m}^2 \text{ (to nearest m}^2\text{)}$

Area of semicircle $= \frac{1}{2}$ area of circle. Diameter of $B = 12$ m, so radius $= 6$ m. Diameter of $C = 9$ m, so radius $= 4.5$ m.

If you round each of the three answers to the nearest whole number and then add them your answer would be $108 + 57 + 32 = 197 \text{ m}^2$ (which is incorrect). You should round at the *end* of the calculation.

Exercise 9E

1 Calculate the area of these circles, leaving your answers as a multiple of π.

   (a) Diameter $= 12$ cm   **(b)** Radius $= 4$ cm

   (c) Radius $= 15$ cm

2 Calculate the area of these circles.

   (a) Diameter $= 23.5$ cm   **(b)** Radius $= 6.7$ cm

   (c) Radius $= 0.84$ m

3 Calculate the area of the three shapes in question 5 in exercise 9D.

4 A circle has area 164 cm². Calculate its diameter, giving your answer to 3 s.f.

5 A circle has area 208 cm².
Find   **(a)** its radius   **(b)** its circumference.

6 Find the shaded area in each of these:

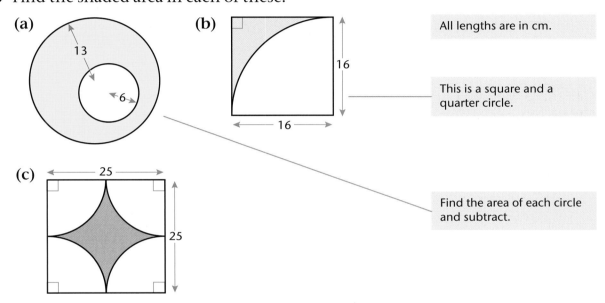

(a)

13

6

(b)

16

16

All lengths are in cm.

This is a square and a quarter circle.

Find the area of each circle and subtract.

(c)

25

25

7 A square of side 14 cm and a circle have the same area.

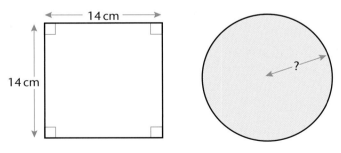

14 cm

14 cm

?

Find the radius of the circle.

8 The circumference of a circle is 85 cm.
Find its area, giving your answer to the nearest cm².

9.3 3-D objects

Isometric drawing

You can draw 3-D objects on isometric paper.
Draw along the printed lines of the paper.
Vertical lines on the paper represent vertical lines
of the object.
The lines at an angle on the paper represent
horizontal lines on the object.

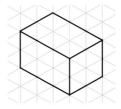

Example 12

This is the cross-section of a
3-D object.
Draw the 3-D object on isometric
paper.

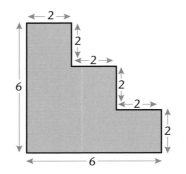

The cross-section is the
shape that runs through
the whole of the solid.

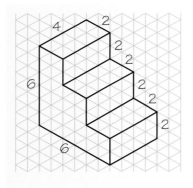

Draw the cross-section first.
Draw the depth of the
object next ... the depth
could be any length, here it
has been drawn 4 units
long.
Join the ends of these lines
to form the other side of
the steps.

Exercise 9F

On isometric paper draw 3-D diagrams of the objects with
these cross-sections.

1

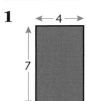

2

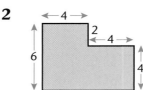

Choose any value for the
depth.

3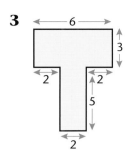

4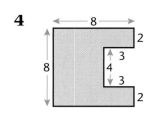

Planes of symmetry

You can also draw 3-D objects on squared paper.
Here are three diagrams of a cuboid. In each one you can
see a **plane of symmetry** (shaded red).

> **A plane of symmetry divides a 3-D object into two
> equal halves where one half is the mirror image of
> the other.**

It is the 3-D equivalent of a
line of symmetry.

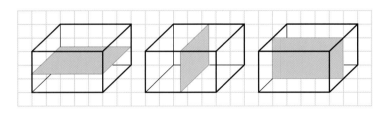

Example 13

Copy this 3-D object on squared paper. Show all its planes
of symmetry. Draw a separate diagram for each plane of
symmetry.

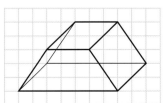

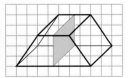 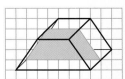

This object has two planes
of symmetry. If you make a
horizontal cut, the top half
would be smaller than the
bottom half, so the cut
would *not* be a plane of
symmetry.

Exercise 9G

Copy these 3-D objects on squared paper.

Show all their planes of symmetry.

Draw separate diagrams for each plane of symmetry.

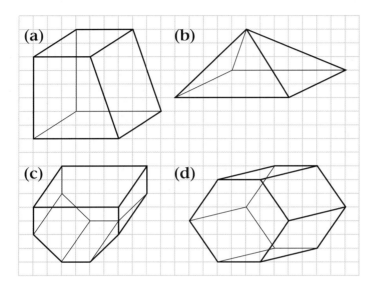

(a) (b) (c) (d)

Nets of 3-D objects

Key words:
net

Here is a closed box in the shape of a cuboid. It has been opened out so that you can see the shape of the card it is made from.

The 2-D shape that you see is called a **net** .

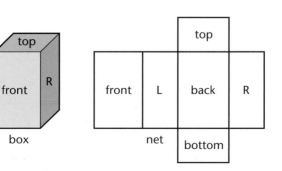

box　　net

The net of a 3-D object is any 2-D shape that folds up to make the 3-D object.

Example 14

Make an accurate drawing of the net of this 3-D object.

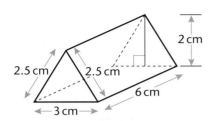

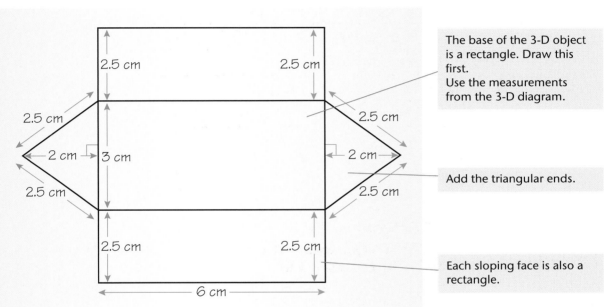

The base of the 3-D object is a rectangle. Draw this first.
Use the measurements from the 3-D diagram.

Add the triangular ends.

Each sloping face is also a rectangle.

Example 15

This is the net of a 3-D object. *Sketch* the object.

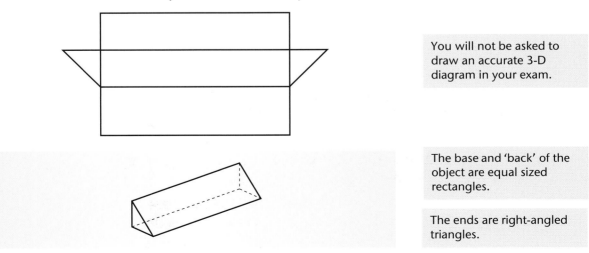

You will not be asked to draw an accurate 3-D diagram in your exam.

The base and 'back' of the object are equal sized rectangles.

The ends are right-angled triangles.

Exercise 9H

1 Make accurate drawings of the nets of these 3-D objects.

(a)

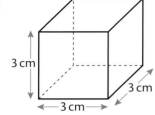

(b)

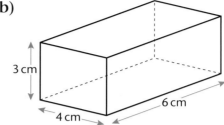

(c)

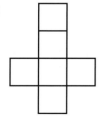

(d)

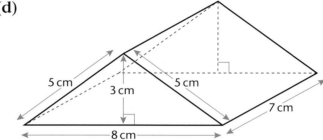

2 Here are some nets of 3-D objects. Sketch each object.

(a)

(b)

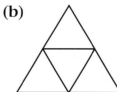

(c)

(d)

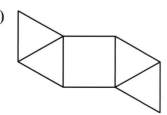

Plans and elevations

You can show a 3-D object in detail by drawing three different 2-D views of it.

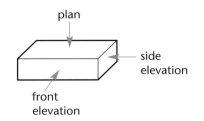

plan

side elevation

front elevation

Key words:
plan
front elevation
side elevation

The **plan** is the view from above.
The **front elevation** is the view from the front.
The **side elevation** is the view from the side.

Example 16

Draw the plan view, the front elevation and the side elevation (from the right-hand side) of this block of six cubes.

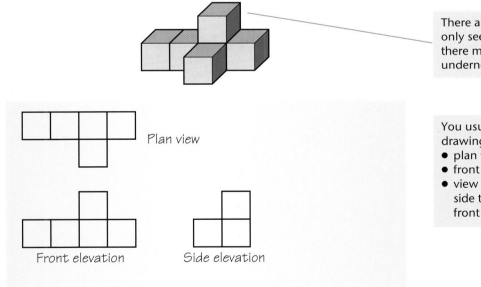

There are 6 cubes. You can only see 5 of them but there must be one underneath this one.

Plan view

Front elevation   Side elevation

You usually set out the drawings like this:
- plan view at the top
- front view beneath it
- view from the right-hand side to the right of the front elevation.

Exercise 9I

Draw a plan view, a front elevation and a side elevation (from the right-hand side) for each block of cubes.

1

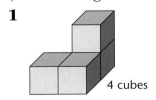

4 cubes

2

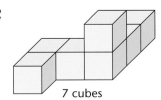

7 cubes

3

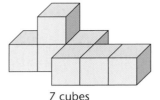

7 cubes

4

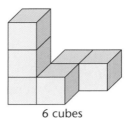

6 cubes

5

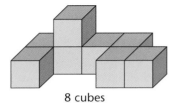

8 cubes

6

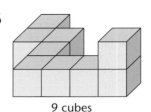

9 cubes

7

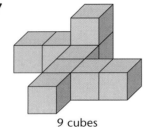

9 cubes

8

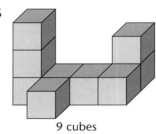

9 cubes

Using plans and elevations

You could be asked to find the volume or surface area of a block of cubes.

In this block, the cubes all have side 1 cm.

So each cube has volume 1 cm^3.

There are 6 cubes, so the volume is $6 \times 1 = 6$ cm^3.

The surface area is the area of all the faces you see if you look at the structure from all sides (including from underneath!).

For these cubes, the area of each face is 1 cm^2.

The surface area of the block is:

number of faces you can see $\times$ 1 cm^2

You can draw the view from the back, the view from the left and the view from underneath, as well as the plan, front and side elevations.

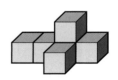

Just count the cubes!

You need to count the faces from *every* direction.

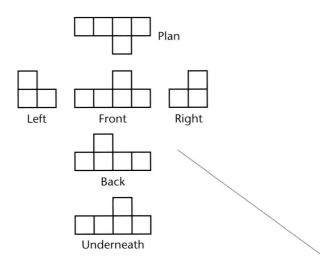

See Example 16 for these.

Can you see that you do not have to draw all six views?
The **back** is a mirror image of the **front**.
The **left** is a mirror image of the **right**.
The **underneath** is a mirror image of the **plan**.

You could count the faces in the plan, front and side elevations and then double your answer.

Each square is a face of a cube.

Counting all the faces you can see will give you the surface area. It is 26 cm².

Exercise 9J

Calculate the volume and surface area for each block of cubes in exercise 9I.

All the cubes have side 1 cm.

9.4 Prisms

Key words:
prism
cross-section

You will need to know:

- **area of a rectangle = length × width**
  area = $l \times w$
- **area of a triangle = $\frac{1}{2}$ × base × height**
  area = $\frac{1}{2} \times b \times h$
- **area of a trapezium = $\frac{1}{2}$ × (sum of parallel sides) × height**
  area = $\frac{1}{2} \times (a + b) \times h$

A **prism** is a 3-D object whose **cross-section** is the same all through its length.

In these prisms the cross-section is shaded.

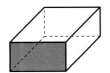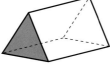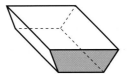

If you cut a 'slice' parallel to the end face, your slice will be the same shape as the end face.

Volume of a prism

A **cuboid** is a prism. Its cross-section is a rectangle.

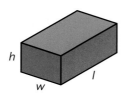

To calculate its volume you use this formula:

> **Volume of cuboid = height × width × length**
> **= $h \times w \times l$**

Another way of writing this formula is

> Volume = area of end face × length

You can use a similar formula to calculate the volume of *any* prism:

> **Volume of prism = area of cross-section × length**

Imagine the cuboid made from cubes, for example:

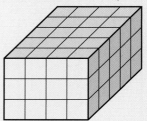

In the end face there are
3 × 4 cubes.
Along the length of the cuboid there are 5 'slices' of 3 × 4 cubes.
So there are 3 × 4 × 5 cubes in total.

Area of end face = $h \times w$

Area of cross-section
= area of end face

Example 17

Find the volume of this cuboid.

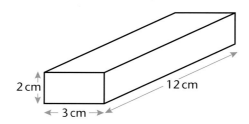

Volume = $h \times w \times l$
  = 2 × 3 × 12
  = 72 cm³

Cubic units for volume.

Example 18

The cross-section of this prism is a trapezium.
Calculate the volume of the prism.

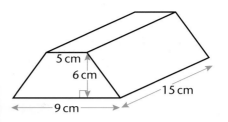

Volume = area of cross-section × length
Area of cross-section = $\frac{1}{2}(a + b) \times h$
$= \frac{1}{2}(5 + 9) \times 6$
$= \frac{1}{2} \times 14 \times 6$
$= 7 \times 6$
$= 42\ cm^2$
Volume of prism = 42 × 15
$= 630\ cm^3$

For this trapezium $a = 5$,
$b = 9$, $h = 6$.

Order of operations:
brackets first.

Example 19

Calculate the volume of this prism.

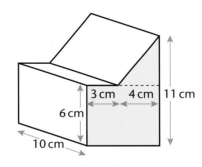

Area of rectangle = 7 × 6 = 42 cm²
Area of triangle = $\frac{1}{2} \times 4 \times 5 = 10\ cm^2$
Area of cross-section = 42 + 10 = 52 cm²
Volume of prism = area of cross-section
    × length
    = 52 × 10 = 520 cm³

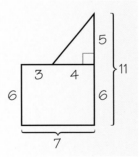

First find the area of the
cross-section (end face).
Split it into a rectangle and
a triangle, as shown.

Then add the two areas
together.

Exercise 9K

1 Find the volume of each of these prisms:

(a)

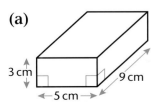

(b)

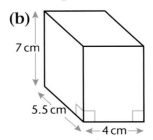

(c)

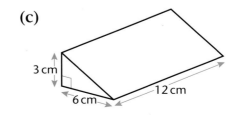

2 Find the volume of each prism:

(a)

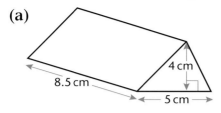

(b)

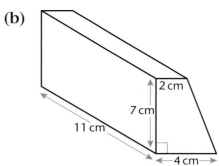

3 Calculate the volume of these prisms:

(a)

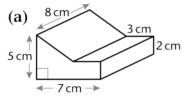

(b)

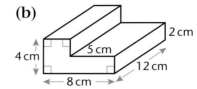

(c)

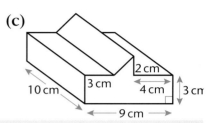

> Split the end face into simpler shapes to calculate its area.

Surface area of a prism

The surface area of a 3-D object is the area of all the faces you can see.
You can use nets to help you calculate surface area.
For the cuboid in Example 17:

> In Section 9.3 you found surface area by counting cubes in plans and elevations.

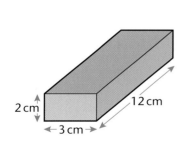   the net is

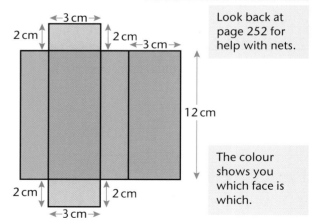

> Look back at page 252 for help with nets.

> The colour shows you which face is which.

The net is the 3-D object 'flattened out', so you can see all the faces at once.

These are 2 green faces,   area 3 × 12
            2 orange faces, area 2 × 12
            2 blue faces,   area 3 × 2

Total surface area $= 2 \times (3 \times 12 + 2 \times 12 + 3 \times 2)$
$$= 2(36 + 24 + 6)$$
$$= 2 \times 66$$
$$= 132 \text{ cm}^2$$

You can use this formula to calculate the surface area of a cuboid:

Surface area $= 2(h \times w + h \times l + w \times l)$
$$= 2(hw + hl + wl)$$

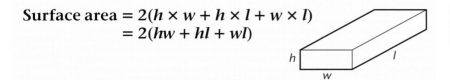

These are the areas of the faces in its net.

Example 20

Calculate the surface area of this cuboid.

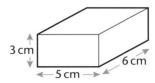

Surface area $= 2(hw + hl + wl)$
$$= 2(3 \times 5 + 3 \times 6 + 5 \times 6)$$
$$= 2(15 + 18 + 30)$$
$$= 2 \times 63$$
$$= 126 \text{ cm}^2$$

$h = 3, w = 5, l = 6$

Example 21

Calculate the surface area of this prism.

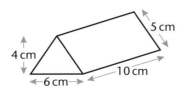

Area of each blue face $= 10 \times 5$
$$= 50 \text{ cm}^2$$
Area of green face $= 6 \times 10$
$$= 60 \text{ cm}^2$$

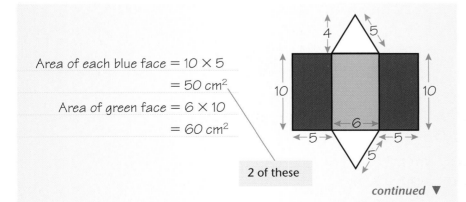

Sketch the net and label the measurements.

2 of these

continued ▼

Area of triangle $= \frac{1}{2} \times h \times b$
$= \frac{1}{2} \times 4 \times 6$
$= 12 \, cm^2$ —————— **2 of these**

Surface area of prism $= 2 \times 50 + 60 + 2 \times 12$
$= 100 + 60 + 24$
$= 184 \, cm^2$

Exercise 9L

1 Calculate the surface area of these cuboids:

(a)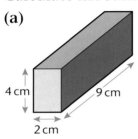

4 cm   9 cm   2 cm

(b)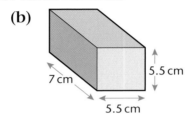

7 cm   5.5 cm   5.5 cm

(c)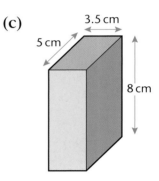

3.5 cm   5 cm   8 cm

2 Calculate the surface area of these prisms:

Sketch and label the net first.

(a)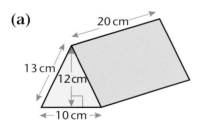

20 cm   13 cm   12 cm   10 cm

(b)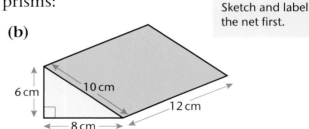

6 cm   10 cm   12 cm   8 cm

9.5 Cylinders

Volume of a cylinder
You will need to know:
- **area of a circle $= \pi r^2$**

A cylinder is a prism whose cross-sectional area is a circle.

Volume of a cylinder = area of cross-section × height
$$V = \pi r^2 h$$

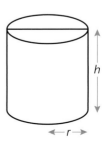

Example 22

Find the volume of a cylinder of radius 9 cm and
height 30 cm.
Give your answer as a multiple of π.

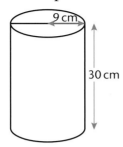

$$Volume = \pi r^2 h$$
$$= \pi \times 9^2 \times 30$$
$$= \pi \times 81 \times 30$$
$$= 2430\pi \ cm^3$$

Example 23

A cylindrical container holds 8.5 litres of liquid.
The height of the container is 24 cm.
Calculate the radius of the base.

$$8.5 \ litres = 8.5 \times 1000 \ cm^3 = 8500 \ cm^3$$
$$Volume = \pi r^2 h$$
$$8500 = \pi \times r^2 \times 24$$
$$\frac{8500}{\pi \times 24} = r^2$$
$$112.73\ldots = r^2$$
$$\sqrt{112.73\ldots} = r$$
$$radius = 10.6 \ cm \ (3 \ s.f.)$$

The volume is 8.5 ℓ. You must use 8500 cm^3

Substitute the values you know.
$V = 8500$, $h = 24$

Rearrange the formula to make r^2 the subject – divide by π and by 24.

Use the $\boxed{\sqrt{}}$ key on your calculator.

Exercise 9M

1 Find the volume of these cylinders. Give your answers
   in terms of π.
   (a) Base radius = 2 cm, height = 8 cm
   (b) Base radius = 3 cm, height = 5 cm
   (c) Base diameter = 14 cm, height = 20 cm
   (d) Base radius = 4 cm, height = 2 m

2 Find the capacity, in litres, of an oil drum of base radius 30 cm and height 80 cm.
Give your answer in terms of π.

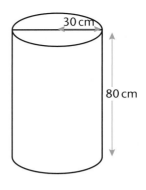

$1000 \text{ cm}^3 = 1\ell$

3 A cylinder has a volume of $180\pi \text{ cm}^3$.
The diameter of the cylinder is 12 cm.
Calculate the height of the cylinder.

You are given the *diameter*.

4 These objects were made by cutting cylinders along axes of symmetry. Find the volume of each object.

(a)

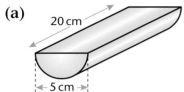

(b)

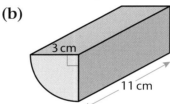

UAM **5** The diagrams show two cylindrical tins of cat food.

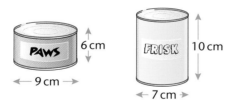

Which tin holds the most food?

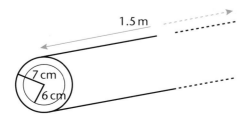

6 A cylinder of height 27 cm holds 3.6 litres of liquid.
What is the radius of the base of the cylinder?

Use Example 23 to help you.

UAM **7** A hollow metal tube, 1.5 m long, is in the form of a cylinder. The external radius is 7 cm and the internal radius is 6 cm.

(a) Calculate the volume of metal in the tube.

(b) The density of the material from which it is made is 3.4 g/cm³.
Calculate the mass of the tube.

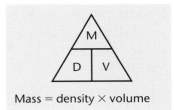

Mass = density × volume

8 A cylindrical metal rod 2 m long has a diameter of 4 cm.
The mass of the rod is 17 kg.
Calculate the density of the metal.

9 Cylindrical tins of diameter 10 cm and height 12 cm are packed *two deep* in a box in the shape of a cuboid 40 cm long, 20 cm wide and 24 cm high.

(a) How many tins will fit into the box?

(b) Calculate the volume of one tin.

(c) What is the total volume of all the tins?

(d) Calculate the volume of the box.

(e) What is the volume of unused space in the box when the tins are packed into it?

Surface area of a cylinder

Key words:
curved surface area
total surface area

You will need to know:

● **circumference of a circle = $2\pi r$**

The net of this cylinder is

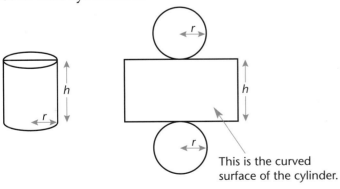

This is the curved surface of the cylinder.

When you peel the label off a tin of beans, the label is rectangular.

The width of the rectangle is the same as the circumference of the circular top.

When you wrap the label round the tin, its top edge makes a circle.

Curved surface area = $2\pi r \times h$ CSA for short.
$$= 2\pi rh$$

The area of each circular end = πr^2

There are 2 of these.

Total surface area of a cylinder = $2\pi rh + 2\pi r^2$
$$= 2\pi r(h + r)$$

TSA for short.

Area of curved surface + area of ends.

Factorising.

Example 24

Calculate the surface area of a cylinder with base radius
12 cm and height 20 cm.
Give your answer as a multiple of π.

TSA $= 2\pi r(h + r)$
$= 2\pi \times 12 \, (20 + 12)$
$= 2\pi \times 12 \times 32$
$= \pi \times 2 \times 12 \times 32$
$= 768\pi \, cm^2$

Numbers before letters.

Square units for area.

Example 25

A thin cylindrical metal rod is 2 metres long and has
diameter 4.6 cm.
Calculate its total surface area.

2 metres $= 2 \times 100$ cm $= 200$ cm

Diameter $= 4.6$ cm, radius $= \frac{1}{2} \times 4.6 = 2.3$ cm

TSA $= 2\pi r(h + r)$
$= 2\pi \times 2.3 \, (200 + 2.3)$
$= 2 \times \pi \times 2.3 \times 202.3$
$= 2923.503...$
$= 2924 \, cm^2$

Mixed units. Change them
all to centimetres.

Substitute $h = 200$, $r = 2.3$

Using the $\boxed{\pi}$ key.

Rounding to a sensible
degree of accuracy.

Exercise 9N

1 Find the total surface area of each cylinder in
Exercise 9M question 1.
Give your answers in terms of π.

2 A cylindrical cushion is 30 cm long.
The radius of each end is 10 cm.
Calculate the area of fabric in the cover for this
cushion.

3 A tin of beans has a diameter of 7.6 cm and a height
of 11 cm.
The label on the tin overlaps by 1.5 cm.
The height of the label is 5 mm less than the height of
the tin.

Beware – mixed units!

Calculate the surface area of the label around the tin.

4 A cylinder of height 10 cm has a curved surface area
of $240\pi\,\text{cm}^2$.
Calculate the volume of this cylinder.
Give your answer as a multiple of π.

5 Calculate the total surface area of each shape in
Exercise 9M question 4.

Sketch the net.

Examination style questions

1 The length of a rectangle is 9.4 cm.
The perimeter of the rectangle is 30.2 cm.

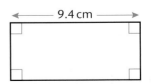

Calculate **(a)** the width of the rectangle
 (b) the area of the rectangle.

(5 marks)

2 Calculate the area of each of these shapes.

(a)

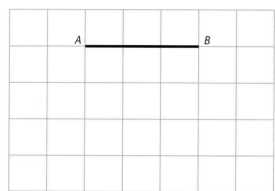

(b)

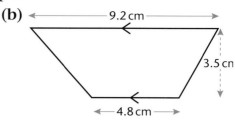

(5 marks)

3 The diagram shows a solid shape made from 8 cubes.
Copy and complete the plan view of the shape on the grid below.

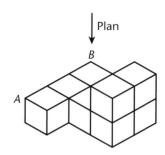

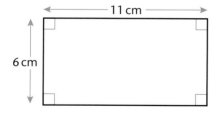

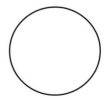

(2 marks)
AQA, Spec A, I, November 2003

4 The diagrams show a rectangle of length 11 cm and width 6 cm and a circle.
The circumference of the circle is equal to the perimeter of the rectangle.

Calculate the area of the circle, giving your answer to the nearest cm^2.

(5 marks)

5 A school hall is in the shape of a cuboid.
(a) The school hall is 30 m long, 12 m wide and 4 m high.
(i) Calculate the volume of the hall.
(ii) Calculate the total area of the **four walls** of the hall.

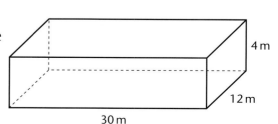

(b) The school buys ten 5 litre tins of paint to paint the hall.
The area to be painted is 279 m^2.
Each tin cover 30 m^2.
Calculate the percentage of paint used.

(8 marks)
AQA, Spec A, I, June 2003

Summary of key points

Perimeters and areas of simple shapes (grade F/E for rectangle, otherwise D)

The perimeter of a shape is the sum of the lengths of all its sides.
The area of a shape is the amount of space inside it.

Rectangle
$\qquad$ Perimeter = $2l + 2w = 2(l + w)$
$\qquad$ Area = $l \times w$

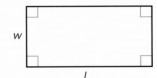

Parallelogram
$\qquad$ Area = $b \times h$
$\qquad$ (h = perpendicular height)

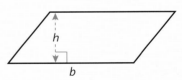

Triangle
$\qquad$ Area = $\frac{1}{2} \times b \times h$

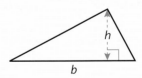

Trapezium
$\qquad$ Area = $\frac{1}{2} \times (a + b) \times h$

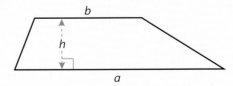

Circle (grade D/C)

Circumference = $2\pi r$ $\quad$ or $\quad$ πd
Area = πr^2

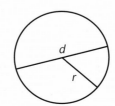

Areas of compound shapes (grade D)

To find the area of a compound shape you split it into simple shapes. Use the formula for area for each shape separately.

For example

Isometric drawing (grade E)

You can draw 3-D objects on isometric paper.
Draw along the printed lines of the paper.

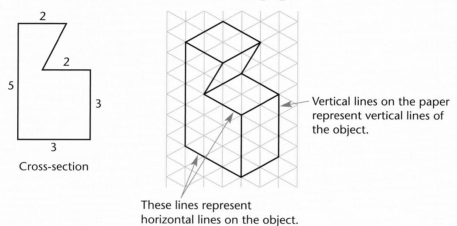

Cross-section

Vertical lines on the paper represent vertical lines of the object.

These lines represent horizontal lines on the object.

Plane of symmetry (grade D)

A plane of symmetry divides a 3-D object into two equal halves where one half is the mirror image of the other.

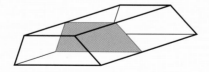

Net of a 3-D object (grade G/F)

The net of a 3-D object is the 2-D shape that folds up to make the 3-D object.

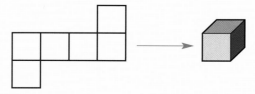

Plans and elevations (grade D)

For any 3-D object:

- The plan is the view from above.
- The front elevation is the view from the front.
- The side elevation is the view from the side.

 Plan view

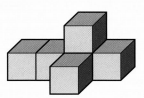

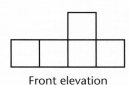

Front elevation   Side elevation

Prisms (grade E/D for cuboids, otherwise C)

Volume of cuboid = $h \times w \times l$

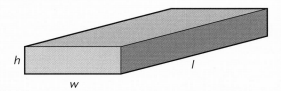

Volume of prism = area of cross-section × length

The cross-section of a prism must be constant.

Surface area of cuboid = $2 \times (h \times w + h \times l + w \times l)$
$$= 2(hw + hl + wl)$$

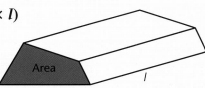

Cylinders (grade C)

Volume of cylinder = $\pi r^2 h$

Curved surface area = $2\pi rh$

Total surface area = $2\pi rh + 2\pi r^2 = 2\pi r(h + r)$

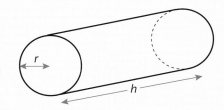

10 Collecting data

This chapter will show you how to:

✔ identify different types of data
✔ construct tally charts or frequency tables for discrete and grouped data
✔ design a questionnaire and a data-capture sheet and select a representative sample
✔ design and use two-way tables for discrete and grouped data

10.1 Types of data

Key words:
primary data
secondary data
qualitative data
quantitative data
discrete data
continuous data

Primary data is information you collect directly yourself, for example from questionnaires.

Secondary data is information that you get from existing records, for example newspapers, magazines, the internet.

Qualitative data contains descriptive words, for example a colour (red, green), or an activity (climbing, sailing), or a location (London, Paris). It is sometimes called categorical data.

Quantitative data contains numbers, such as temperatures, masses, areas, lengths, time, number of TVs or cars.

There are two types of quantitative data:

1 **Discrete data** can only have particular values. Discrete data is 'countable'.
Discrete data examples:
- Shoe size $5, 5\frac{1}{2}, 8$
- Scores on a dice $4, 2, 6$
- Goals scored in a match $0, 2, 3$

> Shoe sizes can only be whole or half numbers.

> You can't score $2\frac{1}{2}$ goals!

2 **Continuous data** can take any value in a particular range.
Continuous data examples:
- Mass 72 kg, 15.3 g, 5 lbs
- Temperature $-4°C, 25°C, 100°C$
- Length 800 m, $300\,000$ km, 2.6 mm

> 5 lbs is measured to the nearest pound.

> 2.6 mm is measured to the nearest tenth of a millimetre.

Continuous data cannot be measured exactly. The accuracy depends on the measuring instrument, for example a ruler, or thermometer.

For more on this, look back at Section 8.7.

Exercise 10A

1 State whether each set of data is quantitative or qualitative.

  (a) Height   **(b)** Age

  (c) Eye colour   **(d)** Place of birth

  (e) Distance   **(f)** Shoe size

2 State whether each set of data is discrete or continuous.

  (a) Cost in pence

  (b) Number of creatures in a rock pool

  (c) Time

  (d) Mass

  (e) Area

  (f) Score on a dartboard

  (g) Hours worked

3 State whether each source will give primary or secondary data.

  (a) Collecting data by observing traffic.

  (b) Downloading data from the internet.

  (c) Looking at data from the 2001 Census.

  (d) Using data found in a newspaper.

  (e) Giving people a questionnaire.

4 (a) How could you collect data on the following:

    (i) car engine sizes and acceleration

    (ii) pet ownership in the UK

    (iii) how people from one company travel to work

    (iv) climate – local and national?

  (b) For each type of data you describe, say whether it is primary or secondary, qualitative or quantitative.

10.2 The data handling cycle

Key words:
data handling cycle
hypothesis
analysis

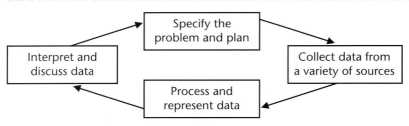

When you carry out an investigation, you will follow the **data handling cycle** . Start with a **hypothesis** . This is simply a statement that you want to test, for example 'the vowels a, e, i, o and u are all equally likely to occur in the English language'.

You will then collect the data and carry out an **analysis** of your data. This means arranging your data in a way that enables you to make sense of it.

Putting data into a bar chart, pie chart or frequency diagram helps you to analyse it. Finally, interpret or draw conclusions from your data – say what you have found out. This interpretation might lead to an alternative hypothesis.

10.3 Data collection

Key words:
tally chart
tally marks
frequency
frequency table
frequency distribution
class interval

If you collect a large amount of data, you can organise it in a table.

Here are the vowels in the first four sentences of a book.

i, a, o, e, a, i, i, e, i, i, o, a, i, a, a, e, a, a, i, e
e, a, i, o, e, i, i, o, a, i, a, a, i, e, o, i, a, i, a, a
i, i, i, i, e, a, e, u, o, o, e, i, a, u, i, e, e, o, o, e
e, u, e, e, a, a, i, a, u, i, e, a, u, i, e

It is difficult to see if the vowels all occur the same number of times.

You could put the results into a **tally chart** like this:

Vowel	Tally	Frequency
a	|||| |||| |||| ||||	20
e	|||| |||| |||| |||	18
i	|||| |||| |||| |||| |||	23
o	|||| ||||	9
u	||||	5

This is the total for each vowel.

Tallies are easier to count if you group them in 5s: ||||

You can now see that i occurred most often.

Work through the data, making a **tally mark** in the correct row for each vowel.

The complete table showing the tally marks and the **frequency** is called a **frequency table** or a **frequency distribution** .

When you have a large set of continuous data, you usually group the results together in equal sized groups. These are classes or **class intervals** .

Example 1

The heights in centimetres of dancers in a musical production were:

There are 31 heights here.

161, 153, 168, 171, 143, 161, 165, 156, 147, 158
161, 160, 180, 173, 149, 155, 164, 167, 173, 158
151, 163, 162, 174, 166, 157, 170, 161, 156, 181
168

Put these heights into a tally chart using class intervals of 140–149, 150–159, etc.

Tally the data into the correct class intervals.

Height (cm)	Tally	Frequency
140–149	|||	3
150–159	⊮ |||	8
160–169	⊮ ⊮ |||	13
170–179	⊮	5
180–189	||	2
	Total	31

Add the tallies to find the frequencies.

Check the total in the frequency column is the same as the original number of values.

Draw the table with the class intervals.

You cannot find individual heights from the table.

Class intervals are groupings of quantitative data. When you group data, you should use between 5 and 10 class intervals.

You can also write class intervals using $<$ and $\leqslant$ notation.

Instead of 140–149 you could write $140 \leqslant h < 150$.

This means the height h is 140 or more, but *less than* 150.

Using this notation, the table in Example 1 is:

Height h (cm)	Tally	Frequency
$140 \leqslant h < 150$	⦀	3
$150 \leqslant h < 160$	⦀⦀⦀ ⦀⦀⦀	8
$160 \leqslant h < 170$	⦀⦀⦀ ⦀⦀⦀ ⦀⦀⦀	13
$170 \leqslant h < 180$	⦀⦀⦀	5
$180 \leqslant h < 190$	⦀⦀	2
	Total	31

This includes all heights *up to* 180 cm, but not 180 cm.

A height of 150 cm is recorded in this class.

Data-capture sheets

Before you collect data, you can prepare a tally chart to record it in. A pre-prepared tally chart to record data is called a **data-capture sheet**.

For example, to collect data on newspapers people buy, you could prepare a chart like this:

Newspaper	Tally	Frequency
The Times The Sun Daily Mail		

Exercise 10B

1 The frequency table shows the type and number of pets a vet treated one week.

Copy and complete the tally marks and the frequency table.

Pet	Tally	Frequency
Dog	⦀⦀⦀ ⦀⦀	
Cat		9
Bird	⦀⦀⦀ ⦀	
Other		
	Total	32

2 The list below shows the calls made by a salesperson over 29 weeks.

```
22   27   18   23   25   19   17   20   23   20
18   22   20   19   24   21   16   19   18   22
14   19   18   21   20   19   25   25   21
```

Copy and complete this frequency table for the data.

Calls per week	Tally	Frequency
1–10		
11–20		
21–30		
	Total	29

3 The masses of 20 people starting a fitness class were recorded to the nearest kilogram:

```
46   55   53   75   62   59   84   63
68   74   78   60   57   61   48   81
73   67   64   77
```

Copy and complete this frequency table for the data. Use five equal class intervals.

Mass m (kg)	Tally	Frequency
$40 \leqslant m < 50$		
$80 \leqslant m < 90$		
	Total	20

4 The times for a cross-country run, in minutes, were:

```
15   18   23   27   17   21   24   19   20   28
33   26   24   19   23   28   25   19   17   31
38   24   28   21   24   26   26   19   18   27
17   26   21   23   26   19   18   26   24   27
26   16   24   28   29   19   18   33   23   29
```

Construct a frequency table for this data, showing the tally marks and frequencies. Use the class intervals 10–14, 15–19, 20–24, 25–29, 30–34, 35–40.

The times are rounded to the nearest minute.

5 The amounts of money collected during a charity week were:

£86 £89 £51 £61 £11 £56 £82 £87 £4 £89
£43 £93 £31 £20 £61 £3 £65 £3 £56 £84
£38 £80 £42 £64 £66 £22 £3 £34

Construct a frequency table using five equal class intervals. Begin with $0 \leqslant m < 20$, where m is the amount of money collected in £s.

6 Construct a data-capture sheet for throwing a fair six-sided dice. Include headings for tally marks and frequency.

7 A car designer is investigating the colours of cars sold by different dealerships. Construct a data-capture sheet to record this information.

10.4 Questionnaires

Key words:
survey
questionnaire
hypothesis
bias
pilot study

A **survey** collects primary data. One way of collecting primary data is to use a **questionnaire** .

Questionnaire guidelines:
- Be clear about what you want to find out. A survey should test your theory or **hypothesis** . Think about what data you need to collect.
- Ask specific questions, for example, what music do you like, how much do you earn, what is your favourite sport?
- Keep your questions short and simple. Tick boxes give a clear choice of replies.

A traffic survey collects data on traffic.

A questionnaire makes sure you ask everyone in the survey for the same information.

Are you:

    male? ☐   female? ☐

This question has a clear choice of two options. They include all possible answers.

What is your favourite sport?

Hockey ☐   Football ☐   Gymnastics ☐

Rugby ☐   Athletics ☐   Other ☐

There are six choices here. 'Other' covers all the choices not in the first five.

- Avoid questions that are vague or could be misunderstood.

> Do you use the internet:
>
> sometimes? ☐   occasionally? ☐   often? ☐

These words mean different things to different people, so avoid using them.

Instead you could ask:

> Do you use the internet:
>
> never? ☐   once a day? ☐   twice a day? ☐   3 times or more a day? ☐

- Avoid using personal or embarassing questions.

> Do you have a boyfriend? ☐   girlfriend? ☐

- Avoid asking questions that are **biased** :

> Do you agree that television is bad for you?
>
> Yes ☐   No ☐

This question suggests that the answer is Yes. A biased or 'leading question' encourages people to give a particular answer.

- Do not ask too many questions.

If your questionnaire is too long, people won't want to answer it.

It is sensible to test a questionnaire on a few people first to find out if it works. This is called a **pilot study** . It may show that your questionnaire needs improving.

Questionnaire tips:
- **Do not ask for names (personal information).**
- **Make sure that the questionnaire is easy to complete quickly (tick boxes are best).**
- **Make sure that the questionnaire can be photocopied easily. You may need lots.**
- **Design a data-capture sheet. You will need to copy the questionnaire answers on to this.**

Exercise 10C

1 Say whether each of these questions is suitable for a questionnaire. If you think not, explain why and show how the question could be improved.

(a)
> Which of these do you like?
>
> Pop ☐   Rock ☐   Soul ☐   Classical ☐   None ☐

(b)
> Do you agree that PE is good for you?
>
> Yes ☐   No ☐

(c)
> How old are you?
>
> ☐

(d)
> What types of books do you read?
>
> Fiction ☐   Non-fiction ☐

(e)
> How often do you use your local sports centre?
>
> Often ☐   Sometimes ☐   Never ☐   Depends ☐

2 Design a questionnaire with up to six questions about a topic of your choice. State your hypothesis clearly.

> A hypothesis is a theory you want to test.

10.5 Sampling

When you carry out a survey it would be too time consuming to ask everyone (in your school, or town). The total number you *could* ask is called the **population**.

Instead of asking everyone, you ask *some* of the population.

> **Key words:**
> population
> sample
> representative
> random
> biased

You select a **sample** of people to ask.

The sample must be **representative** of the population.

For example, in a school there are 1000 pupils. The number in each year group is:

Population = 1000.

Year group	7	8	9	10	11
Number of pupils	250	200	250	200	100

A sample size of 10% is large enough to represent the views of the whole school.

For a school survey you could choose a sample 10% of the population. 10% of 1000 = 100 so you need 100 pupils in your sample.

For more on percentages see Chapter 15.

These 100 pupils should represent each year group and reflect the numbers in each year group.

This means that in the sample there should be:

Years 7 and 9 both have 250 pupils.
Years 8 and 10 both have 200 pupils.

- equal numbers of year 7 and year 9 pupils.
- equal numbers of year 8 and year 10 pupils.
- twice as many year 8 (or year 10) pupils as year 11 pupils.

200 pupils in years 8 and 10 but only 100 pupils in year 11.

If you take 10% from each year group all of these conditions will be met.

Year group	7	8	9	10	11
Number of pupils in sample	25	20	25	20	10

Each year group in the school has roughly equal numbers of boys and girls.
So the sample from each year group should have roughly equal numbers of boys and girls.

It doesn't have to be *exactly* the same number of each but it needs to be quite close.

You choose the sample from each year group using a **random** sampling method.
For example, for year 7 you could put all 250 names into a hat and then draw out 25.
Alternately, you could give each pupil a number between 1 and 250 and then generate random numbers on a computer or calculator to select the 25 for the sample.
Always check that the boys/girls numbers are roughly equal.

If there are twice as many boys as girls in a particular year group then the sample from that year group should have approximately twice as many boys as girls in it.

If you do not follow these rules your sample will not be representative, it will be **biased** .

In a random method, each person is equally likely to be chosen.

Exercise 10D

1 A whole school survey is being conducted in a school. The school has roughly equal numbers of boys and girls. For each of the following sampling methods say whether the results would be a random sample or if it would be biased.

  (a) Taking all surnames on a school list beginning with H.

  (b) Taking only 10 boys in each year group.

  (c) Putting all the names into a box and picking out 50 without looking.

  (d) Asking an equal number of boys and girls from each year group.

  (e) Taking every 20th name from the alphabetical school list.

  (f) Only asking 10 girls and 10 boys in years 7, 9 and 11.

2 'Most students in school carry a mobile phone.' Describe how you would collect data to find out if this statement is true. You should include a questionnaire and explain how you would select your sample.

> This is a hypothesis.

3 'More boys than girls eat fruit during the school day.' Design a questionnaire to find out if this statement is true. Include details of how you would select your sample.

10.6 Two-way tables

> **Key words:**
> two-way tables

Two-way tables are similar to frequency tables. They show two or more types of information at the same time.

> Bus timetables, league tables, school performance tables are all two-way tables.

Example 2

The table shows the type and outcome of matches played by a cricket team.

	Home matches	Away matches
Won	5	2
Drawn	4	6
Lost	4	5

The two types of information are:
1 home or away
2 win, draw, lose.

(a) How many matches were played altogether?

(b) How many matches were lost altogether?

(c) How many matches in total were *not* drawn?

Often the easiest way of answering these questions is to extend the two-way table to include the totals going across → and downwards ↓.

	Home matches	Away matches	Total
Won	5	2	7
Drawn	4	6	10
Lost	4	5	9
Total	13	13	26

The totals across
(13 + 13 = 26) and down
(7 + 10 + 9 = 26) should
be the same.

(a) 26 matches played altogether

(b) 9 matches lost altogether.

(c) 26 − 10 = 16 matches were not drawn.

You could also work out
total lost + total won
= 7 + 9 = 16.

Example 3

In an office survey of 32 staff, 6 women said they walked to work, 10 men came by bus, and 4 men cycled. Of the remaining 11 women, only 1 cycled and the rest came by bus or walked.

(a) Draw a two-way table to show this information.

(b) Complete the table.

(c) How many women went by bus?

(d) How many people walked to work?

(e) What percentage of people went by bus?
Give your answer to the nearest whole number.

(a)

	Walked	Cycled	Bus	Total
Men		4	10	
Women	6	1		17
Total				32

continued ▼

(b)

	Walked	Cycled	Bus	Total
Men	1	4	10	15
Women	6	1	10	17
Total	7	5	20	32

The total number of men is $32 - 17 = 15$.

The number of men that walked is $15 - (4 + 10) = 1$.

The number of women who came by bus is $17 - (6 + 1) = 10$.

Once you have filled in all the values you can calculate the totals.

From the table:

(c) 10 women went by bus

(d) 7 people walked to work.

(e) 20 people out of 32 went by bus.

As a fraction this is $\frac{20}{32}$

As a percentage this is $\frac{20}{32} \times 100\% = 62.5\%$

$= 63\%$ (to the nearest whole number)

For more on percentages see Section 3.5.

Exercise 10E

1 In a class of 30 people, 6 men and 8 women own a bicycle. There were 17 women in the survey. Copy and complete the two-way table to show this information.

	Bicycle	No bicycle	Total
Men	6		
Women	8		17
Total			30

2 In a school survey of 50 boys and 50 girls, 41 boys were right-handed and only 6 girls were left-handed. Copy and complete the two-way table.

	Left-handed	Right-handed	Total
Girls			
Boys			
Total			

Use the table to work out an estimate of the percentage of left-handed pupils in the school.

3 In the 2001 Census the male population of Poynton (Central) was 3522. The number of females in Poynton (West) was 3898. The population of Poynton (Central) was 6792. The total population of Poynton was 13 433.

> Poynton was split into two areas: West and Central.

(a) Construct a two-way table to show this information.

(b) Complete the table.

(c) What percentage of the population of Poynton is female? Give your answer to the nearest whole number.

4 The table gives the KS3 English test results for a local school.

		Level					
		3	4	5	6	7	8
English	Boys	11	28	34	31	15	1
	Girls	4	20	36	43	22	5

(a) Copy the table and extend it to find the totals for each row and column.

(b) How many pupils took the test?

(c) What percentage of boys achieved a level 5 or higher?

(d) What percentage of girls achieved a level 7?

Grouped data in two-way tables

Discrete data (e.g. money) or continuous data (e.g. height) can also be shown in a two-way table.

Example 4

The two-way table shows the heights of people in five athletics teams.

Height h (cm)	Frequency				
	Team A	Team B	Team C	Team D	Team E
$140 \leqslant h < 150$	3	6	4	5	4
$150 \leqslant h < 160$	6	5	7	5	6
$160 \leqslant h < 170$	10	9	11	8	12
$170 \leqslant h < 180$	7	6	4	9	5
$180 \leqslant h < 190$	4	5	4	3	5

(a) How many people were 180 cm or taller?

(b) How many people are there in Team C?

(c) How many people took part in the survey altogether?

(d) What percentage of people were 160 cm or taller, but less than 180 cm? Give your answer to the nearest whole number.

Height h (cm)	Frequency					
	Team A	Team B	Team C	Team D	Team E	Total
$140 \leqslant h < 150$	3	6	4	5	4	22
$150 \leqslant h < 160$	6	5	7	5	6	29
$160 \leqslant h < 170$	10	9	11	8	12	50
$170 \leqslant h < 180$	7	6	4	9	5	31
$180 \leqslant h < 190$	4	5	4	3	5	21
Totals	30	31	30	30	32	153

Draw in the totals column.

(a) 21 people.

(b) 30 people in Team C.

(c) 153 people took part altogether.

(d) 50 + 31 people were in the range $160 \leqslant h < 180$.

As a fraction this is $\frac{81}{153}$.

As a percentage this is $\frac{81}{153} \times 100\% = 52.94\%$.

$$= 53\% \text{ (to the nearest whole number).}$$

For more on percentages see Section 3.5.

Example 5

The times for a fun run were:

Age a (years)	Time t (minutes)			
	$0 \leqslant t < 10$	$10 \leqslant t < 20$	$20 \leqslant t < 30$	$30 \leqslant t < 40$
$5 \leqslant a < 10$	0	1	12	36
$10 \leqslant a < 15$	2	46	59	27
$15 \leqslant a < 20$	7	65	37	13

Two-way table using two types of continuous data: age and time.

(a) How many runners finished in under 20 minutes?

(b) How many runners were aged 15 years or over?

(c) How many runners took part in total?

(d) What percentage of runners completed the fun run in under 10 minutes? Give your answer to the nearest whole number.

Age a (years)	Time t (minutes)				
	$0 \leqslant t < 10$	$10 \leqslant t < 20$	$20 \leqslant t < 30$	$30 \leqslant t < 40$	Totals
$5 \leqslant a < 10$	0	1	12	36	49
$10 \leqslant a < 15$	2	46	59	27	134
$15 \leqslant a < 20$	7	65	37	13	122
Totals	9	112	108	76	305

Add the totals to the table.

(a) $9 + 112 = 121.$

(b) 122 runners were aged 15 years or over.

9 runners finished in under 10 minutes.

(c) 305 runners took part.

(d) $\frac{9}{305} \times 100\% = 2.95\%$

$= 3\%$ (to the nearest whole number).

Exercise 10F

1 University students in five towns were asked how much rent they pay each week.

Rent r (£)	Town				
	Amyngton	Benford	Catson	Durmead	Edbridge
$20 \leqslant r < 40$	4	2	10	12	6
$40 \leqslant r < 60$	17	4	25	20	7
$60 \leqslant r < 80$	17	4	25	14	32
$80 \leqslant r < 100$	25	10	13	25	18
$100 \leqslant r < 120$	6	28	3	7	3
$120 \leqslant r < 140$	2	12	2	1	3

(a) How many students pay £80 or more each week?

(b) How many students were surveyed in Edbridge?

(c) How many students were surveyed altogether?

(d) What percentage of the students surveyed pay between £20 and £80 per week? Give your answer to the nearest whole number.

2 The same university students were asked what percentage they achieved in their last exam and how many hours they studied for it.

Hours h	Percentage, p				
	$0 \leq p < 20$	$20 \leq p < 40$	$40 \leq p < 60$	$60 \leq p < 80$	$80 \leq p < 100$
$0 \leq h < 10$	13	9	5	2	1
$10 \leq h < 20$	9	16	27	16	5
$20 \leq h < 30$	6	29	42	26	14
$30 \leq h < 40$	2	6	24	19	21
$40 \leq h < 50$	1	3	12	31	18

(a) How many students scored 60% or more?
(b) How many students studied for less than 20 hours?
(c) How many students studied for 30 hours or more and scored 80% or more?
(d) What percentage of students scored less than 20%? Give your answer to the nearest whole number.

More two-way tables

Key words: timetables

Two-way tables can show different types of information. Transport **timetables**, calendars, holiday brochure information, statistics from a census and currency conversion tables are all types of two-way table.

Example 6

This rail timetable shows the times that trains arrive at different stations.

For more on timetables see Chapter 8.

Monday to Friday				
Acton	0936	0958	1002	1017
Beeston	0942		1008	
Clapton	0945		1011	
Dunburry	0949	1008	1015	1027
Epton	0953		1019	

Bus, rail and aeroplane timetables all use the 24-hour clock.

A blank space means the train does not stop at this station.

(a) How long does the 1002 from Acton take to reach Dunburry?
(b) If you need to be in Dunburry by a quarter past ten, what time train could you catch from Beeston?

Monday to Friday				
Acton	0936	0958	1002	1017
Beeston	0942		1008	
Clapton	0945		1011	
Dunburry	0949	1008	**1015**	1027
Epton	0953		1019	

The 1002 train from Acton arrives in Dunburry at 1015.

(a) 1015 − 1002 = 13 minutes

(b) The 0942 or 1008 trains from Beeston.

These both arrive in Dunburry at or before 1015.

Example 7

The table shows the monthly rainfall (in mm) and the maximum and minimum temperatures (in °C) for Paris.

The letters stand for the months.

	J	F	M	A	M	J	J	A	S	O	N	D
Rainfall	20	16	18	17	16	14	13	12	14	17	17	19
Max. temperature	6	7	11	14	18	21	24	24	21	15	9	7
Min. temperature	1	1	3	6	9	12	14	14	11	8	4	2

(a) Which month has the most rain?

(b) Which month(s) has the smallest temperature range?

(c) Which month(s) has the largest range?

The range of temperature is maximum − minimum.

From the table:

	J	F	M	A	M	J	J	A	S	O	N	D
Rainfall	20	16	18	17	16	14	13	12	14	17	17	19
Max. temperature	6	7	11	14	18	21	24	24	21	15	9	7
Min. temperature	1	1	3	6	9	12	14	14	11	8	4	2
Temp. range	5	6	8	8	9	9	10	10	10	7	5	5

(a) January has the most rain (20 mm).

(b) January, November, December.

Each of these has temperature range 5 °C.

(c) July, August, September.

Each has temperature range 10°C.

Example 8

The table below shows the cost of a fly-drive holiday. The prices are per person, in £s.

Group	Number of days						Extra night
	2	3	4	5	6	7	
5/6 adults sharing	170	178	185	190	193	196	25
4 adults sharing	173	184	190	197	199	205	25
3 adults sharing	179	192	202	213	220	227	25
2 adults sharing	179	192	202	213	220	227	25
Child	148	148	148	148	148	148	25

You find this type of table in holiday brochures.

Each price is *per person*.

(a) Find the cost of a 3-day holiday for 4 adults and 3 children.

(b) What is the cost of a fly-drive holiday for 2 adults and 2 children for 10 days?

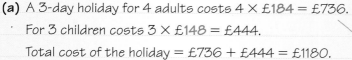
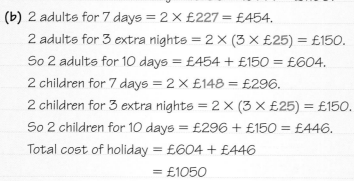

(a) A 3-day holiday for 4 adults costs 4 × £184 = £736.

For 3 children costs 3 × £148 = £444.

Total cost of the holiday = £736 + £444 = £1180.

(b) 2 adults for 7 days = 2 × £227 = £454.

2 adults for 3 extra nights = 2 × (3 × £25) = £150.

So 2 adults for 10 days = £454 + £150 = £604.

2 children for 7 days = 2 × £148 = £296.

2 children for 3 extra nights = 2 × (3 × £25) = £150.

So 2 children for 10 days = £296 + £150 = £446.

Total cost of holiday = £604 + £446

= £1050

Use the prices in the '3 days' column. Read the rows for 4 adults (£184 each) and child (£148 each).

Work out the cost for 7 days then add on **3** extra nights.

Exercise 10G

1 The table shows the distances between some major French cities in kilometres (km).

Bordeaux

870	Calais					
658	855	Grenoble				
649	1067	282	Marseille			
804	1222	334	188	Nice		
579	292	565	776	931	Paris	
244	996	536	405	560	706	Toulouse

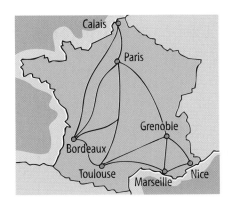

The table shows that the distance between Calais and Paris is 292 km.
Find:

(a) the distance between Bordeaux and Marseille

(b) the distance between Toulouse and Grenoble

(c) the total distance from Paris to Calais to Bordeaux and then back to Paris.

2 Use the two-way table in Example 8 to find the cost of:

(a) a 5-day holiday for 3 adults and 4 children

(b) an 8-day holiday for 6 adults (no children).

Examination style questions

1 Esher asks her friends which type of berry they like best. Their replies are:

Strawberry	Blackberry	Raspberry
NONE	Strawberry	Raspberry
Strawberry	Blackberry	Strawberry
Raspberry	Strawberry	NONE
Strawberry	Strawberry	Raspberry

Copy the table, and complete using tallies.

Berry	Tally	Frequency
Strawberry		
Blackberry		
Raspberry		
NONE		

(2 marks)
AQA, Spec B, 1F, June 2002

2 The distances, to the nearest mile, that 40 people travel to work are recorded below.

```
5    2    4    14   19   6    21   28
14   44   12   18   4    7    32   16
13   9    7    13   11   19   21   4
1    2    14   8    10   38   27   9
5    2    11   6    3    8    2    3
```

(a) Copy and complete the tally and frequency columns in the chart below.

Distances (miles)	Tally	Frequency
1–10		
11–20		
21–30		
31–40		
41–50		

(b) How many people travel between 11 and 20 miles to work? *(3 marks)*

AQA, Spec B, 1F, June 2003

3 The two-way table shows the number of doors and the number of windows in each classroom in a school.

		Number of windows			
		1	2	3	4
Number of doors	1	3	3	3	2
	2	1	5	7	3
	3	0	0	1	2

(a) How many classrooms have 4 windows?

(b) How many classrooms have the same number of windows as doors? *(3 marks)*

AQA, Spec B, 1I, March 2002

4 Emma reads in a magazine that there is a link between the number of children and the number of pets in a family.

   (a) Design a two-way table to record the number of pets and the number of children in a sample of families.

   (b) Complete your two-way table by inventing data for eight families.

        (4 marks)

AQA, Spec B, 1I, June 2003

5 The manager of a cinema wants to find out how often teenagers attend the cinema. He uses a questionnaire.

   (a) Here is part of the questionnaire:

Question	How often do you attend the cinema?
Response	Sometimes ☐ Occasionally ☐
	Regularly ☐

    Write down two criticisms of his response section.

   (b) Explain how the manager could distribute 50 questionnaires randomly to pupils from a school of 1000 pupils.

        (3 marks)

AQA, Spec B, 1I, November 2003

Summary of key points

Types of data

Primary data is information you collect directly yourself.

Secondary data is information that you get from existing records.

Qualitative data contains descriptive words. For example, colour, town name.

Quantitative data contains numbers. For example, measurements, number of shoppers.

Discrete data can only have particular values and is countable. For example, shoe size, number of cars.

Continuous data can have any value in a particular range. For example, measurements of time, length.

Collecting data

You can use tally marks to record data. They are grouped in 5s. ||||

Class intervals are groupings of quantitative data. When you group data, you should use between 5 and 10 class intervals.

You use a data-capture sheet to record data as you collect it.

A survey collects primary data.

A hypothesis is the theory you are testing.

For a pilot study, you try out a questionnaire on a few people. It may show that the questionnaire needs improving.

The population is the total number of people you *could* ask.

A sample is a smaller number of people, chosen from the population.

A random sample is a sample that has been chosen fairly without bias.

Two-way tables (grade E/D)

Two-way tables show two types of information at the same time, for example timetables, numbers of men and women using different transport.

Collecting data is grade G, but becomes grade D when grouped frequency tables are used.

Data-capture sheets, surveys, hypotheses and pilot studies are grade D/C.

This chapter will show you how to:

✔ draw pictograms, frequency diagrams, pie charts, line graphs and stem-and-leaf diagrams
✔ construct scatter graphs and identify correlation
✔ construct frequency polygons
✔ compare two data sets

You record survey data in a tally chart or frequency table. You can then display the data in a picture or diagram as well.

> A diagram can help you to see patterns in the data.

11.1 Pictograms

> **Key words:**
> pictogram
> key

In a **pictogram** a picture or symbol represents an item or number of items.

The table shows the amount of gold produced in tonnes each year, in four different countries.

Country	Gold produced (tonnes)
South Africa	625
USA	325
Australia	250
Canada	150

You can show this information in a pictogram.

South Africa	🔲🔲🔲🔲🔲🔲🔲🔲🔲🔲🔲🔲[
USA	🔲🔲🔲🔲🔲🔲[
Australia	🔲🔲🔲🔲🔲
Canada	🔲🔲🔲

Key: 🔲 represents 50 tonnes

> The **key** tells you what each symbol represents.

> [represents 25 tonnes

Exercise 11A

1 The pictogram shows the number of students in different tutor groups who own mobile phones.

Tutor group	
A	📱 📱 📱 📱
B	
C	📱 📱 📱 📱 📱 ▌
D	

Key: 📱 represents 2 mobile phones

10 students in Group B own mobile phones.
7 students in Group D own mobile phones.

(a) Complete the pictogram to show this information.

(b) How many students own mobile phones in total?

2 Fiona asks her friends what their favourite sport is.
The results are shown below.

Sport	Frequency
Football	4
Netball	6
Riding	8
Other	6

Draw a pictogram to represent Fiona's results. Use the symbol ☺ to represent 4 friends.

3 This table shows the sales from a canteen drinks' machine.

Tea	40
Coffee	47
Chocolate	26
Soup	18
Fruit juice	14

Think how you will display numbers such as 47.

Draw a pictogram to represent these drinks sales.
Use a symbol of a cup to represent 5 drinks.

11.2 Frequency diagrams for discrete data

Bar charts

Frequency diagrams, such as **bar charts** and **vertical line graphs** , can show patterns or trends in data. In a bar chart, the bars can be either vertical or horizontal. They must be of equal width.

Bar charts can be used for quantitative or qualitative data.

Example 1

The table shows the frequency of vowels occurring in the first four lines of a book.

Vowel	a	e	i	o	u
Frequency	20	18	23	9	5

This is qualitative data.

Draw a bar chart for this data.

Choose a sensible scale.

Plot frequency on the vertical axis.

The height of each bar represents the frequency.

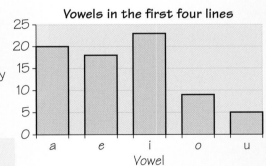

Give your bar chart a title.

Leave gaps between the bars.

Label the axes.

You could plot this bar chart with horizontal bars:

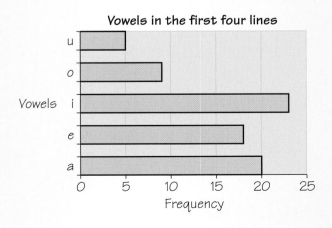

The length of each bar represents the frequency.

There is still a gap between the bars.

Compound bar charts

Key words:
compound bar chart

You can use **compound bar charts** to make comparisons.

Example 2

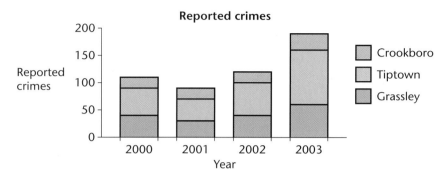

Use the bar chart to say whether these statements are true or false:

(a) Tiptown has the highest number of reported crimes.

(b) Crimes have increased for all three towns in 2003.

(c) Grassley's reported crimes fell by one half from 2000 to 2001.

(d) Crookboro's crime levels stayed roughly the same between 2000 and 2003.

(a) True   **(c)** False

(b) True   **(d)** True

When you draw a bar chart make sure that you:
- label the horizontal and vertical axes clearly
- give the chart a title
- use a sensible scale to show all the information clearly
- shade or colour the bars
- leave spaces between the bars.

Vertical line graphs

Key words:
vertical line graph

Vertical line graphs can also be used to show discrete data. You can construct a vertical line graph in the same way as a bar chart, using a thick line instead of a bar.

Example 3

The number of goals scored in hockey matches one season is recorded below. Draw a vertical line graph for the data.

Number of goals	0	1	2	3	4
Number of matches	5	7	3	2	2

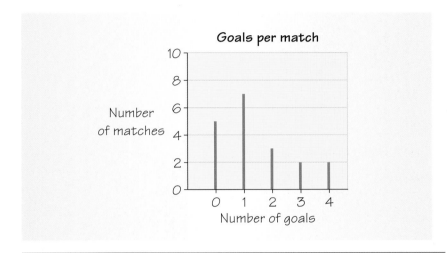

The vertical lines are evenly spread across the graph.

Exercise 11B

1 The table shows the number of cars parked in three hospital car parks at 2 pm on one afternoon.

Car park	Number of cars
Staff (S)	40
Visitors (V)	70
Casualty (C)	65

(a) Draw a bar chart to show this information.

(b) Work out how many more cars were parked in the Visitors car park than in the Staff car park.

2 Emma asks her friends what type of TV programme they like best.

Draw a bar chart to show Emma's results.

Type of TV programme	Frequency
Cartoons	4
Drama	2
Quizzes	1
Soaps	6

3 The following frequency table shows the results of a crisp manufacturer's survey to find the most popular flavour among boys and girls.

	Frequency	
Flavour	**(boys)**	**(girls)**
Plain	6	6
Cheese and Onion	12	4
Smokey Bacon	21	12
Prawn Cocktail	4	9
Salt and Vinegar	5	16
Beef	2	3

(a) Draw a compound bar chart to display this.

(b) How many people took part in the survey?

4 This bar chart represents sales of cars at an auction.

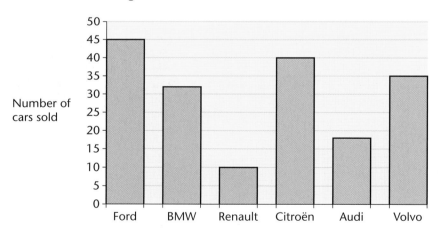

Use the bar chart to answer these questions.

(a) How many cars were sold altogether?

(b) Which make of car totalled exactly 10% of all the cars sold?

(c) Which make of car totalled exactly one quarter of all the cars sold?

(d) Which make of car totalled almost $\frac{1}{5}$ of all the cars sold?

(e) One make of car sold almost four times as many as another. Which makes of cars were these?

5 A survey of the most common birds in the UK gave the
following results:

Bird	Number (millions of pairs)
Blackbird	4.7
Blue tit	3.5
Chaffinch	5.8
Robin	4.5
Sparrow	3.8
Wood pigeon	2.4
Wren	7.6

Draw a vertical line graph to represent this information.

6 At a holiday show, families were asked how many
holidays they had taken last year. Draw a vertical line
graph to show this data.

Number of holidays	Number of families
0	2
1	14
2	17
3	8
4	1

11.3 Frequency diagrams for continuous data

If you collect data that is grouped, you will need to draw a
frequency diagram similar to a bar chart, but with no gaps
between the bars.

Example 4

The heights of 31 sunflowers were measured.

Height (cm)	Frequency
$140 \leqslant h < 150$	3
$150 \leqslant h < 160$	8
$160 \leqslant h < 170$	13
$170 \leqslant h < 180$	5
$180 \leqslant h < 190$	2

Draw a frequency diagram to show this data.

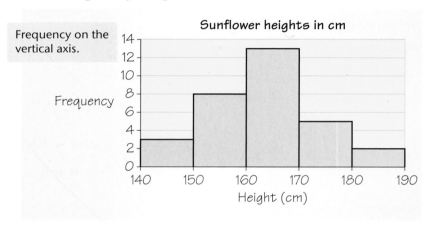

Frequency on the vertical axis.

For continuous data there are no gaps between the bars.

The width of each bar is the same as the class interval.

A frequency diagram often shows the 'spread' of the data.

Exercise 11C

1 Draw a frequency diagram to show the following information.

Length, x (cm)	Frequency
$0 < x \leqslant 5$	6
$5 < x \leqslant 10$	11
$10 < x \leqslant 15$	8
$15 < x \leqslant 20$	5

2 This table gives the age range of the members of a local leisure centre.

Draw a frequency diagram to show the spread of ages.

Age	Frequency
$0 \leqslant \text{age} < 10$	23
$10 \leqslant \text{age} < 20$	45
$20 \leqslant \text{age} < 30$	56
$30 \leqslant \text{age} < 40$	36
$40 \leqslant \text{age} < 50$	49
$50 \leqslant \text{age} < 60$	32
$60 \leqslant \text{age} < 70$	16

11.4 Pie charts

Pie charts show how data is shared or divided.

Interpreting pie charts

The whole pie chart represents the total number of items.
If you know how much the whole pie chart represents, then you
can work out how much each slice represents.
The angle in each slice is proportional to the number of
items in each of the different categories.

The pie chart shows the results
of a Key Stage 3 Science test for
180 pupils in terms of levels.

By measuring the angle of each
slice you can work out how
many pupils each slice
represents.

The chart indicates that the
whole 360° represents 180 pupils.

This means that 1° represents $\dfrac{180}{360} = 0.5$ pupils.

In this example:
140° is the same as $140 \times 0.5 = 70$ pupils, so 70 pupils
achieved level 7; 60° is the same as $60 \times 0.5 = 30$ pupils,
so 60 pupils achieved level 5, and so on.

The answers are only approximate as the value depends
on the accuracy of the angle measured. Pie charts are not
particularly useful for reading off accurate values.

Pie charts are good for showing comparisons. From these
two pie charts, it is easy to spot the difference in spending
patterns of the men and women in the last week.

Most expensive luxury cost in the last week.

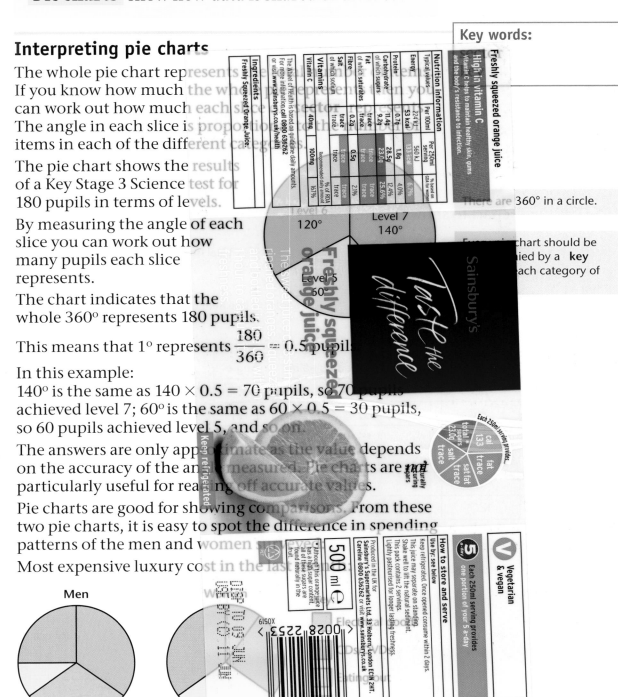

Men

Drawing pie charts

The sections of the pie chart are called sectors.

To draw a pie chart, you first calculate the angle for each sector. You draw the angles using a protractor. Give the pie chart a key to explain what each sector represents.

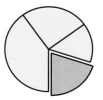

A sector is like a 'slice' of the pie.

For help using a protractor see Section 6.1.

Example 5

In a pet shop survey, people were asked about the pets they owned.

Pet	Dog	Cat	Bird	Fish	Other
Frequency	20	37	15	32	16

(a) How many pets were recorded in total?
(b) Calculate the angle of the pie chart sector for each pet.
(c) Draw a pie chart to show this information.

(a) 20 + 37 + 15 + 32 + 16 = 120

Total frequency = total number of pets.

120 pets were recorded

(b) 360° represents 120 pets

So $\dfrac{360°}{120} = 3°$ represents 1 pet

The angle for one item is always $\dfrac{360°}{\text{total number of items}}$.

Pet	Frequency	Sector angle calculation	Angle
Dog	20	20 × 3°	60°
Cat	37	37 × 3°	111°
Bird	15	15 × 3°	45°
Fish	32	32 × 3°	96°
Other	16	16 × 3°	48°
Total	120	Total angle	360°

Check that the angles add up to 360°.

(c)   Type of pets owned

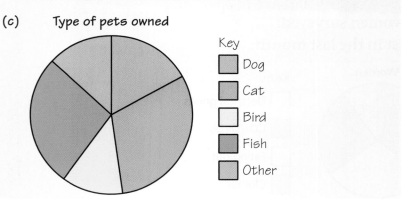

Key
- ▦ Dog
- ▦ Cat
- ☐ Bird
- ▦ Fish
- ▦ Other

The key tells you what each sector represents.

Example 6

Thirty people were asked how they travelled to work. The results were:

Mode of travel	Frequency
Walk	7
Bus	9
Car	13
Cycle	1

Draw a pie chart to show this information.

360° represents 30 people

So $\dfrac{360°}{30}$ = 12° represents 1 person

Mode of travel	Frequency	Sector angle calculation	Sector angle
Walk	7	7 × 12°	84°
Bus	9	9 × 12°	108°
Car	13	13 × 12°	156°
Cycle	1	1 × 12°	12°
Total	30	Total angle	360°

Check that the total frequency equals the number of people and that the angles add up to 360°.

How people travel to work

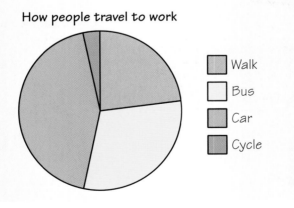

- Walk
- Bus
- Car
- Cycle

Example 7

This pie chart shows how a family spends its money in a week.

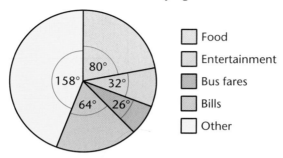

☐ Food
☐ Entertainment
☐ Bus fares
☐ Bills
☐ Other

The amount spent on food is £120.
How much do they spend on each of the other items?

Food has a sector of 80° and the amount for food is £120.	Work out how much an angle of 1° represents.
80° represents £120	
1° represents $\dfrac{£120}{80}$ = £1.50	
Entertainment = 32 × £1.50 = £48	If the sector angles were not labelled, you could measure them using a protractor.
Bus fares = 26 × £1.50 = £39	
Bills = 64 × £1.50 = £96	
Other = 158 × £1.50 = £237	
Check: £120 + £48 + £39 + £96 + £237 = £540	The whole pie chart is 360° so the total of all the amounts of money must be 360 × £1.50.
and 360 × £1.50 = £540	

Exercise 11D

1 In a café, the number of people eating these main courses for lunch was:

Fish pie	16
Sausages	10
Omelette	17
Salad	6
Steak pie	19
Pizza	22

   (a) How many people were in the café?
   (b) What angle will represent 1 person?
   (c) Draw a pie chart to show this information.

2 Just before the General Election, a small survey asked how people intended to vote. Here are the results:

Draw a pie chart to show the results of this survey.

Conservative	60
Green	24
Labour	72
Liberal Democrat	56
Others	28

3 540 pupils were asked which was their favourite school subject.

The results are shown in this pie chart.

Work out the number who voted for each subject.

Use Example 7 to help you.

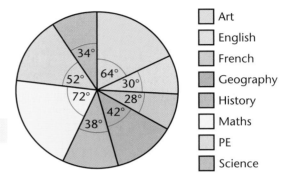

- ☐ Art
- ☐ English
- ☐ French
- ☐ Geography
- ☐ History
- ☐ Maths
- ☐ PE
- ☐ Science

4 This pie chart shows what year 11 pupils did in the year after their GCSE exams.

135 pupils went to college.

Work out the number of pupils in each of the other categories.

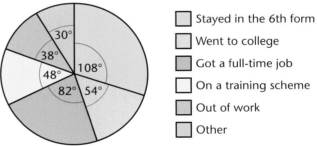

- ☐ Stayed in the 6th form
- ☐ Went to college
- ☐ Got a full-time job
- ☐ On a training scheme
- ☐ Out of work
- ☐ Other

5 A packet of breakfast cereal showed the following nutritional information:

Ingredient	Protein	Carbohydrate	Fat	Fibre
Amount per 100g of cereal	15 g	62.5 g	10 g	12.5 g

(a) If 360° represents 100 g of cereal, what angle represents 1 g?

(b) Calculate the angle of the sector for each ingredient.

(c) Draw a pie chart to show this information.

6 A different cereal brand has ingredients in these proportions:

> 400 g carbohydrate
> 150 g protein
> 120 g fibre
> 50 g fat

(a) Copy and complete the calculation and table below.

$$1 \text{ g of ingredient} = \frac{360°}{\text{total weight of cereal}} = \frac{360}{\boxed{}} \text{ g} = 0.\boxed{}°$$

Ingredient	Amount in g	Sector angle calculation	Angle
Carbohydrate	400 g	400 × 0.☐°	
Protein	150 g	150 × 0.☐°	
Fibre	120 g	120 × 0.☐°	
Fat	50 g	50 × 0.☐°	
Total		Total angle	360°

(b) Draw a pie chart to show this information.

(c) Compare your pie charts for questions 5 and 6. Which brand could claim it has
   (i) less carbohydrate
   (ii) less fat?

11.5 Time series line graphs

Key words:
line graph
time series

In a **line graph**, the points plotted are joined with straight lines.

> **Line graphs show trends in the data. You can only draw line graphs for continuous data.**

Line graphs often show how something changes over time. These are called **time series**.

Example 8

The table below shows the maximum temperature (in °C) in Paris in each month.

Month	J	F	M	A	M	J	J	A	S	O	N	D
Max temp (°C)	6	7	11	14	18	21	24	24	21	15	9	7

Draw a time series line graph to show this data.

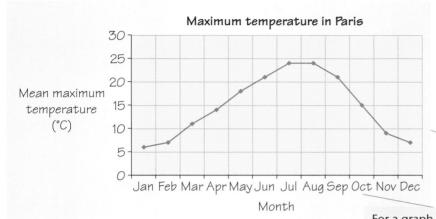

Temperature (°C) and time (months) are continuous data, so we can draw a line graph.

Join the points with straight lines.

This line shows the trend from October to November was a fall in temperature. It does not show actual temperature values.

For a graph showing changes over time, always put the time on the horizontal axis.

Exercise 11E

1 The table shows the cost of second-hand Ford Focus Zetecs (to the nearest hundred pounds).

    (a) Draw a line graph to represent this information.

    (b) What is the trend in the cost of second-hand cars?

Year and registration	Cost (£)
1998 (S)	5700
1999 (T)	6300
1999 (V)	6600
2000 (W)	7300
2000 (X)	7600
2001 (Y)	8500
2001 (Si)	8800

2 The table shows the amount of rainfall during the first week in February.

Draw a line graph to display this data.

Day	Rainfall (mm)
1	1.5
2	2.1
3	3.2
4	1.4
5	2.3
6	0.6
7	0.4

3 The maximum temperature (°C) and amount of
rainfall (mm) recorded over a 12-month period in
Santa Cruz (Bolivia) are shown in the table.

Month	Temperature	Rainfall
Jan	29	200
Feb	30	125
Mar	29	122
Apr	27	120
May	25	95
June	23	80
July	24	65
Aug	26	55
Sept	27	68
Oct	28	105
Nov	30	121
Dec	30	162

Draw two line graphs, one for temperature and one for rainfall, to
show how the temperature and rainfall vary throughout the year.

11.6 Stem-and-leaf diagrams

Key words:
stem-and-leaf

A **stem-and-leaf** diagram keeps the original data
and gives a 'picture' of the spread of the data.

Here are the scores out of 50 for a general knowledge quiz:

32 16 46 31 27 29 36 41 22 44
37 28 34 42 33 25 37 43 20 37
32 41 26 38 46 18 20 30 38

You can write the data using the 'tens' digit as the stem and the 'units'
digit as the leaf.

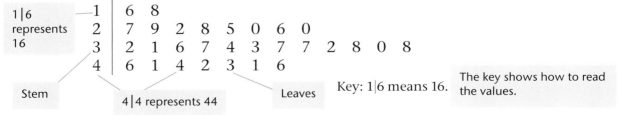

1|6 represents 16

1	6 8
2	7 9 2 8 5 0 6 0
3	2 1 6 7 4 3 7 7 2 8 0 8
4	6 1 4 2 3 1 6

Stem

4|4 represents 44

Leaves

Key: 1|6 means 16.

The key shows how to read
the values.

You need to write the diagram with the leaves in ascending order. Write the leaves in neat columns, keeping all the numbers in line.

```
1 │ 6  8
2 │ 0  0  2  5  6  7  8  9
3 │ 0  1  2  2  3  4  6  7  7  7  8  8
4 │ 1  1  2  3  4  6  6
```

The total number of values here is 29 and this is the same as in the original list.

> Check that you have not missed out any numbers. Make sure the total number of values is the same as in the original list.

Example 9

The length of 50 nails made in a factory were measured to the nearest tenth of a cm. The results were:

15.1 9.7 12.0 10.2 14.9 15.6 10.8 11.0 15.7 10.9
11.4 17.1 10.1 15.2 15.7 10.4 14.7 16.3 15.0 11.2
 9.9 14.6 10.5 15.3 16.2 10.0 16.2 13.9 15.4 15.2
10.6 9.7 15.0 16.1 11.7 10.9 15.2 14.8   8.8 14.6
16.0 15.8 11.0 14.8 15.6   9.5 10.7 14.9 15.2 11.1

(a) Draw a stem-and-leaf diagram for these results.
(b) How many nails were less than 10 cm long?

(a)
```
 8 │ 8
 9 │ 5  7  7  9
10 │ 0  1  2  4  5  6  7  8  9  9
11 │ 0  0  1  2  4  7
12 │ 0
13 │ 9
14 │ 6  6  7  8  8  9  9
15 │ 0  0  1  2  2  2  2  3  4  6  6  7  7  8
16 │ 0  1  2  2  3
17 │ 1
```

Key: 8│8 represents 8.8

(b) 5 nails were less than 10 cm long.

> Use the 'whole numbers' as the stem and the 'decimal parts' as the leaves.

> The 'leaves' have been written in ascending order.

> Check: there are 50 pieces of data in the diagram, which is the same number as in the original list.

> Count the leaves on the 8 and 9 stems.

Exercise 11F

1 The temperature in °C is recorded in 20 towns on one day.

    23   16   18   21   15   24   21   19   19   11
    17   18   14   22   23   9   12   17   20   19

Copy and complete the stem-and-leaf diagram for this data.

    0 |
    1 |
    2 |

Key: 1│3 represents 13°C.

2 Here are pupils' test scores:

     9   43   33   26   37   12   18   19   25   32
    14   29   43   33   37   31   29   40   17

(a) Draw a stem-and-leaf diagram for these scores.

(b) How many pupils scored less than 13?

Find where 13 would be on the diagram.

3 A number of people were asked how many driving lessons they had taken before passing their test. The results are shown below.

     8   21   21   14   18   41   35   12   17
    14   32   29   38   25   20   34   13   19

(a) How many people were asked?

(b) Draw a stem-and-leaf diagram for this data.

4 The following list gives the number of litres of fuel (given to the nearest litre) sold at a garage one morning.

    26   31   18   44   37   30   29   32   35   40   20
    51   15   36   30   25   40   30   34   27   20   35
    45   38   40   42   28   12   39   44   35   30   24
    36   33   43   23   46   38   42   50

(a) Draw a stem-and-leaf diagram for these results.

(b) How many customers bought more than 40 litres of fuel?

11.7 Scatter graphs

Scatter graphs help you to compare two sets of data. They show if there is a connection or relationship called the **correlation** , between the two quantities plotted.

Sometimes scatter graphs are called *scatter diagrams* or *scattergrams*.

The following table shows the masses and heights of 10 men registered at a gym.

Height (cm)	Mass (kg)
166	65
169	73
172	67
161	62
177	75
171	72
168	66
165	67
170	70
176	75

For a scatter graph, plot the height along the *x*-axis (horizontal) and the mass along the *y*-axis (vertical).

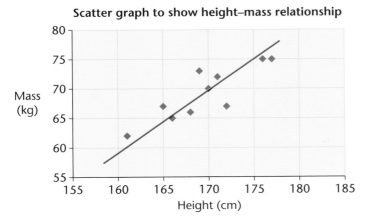

Scatter graph to show height–mass relationship

You do not need to start the axes at zero. Find the smallest and largest values in each set of data to help you decide on the scale. The straight line shows the line of best fit.

The scatter graph suggests that the taller you are the heavier you are.

A good way to show this is by drawing a straight line through, or as close to, as many points as possible. This line is called the **line of best fit** .

You draw it 'by eye', using a ruler.

Here the line of best fit slopes from bottom left to upper right. This is called a **positive correlation** .

Try to have equal numbers of points above and below the line of best fit.

Positive correlation means as one variable gets bigger, so does the other.

Example 10

The following table shows how the fuel consumption (in litres per 100 km) changes as the speed of a car increases.

Speed (kph)	Fuel consumption (litres/100 km)
20	9.6
30	8.7
40	8.0
50	6.8
60	6.0
70	5.5
80	4.5

(a) Plot a scatter graph for this data.

(b) Draw in a line of best fit.

(c) Comment on the correlation.

(a)(b)

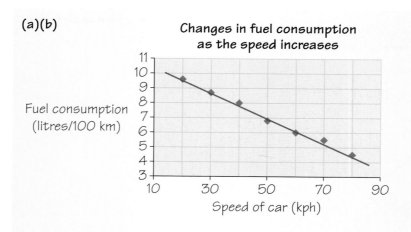

Changes in fuel consumption as the speed increases

As the speed *increases* the fuel consumption *decreases*.

(c) The line of best fit shows **negative correlation** .

Example 11

This table shows the results of a recent Geography test (as a percentage) and pupil's hand span measurements (in cm).

(a) Plot a scatter graph for this data.

(b) Comment on the correlation shown.

Test result (%)	Hand span (cm)
32	15
85	19
54	21
47	16
41	23
36	14
29	18
57	17
67	20
60	21

(a)

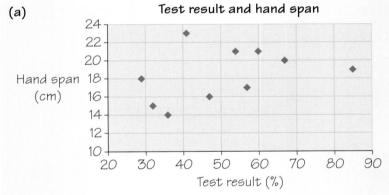

The points are randomly spread out, so you cannot draw a line of best fit.

(b) There is no linear connection between the test results and the hand span measurement. There is **no correlation**.

There is no *linear* relationship because the points do not lie on or near a *line*.

Exercise 11G

1 Here are four sketches of scatter diagrams.

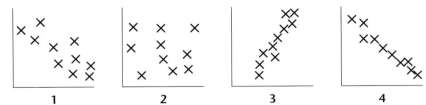

Which ones show:

(a) positive correlation **(b)** negative correlation

(c) zero correlation?

No correlation is sometimes called zero correlation.

2 What type of correlation would you expect if you drew a scatter graph of the following?

   (a) The football league position (where top of the league = 1) against the number of goals conceded.

   (b) A person's earnings against their height.

   (c) The number of ice creams sold against the temperature during the day.

   (d) The marks gained in a mock GCSE exam against those gained in the actual GCSE exam.

   (e) The size of a car engine against the amount of fuel used by that engine.

3 In a science experiment, one end of a metal bar is heated. The results show the temperature at different points along the metal bar.

Position (cm)	Temperature (°C)
1	15.6
2	17.5
3	36.6
4	43.8
5	58.2
6	61.6
7	64.2
8	70.4
9	98.8

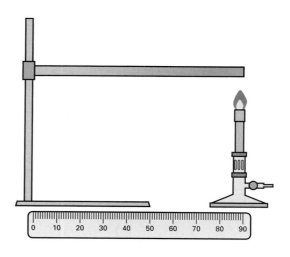

   (a) Draw a scatter graph to show this data. Let the *x*-axis represent the position between 0 and 10 cm, and the *y*-axis represent the temperature between 0 and 100°C.

   (b) Draw the line of best fit.

   (c) Comment on the correlation shown.

4 The table shows the number of female competitors taking part in each Olympic Games from 1948 to 1984.

Year	Female competitors
1948	385
1952	518
1956	384
1960	610
1964	683
1968	781
1972	1070
1976	1251
1980	1088
1984	1620

Put 'Year' on the horizontal axis.

(a) Plot a scatter graph for this data.
(b) Draw the line of best fit.
(c) State what type of correlation you see.
(d) Describe the trend or pattern.

5 The table below shows the mean annual temperature (in °C) for 10 major cities and their latitude.

City	Mean annual temperature (°C)	Latitude (degrees)
Mumbai	32	19
Kolkata	26	22
Dublin	12	53
Hong Kong	26	22
Istanbul	18	41
London	12	51
Oslo	10	60
New Orleans	21	30
Paris	15	49
St Petersburg	7	60

Latitude describes position on the globe, as degrees North from the Equator.

(a) Plot the latitude along the *x*-axis (horizontal) from 0 to 70 degrees.
Plot the mean temperature along the *y*-axis (vertical) between 0 and 40°C.

(b) What type of correlation does the graph show?

(c) What happens to the temperature as you move further North from the Equator?

6 The table below shows the exam results from twelve pupils in maths and science.

Maths mark	Science mark
42	38
83	58
29	23
34	17
45	30
47	35
55	47
74	55
61	36
59	50
53	37
77	63

(a) Draw a scatter graph to show this information.

(b) Draw the line of best fit.

(c) What type of correlation is it?

(d) Estimate the science mark for a pupil who gained 38 marks in the maths exam.

Put the maths mark along the *x*-axis from 10 to 100.

11.8 Frequency polygons

Key words:
frequency polygon
mid-point values

You can display grouped data in a **frequency polygon** . You draw a frequency diagram for continuous data and then join the middle values (**mid-point values**) of the top of the bars for each class interval using straight lines.

A frequency polygon shows patterns or trends in the data more clearly than a frequency diagram.

Example 12

The frequency table shows the heights of 50 boys and 50 girls.

Height (cm)	Frequency (boys)	Frequency (girls)
$155 \leqslant h < 160$	2	5
$160 \leqslant h < 165$	6	9
$165 \leqslant h < 170$	14	22
$170 \leqslant h < 175$	19	11
$175 \leqslant h < 180$	8	3
$180 \leqslant h < 185$	1	0

Draw a frequency polygon for the boys' heights.

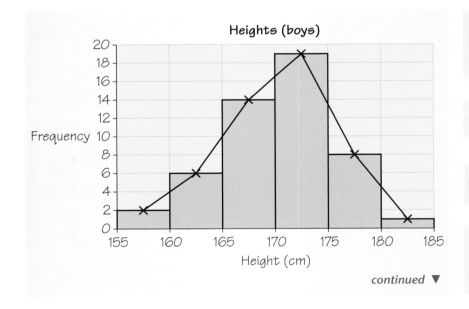

The frequency diagram for the boys' data is shown. Join the mid-points of each bar with straight lines.

Frequency always goes on the vertical axis.

The straight lines form the frequency polygon.

You could draw the frequency polygon without drawing the frequency diagram first.

continued ▼

Height (cm)	Frequency (boys)	Frequency (girls)	Mid-point values
$155 \leqslant h < 160$	2	5	157.5
$160 \leqslant h < 165$	6	9	162.5
$165 \leqslant h < 170$	14	22	167.5
$170 \leqslant h < 175$	19	11	172.5
$175 \leqslant h < 180$	8	3	177.5
$180 \leqslant h < 185$	1	0	182.5

First you extend the table and work out the mid-point values.

For each class interval, add together the boundary values and divide by 2. The mid-point value of $155 \leqslant h < 160$ is $\dfrac{155 + 160}{2} = 157.5$.

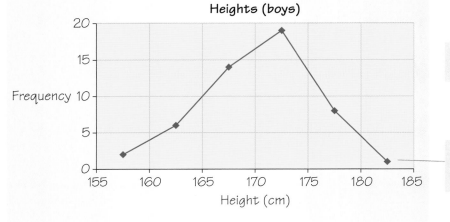

Plot each frequency against its mid-point value.

This is the point (182.5, 1). It represents 1 boy in the range $180 \leqslant h < 185$.

You could draw a frequency diagram and frequency polygon for the girls' data in the same way.

Exercise 11H

1 Work out the mid-point value for each of these class intervals:
- **(a)** 160 cm – 170 cm
- **(b)** $16 \text{ kg} \leqslant m < 18 \text{ kg}$
- **(c)** $3s \leqslant t < 7s$
- **(d)** $15 \text{ m}^2 – 20 \text{ m}^2$
- **(e)** $100 \text{ m}\ell < v \leqslant 200 \text{ m}\ell$

2 (a) Draw a frequency polygon for the girls' data in Example 12.

(b) Compare your polygon with the boys' polygon. What do you notice?

Copy the axes for the boys' data.

3 The table shows the total catches for 32 anglers
in a fishing competition.

Catch, c (kg)	Frequency
$0.05 < c \leqslant 0.55$	6
$0.55 < c \leqslant 1.05$	6
$1.05 < c \leqslant 1.55$	9
$1.55 < c \leqslant 2.05$	4
$2.05 < c \leqslant 2.55$	3
$2.55 < c \leqslant 3.05$	4

(a) Copy and complete this extended table to show
the mid-point values.

Catch, c (kg)	Frequency	Mid-point values
$0.05 < c \leqslant 0.55$	6	0.3
$0.55 < c \leqslant 1.05$	6	0.8
$1.05 < c \leqslant 1.55$	9	
$1.55 < c \leqslant 2.05$	4	
$2.05 < c \leqslant 2.55$	3	
$2.55 < c \leqslant 3.05$	4	

(b) Draw a frequency polygon for this data.

4 The temperature was recorded over a period of 60 days.

Temperature, t (°C)	Frequency
$0 < t \leqslant 4$	6
$4 < t \leqslant 8$	14
$8 < t \leqslant 12$	11
$12 < t \leqslant 16$	15
$16 < t \leqslant 20$	8
$20 < t \leqslant 24$	3
$24 < t \leqslant 28$	3

(a) Copy and extend the table with a column showing
the mid-point values.

(b) Draw a frequency polygon for this data.

5 A box holds 100 tomatoes. The mass of each tomato was measured and recorded in this frequency table.

Mass of tomato m (g)	Frequency
$35 < m \leqslant 40$	7
$40 < m \leqslant 45$	23
$45 < m \leqslant 50$	30
$50 < m \leqslant 55$	26
$55 < m \leqslant 60$	14

Draw a frequency polygon to show this data.

Examination style questions

1 Nick asked his friends what their favourite colour was. His results are shown in the table.

Colour	Tally	Frequency
Blue	|||| |||| ||||	
Red	||||	
Green	|||| ||||	
Black	|||| |||	

(a) (i) Copy the table and complete the frequency column.

(ii) How many friends did Nick ask altogether?

(b) Copy and complete the pictogram to show Nick's results. Use the symbol ☺ to represent 4 friends.

Blue	
Red	
Green	
Black	

(4 marks)
AQA, Spec B, 1F, November 2002

2 Dave and Todd go fishing one day. The table shows the total number of each type of fish they caught.

Type of fish	Total number caught
Perch	6
Roach	8
Eel	5

(a) Draw and label a bar chart to show this information.

(b) Dave caught 4 perch, 5 roach and no eels.
 Work out the number of each type of fish Todd caught. *(5 marks)*

AQA, Spec B, 1F, March 2003

3 Billy asks 40 students how they travel to college.
The table shows the results.

Method of travel	Frequency
Car	20
Bus	10
Walk	6
Other	4

(a) Draw and label a pie chart to represent the information in the table.

(b) Explain why Billy's results may not show the correct proportions for the whole college. *(5 marks)*

AQA, Spec B, 1F, November 2002

4 The number of cars passing through a set of traffic lights each time they show green is recorded.

       12   15   23   20   18   16   27   9   10
       19   22   26   14   11   8   4   12   23

Copy and complete the stem-and-leaf diagram, including the key, to represent the data.

       0 | 8
       1 |
       2 |

(3 marks)

AQA, Spec B, 1F, March 2003

5 The times, in minutes, that seven teenagers spent using their computer and watching TV on one day is recorded in the table.

Time spent using computer (minutes)	10	20	30	40	45	55	60
Time spent watching TV (minutes)	50	40	45	40	30	30	20

    (a) Plot these data on a scatter graph.

    (b) Draw a line of best fit on your scatter graph.

    (c) Describe the relationship shown in the scatter graph.   *(4 marks)*

                                                  AQA, Spec B, 1F, June 2003

6 Carla is a telephone sales assistant. The length and frequency of telephone calls made by Carla during one day are shown in the table.

Length, t (minutes)	Frequency
$0 < t \leqslant 2$	29
$2 < t \leqslant 4$	36
$4 < t \leqslant 6$	26
$6 < t \leqslant 8$	15
$8 < t \leqslant 10$	8

Draw a frequency polygon for this data.   *(2 marks)*

                                   AQA, Spec B, 1I, November 2003

Summary of key points

Pictograms (grade G)

In a pictogram a picture or symbol represents an item or number of items. The key tells you what each symbol represents.

Frequency diagrams (grades G to D)

Frequency diagrams can show any patterns or trends in the data. You can draw bar charts or vertical line graphs, with gaps between the bars, for discrete or qualitative data, or frequency diagrams with no gaps between the bars, for continuous data. You can draw the bars horizontally or vertically.

Frequency polygons (grade C)

A frequency polygon joins the mid-points of the tops of the bars for a frequency diagram. It shows trends in the data. It is usually used for grouped continuous data.

Pie charts (grade F/E)

Pie charts show how data is shared or divided. The angles in the sectors add up to 360°. The key tells you what each sector represents.

Time series line graphs (grade E)

Time series line graphs show trends in the data. You plot the points and join them with straight lines. You can only draw line graphs for continuous data.

Stem-and-leaf diagrams (grade E/D)

A stem-and-leaf diagram keeps the original data and gives a 'picture' of the spread of the data. You need a key to show you how to read the values.

Scatter graphs and correlation (grade D/C)

Scatter graphs help you to compare two sets of data. They show if there is a connection, or relationship, called the correlation, between the two. Draw a line of best fit 'by eye', using a ruler. It should pass through or close to as many points as possible.

- Positive correlation is when one quantity increases as the other quantity also increases.

- Negative correlation is when one quantity decreases as the other quantity increases.

- If the points are randomly scattered, you cannot draw a line of best fit and there is no correlation between the two quantities.

12 Ratio and proportion

This chapter will show you how to:
- ✔ write, use and simplify ratios
- ✔ write a ratio as a fraction and in the form 1 : *n* or *n* : 1
- ✔ divide quantities in a given ratio
- ✔ solve problems using direct and inverse proportion
- ✔ apply ratios to problems involving scales and maps

12.1 Ratio

Writing and using ratios

A **ratio** compares two or more quantities.

Ratios occur often in real life.
A scale model of an Aston Martin Vanquish car is labelled 'Scale 1 : 18'.

This means that the length of any part of the real car is 18 times longer than the corresponding part on the model.

Key words:
ratio

For example:

Part	Length on model	Length on actual car
Length of car door	7 cm	$7 \times 18 = 126$ cm
Diameter of wheels	3 cm	$3 \times 18 = 54$ cm
Length of car	26 cm	$26 \times 18 = 468$ cm

You write ratio with a : between the numbers. For 1 : 18 you say '1 to 18'.

1 : 18 is an example of a ratio.

Example 1

To make 12 toffee cupcakes you need 150 g of flour and 30 g of melted butter.

Grace is making 36 cupcakes. How much of each ingredient does she need?

> $36 \div 12 = 3$ so she will need three times as much of each of the ingredients.
>
> Flour: $3 \times 150\,g = 450\,g$ Melted butter: $3 \times 30\,g = 90\,g$

The multiplier is 3 so multiply the quantity of each ingredient by 3.

Example 2

Elderflower cordial needs to be diluted with water in the ratio 1 : 10. If you put 25 m*l* of cordial in a glass, how much elderflower drink will you make?

A ratio of 1 : 10 means that you need 10 times as much water as cordial.

Amount of water needed = 10 × 25 m*l*

$\qquad\qquad\qquad\qquad$ = 250 m*l*

Amount of elderflower drink = 25 m*l* + 250 m*l*

$\qquad\qquad\qquad\qquad\qquad$ = 275 m*l*

In this case, the multiplier is 10.

Total amount = amount of cordial + amount of water.

275 m*l* is about $\frac{1}{2}$ pint.

Example 3

A recipe for Fudge and Walnut Pudding for 6 people uses

$\qquad$ 6 ounces of butter
$\qquad$ 5 ounces of brown sugar
$\qquad$ 4 ounces of chopped walnuts

(a) How much sugar and chopped walnuts would you need if you used 9 ounces of butter?

(b) How many people would this pudding serve?

(a) 9 ÷ 6 = 1.5

$\qquad$ Amount of sugar = 1.5 × 5 = 7.5 ounces

$\qquad$ Amount of chopped walnuts = 1.5 × 4 = 6 ounces

(b) Number of people served = 1.5 × 6 = 9 people

The multiplier is 1.5. Use it to calculate the quantities of the other ingredients.

Exercise 12A

1 To make 12 fruit scones you need 200 g of flour, 100 g of margarine and 80 g of dried fruit.
How much of each ingredient will you need to make:
  (a) 24 scones **(b)** 6 scones **(c)** 15 scones?

2 A builder makes concrete with 7 bags of sand and 3 bags of cement.
  (a) If he has 28 bags of sand, how many bags of cement will he need to make the concrete?
  (b) The builder has 15 bags of cement. How many bags of sand should he order for the next day?

3 A Sunrise Smile cocktail uses cranberry juice, orange juice and tonic water in the ratio 5 : 3 : 1.

    (a) If I have 25 mℓ of tonic water, how much of the other ingredients will I need?

    (b) How much Sunrise Smile will I have altogether?

> Find the total of all the ingredients.

 4 The ratio of cats to dogs seeing a vet is 4 : 3.

    (a) How many dogs would you expect if there were 36 cats?

    (b) How many cats would you expect if there were 15 dogs?

> Make sure your answer is sensible!

 5 A fruit stall sells 5 apples for every 3 oranges.

    (a) Apples come in boxes of 60. The stall has 4 boxes. How many oranges do they need?

    (b) One day the stall sells 42 oranges. How many apples do they sell?

 6 Cupro-nickel is used to make key rings. It is made from mixing copper and nickel in the ratio 5 : 2.

    (a) How much copper would you need to mix with 3 kg of nickel?

    (b) How much nickel would you need to mix with 2 kg of copper?

 7 Potting compost is made by mixing 8 kg of peat with 3 kg of sand.

    (a) How much sand would be needed to mix with 20 kg of peat?

    (b) If I buy a 33 kg bag of compost, how much peat and sand will be in the bag?

 8 A recipe for 475 mℓ of spaghetti sauce includes 150 g of minced meat, 250 g of tomatoes and 75 g of mushrooms.

    (a) If I have only 200 g of tomatoes, how much of each other ingredient will I need?

    (b) How many millilitres of sauce would this make?

Simplifying ratios

You will need to know:
- **how to simplify fractions**

The method for simplifying a ratio is similar to simplifying fractions.

In Example 1 the amounts of flour and melted butter were 150 g and 30 g.

You could write this as a ratio $150 : 30$
Dividing by 10 gives $15 : 3$
Dividing by 3 gives $5 : 1$

This means that you need 5 times as much flour as melted butter.

The three ratios $150 : 30$, $15 : 3$ and $5 : 1$ are equivalent.

The **simplest form** is $5 : 1$.

> You can divide or multiply ratios to get them into their simplest form.

> When a ratio is in its simplest form you cannot divide the numbers any further.

Key words:
simplest form
lowest terms

Simplifying fractions was covered in Section 3.1.

You can compare two quantities as long as they are in the same units.

You cannot divide this any further.

You could also say 'the ratio $5 : 1$ is in its **lowest terms** '.

These rules are the same as for finding equivalent fractions.

Example 4

Write these ratios in their simplest form:
(a) $8 : 12$ **(b)** $18 : 24 : 6$ **(c)** $40\,\text{cm} : 1\,\text{m}$
(d) $1\,\text{kg} : 350\,\text{g}$ **(e)** $\frac{3}{4} : \frac{1}{3}$ **(f)** $2.7 : 3.6$

(a) $8 : 12 = 2 : 3$ (dividing by 4)
(b) $18 : 24 : 6 = 3 : 4 : 1$ (dividing by 6)
(c) $40\,\text{cm} : 1\,\text{m} = 40\,\text{cm} : 100\,\text{cm}$ Units not the same,
    $= 40 : 100$ change m to cm.
    $= 4 : 10$ (dividing by 10)
    $= 2 : 5$ (dividing by 2)

continued ▼

Divide by a common factor.

This ratio has 3 parts.

The units *must* be the same before you can simplify. This applies to (c) and (d).

(d) $1\,kg:350\,g = 1000\,g:350\,g$ Units not the same,

$= 1000:350$ change kg to g.

$= 100:35$ (dividing by 10)

$= 20:7$ (dividing by 5)

(e) $\frac{3}{4}:\frac{1}{3} = (\frac{3}{4} \times 12):(\frac{1}{3} \times 12)$ (multiplying by 12)

$= 9:4$

(f) $2.7:3.6 = 27:36$ (multiplying by 10)

$= 3:4$ (dividing by 9)

> To be in its simplest form, all parts of the ratio must be whole numbers.

> Find a number that both denominators divide into exactly.
> Multiply both fractions by this number.

> Multiply by 10 to 'get rid' of the decimal places.
> Divide by 9 to simplify to its simplest form.

Exercise 12B

1 Write these ratios in their lowest terms.

   (a) $5:10$ **(b)** $12:8$ **(c)** $6:30$

   (d) $9:15$ **(e)** $40:25$ **(f)** $48:36$

   (g) $120:70$ **(h)** $28:49$ **(i)** $1000:250$

2 Write these ratios in their simplest form.

   (a) £1:20p   **(b)** $6\,mm:3\,m$ **(c)** $5\,\ell:250\,m\ell$

   (d) $300\,m:2\,km$ **(e)** 75p:£5   **(f)** $350\,g:2\,kg$

   (g) £3.50:£1.25 **(h)** $45\,cm:2\,m$ **(i)** $50\,mm:1\,m$

> Change each part into the same units first.

3 Change these ratios into their simplest form.

   (a) $\frac{1}{2}:\frac{1}{4}$ **(b)** $\frac{2}{3}:\frac{3}{4}$ **(c)** $\frac{3}{5}:\frac{7}{10}$

   (d) $3.2:1.6$ **(e)** $5.4:1.8$ **(f)** $4.8:1.2$

4 In a school there are 480 girls and 560 boys. Write the ratio of boys to girls in its lowest terms.

5 In an Army corps there are 480 privates, 96 corporals, and 12 captains. Write the ratio of privates to corporals to captains in its simplest form.

6 A drink was made with $1\frac{1}{2}$ cups of lemonade and $\frac{1}{4}$ cup of blackcurrant cordial. Write the ratio of lemonade to blackcurrant in its lowest terms.

7 The cost of materials for a soft toy were: fabric £3.20, stuffing £1.20 and others 60p. Write the ratio of the cost of fabric to stuffing to others in its lowest terms.

Writing a ratio as a fraction

> You can write a ratio as a fraction (or vice versa).

You can write 2 : 3 as the fraction $\frac{2}{3}$.

This gives you an alternative method of solving some simple problems.

Example 5

In a school the ratio of boys to girls is 5 : 7.
There are 265 boys.
How many girls are there?

Method A (using 'multiplying')

Suppose there are x girls.

The ratio 5 : 7 has to be the same as 265 : x

$265 \div 5 = 53$ so multiply both sides of the ratio by 53

Boys: $5 \times 53 = 265$

Girls: $7 \times 53 = 371$

There are 371 girls in the school.

> Multiplying a ratio by 2, 3, 4, 5 ... etc gives an equivalent ratio.
> Here the multiplier is 53.

Method B (using fractions)

Suppose there are x girls.

The ratio 5 : 7 has to be the same as 265 : x

Using fractions, $\dfrac{5}{7} = \dfrac{265}{x}$

This can also be written as $\dfrac{7}{5} = \dfrac{x}{265}$

Multiplying both sides by 265, $\dfrac{7 \times 265}{5} = x$

This gives $x = 371$, so there are 371 girls in the school.

> When the unknown (x) is on the top, the equation is easier to solve.

> See Section 3.2, Example 5 for how to solve this kind of equation.

Exercise 12C

> Use the fraction method.

1 In a zoo the ratio of penguins to flamingos is 2 : 5. There are 235 flamingos. How many penguins are there?

2 Adam and June share an inheritance from their aunt in the ratio of 4 : 3. If Adam received £480, how much did June get?

3 At a football match the ratio of children to adults is 3 : 8. If there were 630 children, how many adults were there?

4 A piece of wood is cut into two pieces in the ratio 4 : 7.
   (a) If the shortest piece is 140 cm, how long is the longer piece?
   (b) How long was the original piece of wood?

5 A recipe for jam uses 125 g of fruit and 100 g of sugar.
   (a) What is the ratio of fruit to sugar in its lowest terms?
   (b) If I had 300 g of fruit how much sugar would I need?
   (c) What would be the total mass of the jam in part **(b)**?

6 In a box of strawberries the ratio of good to damaged fruit is 16 : 3.
   (a) If I found 18 damaged fruits, how many good ones did I have?
   (b) How many strawberries are in the box?

7 A celebrity magazine has pictures to text in the ratio of 3 : 2. If a magazine has 75 sections of pictures how many sections of text does it have?

8 The ratio of sand to cement to make mortar is 5 : 2.
   (a) If I use 3 buckets of cement, how many buckets of sand will I need?
   (b) A bucket of sand weighs 11 kg and a bucket of cement weighs 17 kg. How much will the mortar weigh?

Writing a ratio in the form 1 : *n* or *n* : 1

In Example 1, 150 : 30 simplified to 5 : 1.

Not all ratios simplify so that one value is 1. For example, 12 : 8 simplifies to 3 : 2.

Dividing by 4.

If you are asked to write this ratio in the form *n* : 1 you cannot leave the answer as 3 : 2.

Look at which number has to be 1. Divide both sides of the ratio by that number.

$$12 : 8 = (12 \div 8) : (8 \div 8)$$
$$= 1.5 : 1$$

You can write all ratios in the form 1 : n or n : 1.

> For n : 1 the number on the right-hand side has to be 1.

> The '8' has to be 1. Divide by 8.

> For a ratio in the form n : 1 or 1 : n, the n value does not have to be a whole number.

Example 6

Write these ratios in the form n : 1.

(a) 27 : 4   **(b)** 6 cm : 25 mm   **(c)** £2.38 : 85p

(a) 27 : 4 = 6.75 : 1

> Divide both sides by 4.

(b) 6 cm : 25 mm = 60 mm : 25 mm
$$= 60 : 25$$
$$= 2.4 : 1$$

> Divide both sides by 25.

> Change of units needed in (b) and (c).

(c) £2.38 : 85p = 238p : 85p
$$= 238 : 85$$
$$= 2.8 : 1$$

> Divide both sides by 85.

Example 7

Write these ratios in the form 1 : n.

(a) 5 : 9   **(b)** 400 g : 1.3 kg   **(c)** 40 minutes : $1\frac{1}{4}$ hours

(a) 5 : 9 = 1 : 1.8

> Divide both sides by 5.

(b) 400 g : 1.3 kg = 400 g : 1300 g
$$= 400 : 1300$$
$$= 1 : 3.25$$

> Divide both sides by 400.

> Change of units needed in (b) and (c).

(c) 40 mins : $1\frac{1}{4}$ hours = 40 mins : 75 mins
$$= 40 : 75$$
$$= 1 : 1.875$$

> Divide both sides by 40.

Exercise 12D

1 Write the following ratios in the form $n : 1$.

(a) $15 : 3$ (b) $7 : 2$

(c) $9 : 4$ (d) $12 : 5$

(e) $3 \text{ kg} : 200 \text{ g}$ (f) $1 \text{ hour} : 40 \text{ min}$

(g) $5 \text{ m} : 40 \text{ cm}$ (h) $1 \text{ day} : 10 \text{ hours}$

(i) $2.5 \ \ell : 500 \text{ m}\ell$

2 Write the following ratios in the form $1 : n$.

(a) $16 : 24$ (b) $45 : 54$

(c) $36 : 96$ (d) $15 \text{ hours} : 1 \text{ day}$

(e) $5 \text{ days} : 4 \text{ weeks}$ (f) $6 \text{ mm} : 2.7 \text{ cm}$

(g) $18 \ \ell : 4500 \text{ m}\ell$ (h) $5 \text{ cm} : 15 \text{ mm}$

Dividing quantities in a given ratio

Ian and Simon share £60 so that Ian receives twice as much as Simon.

You can think of this as, Ian : Simon = 2 : 1.

Ian will get this much: Simon will get this much:

where these bags of money contain equal amounts.

To work out each man's share, you need to divide the money into 3:

$(2 + 1)$ equal **parts** (or **shares**).

$$1 \text{ part} = £60 \div 3 = £20$$

Ian receives, $2 \times £20 = £40$ 2 parts

Simon receives, $1 \times £20 = £20$ 1 part

Check: $£20 + £40 = £60$ ✓

> **To share quantities in a given ratio:**
> 1 **Work out the total number of equal parts.**
> 2 **Work out the amount in 1 part.**
> 3 **Work out the value of each share.**

You can use this method to share amounts between more than two people.

Key words:
part
share

Remember, a ratio compares quantities.

Example 8

The alloy *Alnico* is made from five metals: Cobalt, Iron, Nickel, Aluminium and Titanium in the ratio 6 : 5 : 3 : 1 : 1.

How many grams of each metal are there in 544 g of this alloy?

This fender guitar uses Alnico magnet pickups.

```
Total number of parts = 6 + 5 + 3 + 1 + 1 = 16 parts
              1 part = 544 g ÷ 16 = 34 g
Cobalt      = 6 × 34 g = 204 g     6 parts
Iron        = 5 × 34 g = 170 g     5 parts
Nickel      = 3 × 34 g = 102 g     3 parts
Aluminium   = 1 × 34 g = 34 g      1 part
Titanium    = 1 × 34 g = 34 g      1 part
Check:      204 g + 170 g + 102 g + 34 g + 34 g = 544 g  ✓
```

Check that your final answers add up to the original amount.

Exercise 12E

1 Divide these amounts in the ratio given:
  (a) 50 in the ratio 3 : 2   **(b)** 30 in the ratio 3 : 7
  (c) 45 in the ratio 5 : 4   **(d)** 64 in the ratio 3 : 5

2 Divide each amount in the ratio given:
  (a) £200 in the ratio 9 : 1
  (b) 125 g in the ratio 2 : 3
  (c) 600 m in the ratio 7 : 5
  (d) 250 ℓ in the ratio 16 : 9

3 Divide £120 in the ratio:
  (a) 7 : 3   **(b)** 1 : 3   **(c)** 3 : 2   **(d)** 5 : 7

4 The ratio of pupils to teachers for a school trip is 13 : 2. If 135 people go on the trip, how many of them are teachers?

5 At a farm the ratio of black lambs to white lambs born is 2 : 9. If 132 lambs are born one year, how many of them are black?

6 A cocktail is made from orange juice, sparkling wine and lime juice in the ratio 5 : 6 : 1. If I make 960 mℓ of cocktail, how much of each ingredient will I need?

7 Ali puts 50p towards a raffle ticket, Sam puts 30p and Joe puts 20p. They agree to share any winnings in the ratio of the amount they put in.

   (a) Write the ratio of these three amounts in its lowest terms

   (b) They win £130. How much will each of them receive?

8 One Saturday Jasmine decides to spend her revision time on Maths, Geography and French in the ratio 7 : 5 : 3. If she she spends a total of 4 hours revising, how much time will she spend on each subject?

> Work in minutes.

9 Concrete is made from cement, sand and gravel in the ratio 1 : 3 : 4. If I want to make 2 tonnes of concrete, how much of each ingredient do I need?

10 Paul is 18 years old, John is 15 years old and Sarah is 12 years old. Their grandfather leaves them £255 to be divided between them in the ratio of their ages. How much will each of them receive?

> Simplify the ratio first.

12.2 Proportion

Direct proportion

> **Key words:**
> direct proportion

> Two quantities are in **direct proportion** if their ratio stays the same as they increase or decrease.

For example, a mass of 5 kg attached to a spring stretches it by 40 cm.

A mass of 15 kg attached to the same spring will stretch it 120 cm.

The ratio mass : extension is
$$5 : 40 \ = 1 : 8 \text{ (dividing by 5)}$$
$$15 : 120 = 1 : 8 \text{ (dividing by 15)}$$

The ratio mass : extension is the same so the two quantities are in direct proportion.

You can use direct proportion to solve problems.

Example 9

The cost of 5 apples is £1.40.
How much would 8 apples cost?

Suppose that 8 apples cost x pence.

The ratio 'number of apples' : 'cost' must stay the same.

So $5 : 140 = 8 : x$

Using fractions, $\dfrac{140}{5} = \dfrac{x}{8}$

Multiplying both sides by 8, $\dfrac{140 \times 8}{5} = x$

  $x = 224$

So 8 apples cost £2.24.

Working in pence for both costs.

Using the fractions method as in Example 5.

Unitary method

There is an easier way of solving Example 9.
First find the cost of 1 apple.

5 apples cost £1.40 or 140p

1 apple costs $\dfrac{140}{5} = 28\text{p}$

8 apples cost $28 \times 8, = 224\text{p}$ or £2.24

This is called the   **unitary method** .

Key words:
unitary method

This is easier than the method in Example 9. It only takes three lines.

In the unitary method, you find the value of *1 unit* of a quantity.

Example 10

Susan has a part-time job.
She works for 15 hours a week and is paid £93.
Her employer wants her to work for 25 hours a week.
How much will Susan be paid now?

For 15 hours' work she is paid £93

For 1 hour's work she is paid $\dfrac{£93}{15} = £6.20$

For 25 hours' work she will get $25 \times £6.20 = £155$

Find how much she is paid for 1 hour.

To solve *direct* proportion problems you always do a *division* followed by a *multiplication*.

Exercise 12F

1 Find the cost of 1 unit for each of the following.

    (a) 5 bars of chocolate cost £2.25

    (b) 3 m of wood costs £3.36

    (c) 9 kg of carrots costs £4.05

2 If 8 pens cost £2.96, how much will 3 pens cost?

3 40 litres of petrol will take me 200 miles.
How far can I go on 15 litres?

4 Five books weigh 450 g. How much would 18 books weigh?

5 Alice sews buttons onto trousers. She is paid £30 for sewing 200 buttons.
How much would she get for sewing 35 buttons?

6 6 packets of soap powder cost £7.38.
How much will 10 packets cost?

7 Kim earns £67.20 for 12 hours' work. One week she works 17 hours. How much does she earn for that week?

8 A chocolate cake for 8 people uses 120 g of sugar. How much sugar would you need for a cake for 15 people?

9 Omar walks 500 m to school and it takes him 8 minutes. If he continues to walk at the same speed,

    (a) how far would he walk in 1 hour?

    (b) how long would it take him to walk 2.25 km?

10 My cat eats 2 tins of food every 3 days.
How much does he eat in a week?

Key words:
inverse proportion

Inverse proportion

When two quantities are in direct proportion:
- as one increases, so does the other
- as one decreases, so does the other.

Suppose you travel from Leeds to Newcastle, a distance of 100 miles.

If you travel at an average speed of 50 mph it will take you 2 hours.

If you only average 40 mph it will take you $2\frac{1}{2}$ hours.

The slower you travel, the more time it takes:
- as the speed decreases the time increases
- as the speed increases the time decreases.

See Section 8.8

$$\text{Time} = \frac{\text{distance}}{\text{speed}}$$

$$\frac{100}{50} = 2$$

$$\frac{100}{40} = 2\frac{1}{2}$$

If you *halve* the speed to 25 mph you *double* the time to 4 hours.

> When two quantities are in **inverse proportion**, one quantity increases at the same rate as the other quantity decreases.

The best way to solve problems involving inverse proportion is to use the unitary method.

Example 11

Two people take 6 hours to paint a fence.
How long will it take 3 people?

> 2 people take 6 hours
> 1 person takes 6 × 2 = 12 hours
> 3 people will take 12 ÷ 3 = 4 hours

Read the problem first and decide whether it is *direct* or *inverse* proportion. Use your common sense! The more people painting, the less time it takes inverse proportion.

The unitary method: Work out how long it will take 1 person. 1 person takes twice as long as 2 people.

Example 12

Tarik is repaying a loan from a friend.

He agrees to pay £84 per month for 30 months.

If he can afford to pay £120 per month, how many months will it take to repay?

Repaying *more* each month will take *less* time, so this is inverse proportion.

Paying £84 per month takes 30 months

Paying £1 per month takes (84 × 30) months

Paying £120 per month will take $\frac{84 \times 30}{120}$ months = 21 months

210 years !!!!!

To solve *inverse* proportion questions you always do a *multiplication* followed by a *division*.

Exercise 12G

1 It takes 3 people 4 days to paint a shop. How long would it take 1 person?

2 It takes 2 people 6 hours to make a suit. How long would it take 3 people?

3 Eight horses need a trailer of hay to feed them for a week. How many horses could this feed for 4 days?

4 It takes 5 bricklayers 4 days to build a house. How long would it take 2 bricklayers?

5 It takes 3 pumps 15 hours to fill a swimming pool. How many pumps would be needed to fill the pool in 9 hours?

6 A refugee camp has enough food for 400 people for 25 days. If the food lasts only 20 days, how many people must be in the camp?

7 It takes 7 days for 6 people to dig a trench. How long would it take 14 people?

8 In a library 48 paperback books 20 mm wide fit on a shelf. How many books 24 mm wide would fit on the same shelf?

9 It takes 8 hours to fly to Chicago at a speed of 300 mph.
   (a) How long would it take to fly to Chicago at a speed of 400 mph?
   (b) If the journey took 12 hours, what speed was I travelling?

10 It takes 6 window cleaners 8 days to clean the windows at Buckingham Palace.
   (a) How long would it take 4 window cleaners?
   (b) The windows need to be cleaned in 3 days for a special event. How many window cleaners are needed?

Exercise 12H Mixed questions

1 If 10 metres of material cost £23.50, what will 7 metres cost?

2 A man cuts a hedge in 45 minutes using a cutter with a blade 36 cm wide. How long would it take if the blade was only 15 cm wide?

3 It takes 200 tiles to tile a room when each tile covers 36 cm². If I use tiles which cover 25 cm², how many tiles would I need?

4 Alan works for 9 hours a week in a shop. He is paid £45. He increases his hours to 15. How much will he be paid?

5 It takes $2\frac{1}{2}$ hours to travel to London by train at an average speed of 80 mph. A new train travels at an average speed of 100 mph. How long will the journey to London take on the new train?

6 If 14 kg of potatoes cost £2.38, how much will 6 kg cost?

7 One load of feed lasts 40 sheep 9 days.
   (a) How many sheep could this load feed for 5 days?
   (b) For how many days could it feed 60 sheep?

8 In a hotel it takes a maid 24 minutes to clean 3 rooms.
   (a) How long will it take her to clean 16 rooms?
   (b) How many rooms can she clean in 4 hours?

12.3 Map scales

Key words:
scale

You will need to know:

- **how to convert between metric units of length**

Map **scales** are written as ratios.

A scale of 1 : 50 000 means that 1 cm on the map represents 50 000 cm on the ground.

A scale of 1 : 25 000 means that 1 cm on the map represents 25 000 cm on the ground.

When you answer questions involving map scales you need to:

- use the scale of the map
- convert between metric units of length so that your answer is in sensible units.

Always look carefully to see what scale is being used.

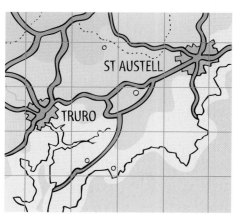

Example 13

The scale of a map is 1 : 50 000. The distance between Truro and St. Austell on the map is 40 cm. What is the actual distance between Truro and St. Austell? Give your answer in kilometres.

Distance on map = 40 cm

Distance on the ground = 40 × 50 000 cm

$\qquad$ = 2 000 000 cm

$\qquad$ = 2 000 000 ÷ 100 m

$\qquad$ = 20 000 m

$\qquad$ = 20 000 ÷ 1000 km

$\qquad$ = 20 km

Work out the real distance in cm then convert to km.

Each 1 cm on the map is 50 000 on the ground.

1 m = 100 cm
1 km = 1000 m

Example 14

The distance between two towns is 24 km. How far apart will they be on a map of scale 1 : 180 000?

Distance on the ground = 24 km

$$= 24 \times 1000 \text{ m}$$

$$= 24\,000 \text{ m}$$

$$= 24\,000 \times 100 \text{ cm}$$

$$= 2\,400\,000 \text{ cm}$$

Distance on map $= \dfrac{2\,400\,000}{180\,000} \text{ cm} = 13.3 \text{ cm (to 1 d.p.)}$

> Convert the real distance to cm before you divide by the scale of the map.

Exercise 12I

1 The scale of a map is 1 : 25 000. Find the actual distance represented by these measurements on the map.

   (a) 4 cm   **(b)** 7 cm   **(c)** 8 mm   **(d)** 12.5 cm

2 A map has scale 1 : 200 000. What measurement on the map will represent:

   (a) 4 km   **(b)** 20 km   **(c)** 15 km   **(d)** 12.5 km?

3 This motorway map has a scale of 1 : 3 000 000.

   (a) The distance on the map from Carlisle to Kendal is 2.5 cm. How far is the actual distance?

   (b) Measure in a straight line the distance from Carlisle to Penrith. How far is the actual distance?

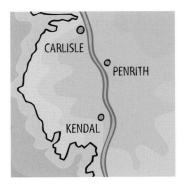

4 A map of Ireland has a scale of 1 : 500 000. What measurement on the map would represent these distances:

   (a) Waterford to Dundalk, 400 km

   (b) Wicklow to Carrigart, 320 km

   (c) Dublin to Tralee, 160 km?

5 On a map with a scale of 1 : 2 500 000 the distance from Leeds to Edinburgh is 13 cm.

   (a) What is the actual distance from Leeds to Edinburgh?

   (b) Newcastle is 150 km from Leeds. How far is this on the map?

6 This map has a scale of 1 : 50 000. Use the map to work out the following distances.

(a) Between the ends of the 2 piers at Tynemouth.

(b) From Sharpness Point to Smuggler's Cave.

(c) The Ferry crossing of the Tyne.

(d) The length of both piers.

(e) From the Coast Guard Station (CG Sta) to the Coast Guard Lookout (CG Lookout).

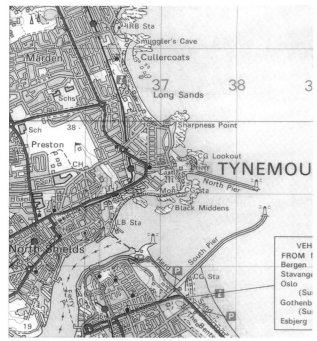

7 A model train is made to a scale of 1 : 40.

(a) The height of the model is 20 cm. How high is the real train?

(b) The real train is 7.4 m long. How long is the model?

8 A designer draws a plan of a garden using a scale of 1 : 50.

(a) The garden is 5.5 m long and 4.2 m wide. What are the measurements of the plan?

(b) She makes some scaled cut-outs of a shed, barbecue and water feature so she can try different layouts. Change these scaled measurements to find the real size of the items.

shed 4 cm × 2 cm
barbecue 1.5 cm × 1 cm
water feature 3 cm × 1 cm

Examination style questions

1 In a class of 28 pupils, the ratio of boys to girls is 3 : 4.
How many of the pupils are

(a) girls?

(b) boys?

(3 marks)
AQA, Spec B, 3I, March 2003

2 While in the USA, John pays $30 for a pair of trainers.
The exchange rate is $1.50 to £1.
Calculate the cost of the pair of trainers in £.

(2 marks)
AQA, Spec B, 3F, June 2004

3 Four melons cost £5.28.
How much will five melons cost?

(3 marks)
AQA, Spec B, 3F, June 2004

4 A pile of 20 sheets of card is 1 cm high.
How high is a pile of 50 sheets of card?

(2 marks)
AQA, Spec B, 3F, November 2003

5 Two friends agree to share £42 in the ratio 3 : 4.
How much is the smaller share?

(2 marks)
AQA, Spec B, 3I, November 2003

6 Pauline goes on holiday.
She changes £200 into Euros (€).
The exchange rate is £1 = €1.7.

(a) How many Euros does Pauline receive for £200?

(b) Pauline spends one quarter of her Euros on presents.
How many Euros does she spend on presents?

(c) A bag is priced at €68.
In a sale, this price is reduced by 25%.
Pauline buys the bag.
How much does she save?
Give your answer in pounds.

(8 marks)
AQA, Spec B, 3F, March 2003

7 Sarmad swims 80 lengths of a swimming pool every day.
The swimming pool is 20 metres long.
- **(a)** **(i)** How far does Sarmad swim each day?
  (ii) What is this distance in kilometres?
- **(b)** Sarmad swims lengths of front crawl and lengths of
  backstroke in the ratio 1 : 3.
  How many lengths of front crawl does he swim each
  day?

  (5 marks)
  AQA, Spec B, 3F, March 2003

8 Bill changes £27 into Swiss francs.
The exchange rate is £1 to 1.55 Swiss francs.
How many Swiss francs does he receive?

  (2 marks)
  AQA, Spec B, 3F, November 2003

Summary of key points

Ratio

A ratio compares two or more quantities.
For example, a ratio of 3 : 1 means that one quantity is
3 times larger than the other.

You can divide or multiply ratios to get them into their
simplest form.
For example, 12 : 3 = 4 : 1 (dividing by 3)
            2.5 : 6 = 5 : 12 (multiplying by 2).

When a ratio is in its simplest form (or lowest terms) you
cannot divide the numbers any further.

You can write a ratio as a fraction. For example 4 : 5 = $\frac{4}{5}$.
You can use ratios as fractions to solve problems.

You can write all ratios in the form 1 : n or n : 1.
For example, 4 : 5 = 1 : $1\frac{1}{4}$ (dividing by 4)
            3 : 2 = $1\frac{1}{2}$: 1 (dividing by 2).

Sharing ratios

To share quantities in a given ratio:
1 Work out the total number of equal parts.
2 Work out the amount in 1 part.
3 Work out the values of each share.

Proportion

Two quantities are in direct proportion if their ratio stays the same as they increase or decrease.

When two quantities are in inverse proportion, one increases at the same rate as the other decreases.

Unitary method

In the unitary method you find the value of 1 unit of a quantity.

It is the best method for solving direct and inverse proportion problems.

Map scales

Map scales are written as ratios.
For example a ratio of 1 : 200 000 means that 1 cm on the map represents 200 000 cm on the ground.
Give answers in sensible units, for example,
200 000 cm = 2000 m = 2 km.

This work ranges from grade F (simple problems on unitary method) through grade D (simple direct proportion problems or, for example, sharing £140 in the ratio 6 : 1) to grade C (harder proportion problems or, for example, sharing £5000 in the ratio 10 : 8 : 7).

This chapter will show you how to:
- ✔ write a probability using the values it can take
- ✔ work out relative frequencies and estimate probabilities
- ✔ list outcomes and draw sample space diagrams
- ✔ deal with mutually exclusive events and independent events

13.1 Describing probability

Key words:
chance	event
certain	likely
even chance	unlikely
impossible	outcome
probability scale	

Probability uses numbers to represent the **chance** that something (an **event**) will happen.

You can describe the chance of an event happening using words such as **certain**, **likely**, **even chance**, **unlikely** and **impossible**.

You can use numbers to represent these based on the **probability scale**. All probability values lie between 0 and 1.

$0 \leqslant$ probability $\leqslant 1$.

Example 1

Describe the likelihood of these events using words:
(a) The sun will set today.
(b) You were born 30 years before your father.
(c) Getting a Head when you throw a coin.
(d) It will rain tomorrow.
(e) It will snow in summer.

(a) Certain	**(b)** Impossible
(c) Even chance	**(d)** Likely
(e) Unlikely	

You can write a probability as a fraction, decimal or percentage.

probability	fraction	decimal	percentage
impossible	0	0	0%
certain	1	1.0	100%
even chance	$\frac{1}{2}$	0.5	50%
likely	e.g. $\frac{8}{10}$	0.8	80%
unlikely	e.g. $\frac{1}{5}$	0.2	20%

Converting between fractions, decimals and percentages was covered in Chapter 3.

Example 2

Write the probability of obtaining a Head when you toss a coin. Give the probability as a fraction.

> The chances of throwing a Head (or a Tail) are even.
>
> The probability $P(H) = \frac{1}{2}$.

There are two possible **outcomes** – a Head or a Tail. They are both equally likely.

P(H) is shorthand for 'probability of throwing a Head'.

Example 3

Write the probability of throwing a 6 on an ordinary six-sided dice. Write the probability as a percentage.

> The chances of throwing a 6 is 1 out of a possible 6 numbers.
>
> The probability $P(6) = \frac{1}{6} = 0.1666\ldots = 16.7\%$ (3 s.f.)

An ordinary six-sided dice has six faces with numbers 1, 2, 3, 4, 5 and 6. There are 6 possible outcomes.

In Examples 2 and 3 you could obtain the probability of an event as the fraction.

$$\text{probability} = \frac{\text{number of successful outcomes}}{\text{total number of possible outcomes}}$$

Remember this, it is an important formula.

Example 4

An eight-sided (octahedral) fair dice numbered 1 to 8 is thrown. Find:

(a) the probability of obtaining a 3

(b) the probability of obtaining a prime number.

(a) The value 3 only occurs once on the dice, so there is only 1 successful outcome.

The total number of possible outcomes is 8.

$P(3) = \frac{1}{8}$

(b) The prime number values are 2, 3, 5 and 7, so there are 4 successful outcomes.

$P(prime) = \frac{4}{8} = \frac{1}{2}$

> The possible outcomes are: 1, 2, 3, 4, 5, 6, 7, 8.

> number of successful outcomes
> number of possible outcomes

> The number of possible outcomes is still 8.

> Give the fraction in its lowest terms.

Exercise 13A

1 Describe the likelihood of these events using words.
  (a) The sun will rise tomorrow.
  (b) England will win the next World Cup.
  (c) You will eat five portions of fruit or vegetables today.
  (d) You will see the Prime Minister on your way home from school.
  (e) You will throw an even number on an ordinary six-sided dice.

2 Copy the probability scale. Label each arrow with an event from the list below.

0

On an ordinary six-sided dice:
  (a) throwing a 1
  (b) obtaining an even number
  (c) getting a number >7
  (d) obtaining a number less than or equal to 4.

3 Copy the probability scale from question 2. Draw and label arrows to show these probabilities.
  (a) It will rain tomorrow.
  (b) You throw a 20p coin and get a 'Head'.
  (c) A dropped drawing pin lands point-down.
  (d) You throw a 7 on an ordinary six-sided dice.
  (e) You buy a new music CD next week.

4 Write these percentages using the notation
P(event) = ☐ and in the form specified.

(a) The probability that you will be given homework is
40% (as a decimal).

(b) The probability that it will snow tomorrow is 0.01
(as a percentage).

(c) The probability of throwing two consecutive sixes
with a dice is $\frac{1}{36}$ (as a decimal).

5 An ordinary six-sided dice is thrown.
Find the probabilities of these events:

(a) throwing a 2

(b) obtaining an even number

(c) throwing a number less than 4

(d) landing on a 3 or a 5

(e) not obtaining a six.

> Use the notation P(2).

6 In a raffle one hundred tickets are sold. What is the
probability of winning if you buy

(a) 3 tickets   (b) 11 tickets?

7 In a bag there are 11 blue marbles and 5 red marbles.
What is the probability of picking out

(a) a blue marble   (b) a red marble

(c) a green marble   (d) a blue or red marble?

13.2 Relative frequency

> **Key words:**
> successful trials
> estimated probability
> relative frequency

The probability of obtaining a Head when a coin is tossed
is P(H) = $\frac{1}{2}$.

In reality, if you toss a coin 10 times you may not get 5
Heads.

However, the number of Heads you get (**successful trials**)
will get closer to $\frac{1}{2}$ the total number of throws the more
times you toss the coin (total number of trials).

When you carry out a probability experiment, the
experimental probability is called the **relative
frequency** .

$$\text{Relative frequency} = \frac{\text{number of successful trials}}{\text{total number of trials}}$$

Remember this – it is an important formula.

This is also called the **estimated probability** , or experimental probability.

Example 5

A 2p coin is tossed 50 times:
Estimate the probability of
(a) a Head
(b) a Tail.

	Tally	Frequency
Head (H)	~~IIII~~ ~~IIII~~ ~~IIII~~ ~~IIII~~ ~~IIII~~ III	28
Tail (T)	~~IIII~~ ~~IIII~~ ~~IIII~~ ~~IIII~~ II	22
	Total	50

(a) Relative frequency $= \dfrac{number\ of\ successful\ trials}{total\ number\ of\ trials}$

$P(H) = \dfrac{number\ of\ Heads}{total\ number\ of\ throws}$

$= \frac{28}{50}$ or 0.56 or 56%

(b) $P(T) = \dfrac{number\ of\ Tails}{total\ number\ of\ throws}$

$= \frac{22}{50}$ or 0.44 or 44%

Check: $0.56 + 0.44 = 1.00$ ✓

Check that the probabilities for all the events add up to 1.

Frequency = total number of trials × relative frequency

Remember this – it is an important formula.

Example 6

An ordinary six-sided dice is thrown 100 times.
The scores are recorded in a frequency table.

Score	1	2	3	4	5	6	Total
Frequency	13	18	19	14	15	21	100

(a) Estimate the relative frequency of each score.
(b) Compare the relative frequency of throwing a 5 with the theoretical probability of P(5).
(c) How could you get a better estimate for the probability?
(d) How many times would you expect to throw a 6 if you rolled the dice a total of 300 times?

Extend the table.

(a)

Score	Frequency	Relative frequency
1	13	$\frac{13}{100} = 0.13$
2	18	$\frac{18}{100} = 0.18$
3	19	$\frac{19}{100} = 0.19$
4	14	$\frac{14}{100} = 0.14$
5	15	$\frac{15}{100} = 0.15$
6	21	$\frac{21}{100} = 0.21$
Total	100	$\frac{100}{100} = 1.00$

$$\frac{\text{number of times 1 is thrown}}{\text{total number of throws}}$$

(b) The theoretical probability P(5) is

$$\frac{\text{number of successful outcomes}}{\text{total number of outcomes}} = \frac{1}{6} = 0.167$$

The relative frequency for P(5) is 0.15, which is slightly

lower.

(c) To improve the estimate, you could carry out more trials.

(d) Number of 6s $= 300 \times \frac{21}{100}$

$$= 63$$

As the number of trials increases, the relative frequency approaches the theoretical probability.

Frequency
$$= \frac{\text{no. of}}{\text{trials}} \times \frac{\text{relative}}{\text{frequency}}$$

Exercise 13B

1 (a) Copy the tally chart shown.

	Tally	Frequency	Relative frequencies
Head (H)			
Tail (T)			
	Total	100	

(b) Toss a coin 100 times and record your results in the chart.

(c) Work out the relative frequency for Head and Tail from your results. Write them in the table.

(d) Compare your results with the theoretical probabilities for P(H) and P(T). How could you get a better estimate for the probability?

2 Copy the tally chart.

Type of number obtained	Tally	Frequency
An even number		
A square number		
A prime number		
A number greater than 4		
	Total	

Some values may appear in more than one column at the same time.

(a) Throw an ordinary six-sided dice 100 times and record your results in the chart.
(b) Calculate the relative frequency for each event.
(c) Compare your answers to the theoretical probabilities for P(even), P(square), P(prime) and P($>$4).

3 A bag contains 7 red counters and 3 blue counters. Pick a counter from the bag, record its colour, then replace it. Repeat 50 times.
(a) Draw a tally chart to record this information.
(b) Work out the relative frequencies for picking a red and blue counter.
(c) Compare your results with the theoretical values for P(R) and P(B).

13.3 Mutually exclusive events

Key words:
mutually exclusive event
exhaustive

When you toss a coin you can get either a Head *or* Tail but not both at the same time. Head and Tail are called **mutually exclusive events** because if you get one, you cannot get the other outcome.

The word *or* is important here.

The outcomes of throwing a dice are all equally likely. They are mutually exclusive – you could throw 1 *or* 2 *or* 3 *or* 4 *or* 5 *or* 6.

One outcome excludes the others.

Example 7

(a) For an ordinary six-sided dice work out the probability of:
   (i) throwing a 6   (ii) throwing an even number.
(b) Are the events mutually exclusive? Explain your answer.

(i) The probability of throwing a 6 is 1 chance in 6 or $\frac{1}{6}$.

(ii) The probability of throwing an even number is 3 chances in 6, or $\frac{3}{6} = \frac{1}{2}$.

Even numbers on a dice = 2, 4, 6.

(b) No, throwing an even number does not exclude throwing a 6.

If there are n mutually exclusive events, all equally likely, then the probability of each event is $\dfrac{1}{n}$.

For a dice,
$P(1) = P(2) = P(3) = P(4) = P(5) = P(6) = \frac{1}{6}$.

If there are n mutually exclusive events and m successful outcomes, then the probability of a successful outcome is $\dfrac{m}{n}$.

For a dice, $P(\text{odd}) = \frac{3}{6}$. There are 3 successful outcomes (1, 3, 5).

When throwing a dice, the events (even) and (odd) are **exhaustive** . This means there is no other possible outcome – you can only get an even or an odd number.

Example 8

Work out the probability of obtaining a Head (H) *or* a Tail (T) when tossing a coin.

$P(H) = \frac{1}{2}$ and $P(T) = \frac{1}{2}$.
$P(H \text{ or } T) = P(H) + P(T) = \frac{1}{2} + \frac{1}{2} = 1$.

This is common sense. You are *certain* to get one *or* the other.

For any two events, say A and B, that are mutually exclusive and exhaustive then,

$$P(A \text{ or } B) = P(A) + P(B) = 1$$

Remember this, it is an important formula.

This is also true for more than two mutually exclusive events.

Example 9

A spinner is divided into three equal sections coloured red, green and blue.
(a) Find the probability that the spinner will land on red.
(b) Find the probability that the spinner will *not* land on red.

(a) The sectors are equal, so all the results are equally likely.

The events are mutually exclusive.

So $P(R) = \frac{1}{3}$.

As the events are equally likely, $P(G) = \frac{1}{3}$ and $P(B) = \frac{1}{3}$ too.

(b) The probability that the spinner will not land on red is the same as the probability of landing on green or blue. So

$P(\text{not } R) = P(G \text{ or } B)$.

$P(G \text{ or } B) = P(G) + P(B) = \frac{1}{3} + \frac{1}{3} = \frac{2}{3} = P(\text{not } R)$.

$P(A + B) = P(A) + P(B)$ for mutually exclusive events.

In Example 9, $P(\text{not } R) = \frac{2}{3} = 1 - P(R)$.

$P(R) = \frac{1}{3}$.

For an event A, the probability of the event A _not_ happening is given by

$$P(\text{not } A) = 1 - P(A)$$

Remember this, it is an important formula.

Exercise 13C

1 A letter of the alphabet is chosen at random. What is the probability that it will be a vowel? What is the probability it will be a consonant?

2 One of the longest rivers in the world is called the MISSISSIPPI. If you choose one of these letters at random, what is the probability of choosing:

(a) the letter I

(b) the letter S

(c) M or I or S or P?

3 An ordinary six-sided dice is thrown. Work out:

(a) P(3), the probability of throwing a 3

(b) P(1), P(2), P(4), P(5) and P(6)

(c) the probability of throwing a 3 _or_ a 4

(d) the probability of throwing _not_ a 2

(e) the probability of throwing _not_ a 2 _or_ a 3.

4 A set of cards are numbered 1 to 30. What is the probability of
   (a) choosing a square number
   (b) choosing a prime number
   (c) choosing a number > 10
   (d) choosing a multiple of 3
   (e) choosing a factor of 24
   (f) choosing a card containing the number 2?

5 A bag contains 5 green counters, 4 blue counters, 2 red counters and 3 yellow counters.
   What is the probability of
   (a) picking out a red counter
   (b) picking a blue or green counter
   (c) picking a counter that is not yellow
   (d) picking a counter that is not green or red?

6 A bag contains 16 marbles of three different colours – red, green and yellow.
   The probability of picking a green marble is P(G) = $\frac{1}{4}$.
   The probability of picking a red marble is P(R) = $\frac{3}{8}$.
   (a) Work out the probability of picking a yellow marble.
   (b) Work out the probability of picking a marble that is not red or yellow.
   (c) How many marbles of each colour are in the bag?

13.4 Listing outcomes

Key words:
sample space diagram

To work out probabilities you need to know the outcomes of an event.

When you list outcomes, work systematically to make sure you don't miss any out.

Example 10

A 10p coin and 5p coin are tossed at the same time.
List all the possible outcomes.

Let H mean Heads and T mean Tails.

The outcomes are: HH HT TH TT.

There are 4 possible outcomes in total.

HT and TH are different.

HT means H on 10p, T on 5p.
TH means T on 10p, H on 5p.

Example 11

A four-sided spinner has 4 equal areas coloured red, blue, green and yellow. The spinner is spun and a coin is tossed at the same time. List all the possible outcomes.

Let R, B, G and Y stand for red, blue, green and yellow.

The possible outcomes are

(R,H) (R,T)

(B,H) (B,T)

(G,H) (G,T)

(Y,H) (Y,T).

There are 8 possible outcomes.

All the outcomes are equally likely.

List them systematically: possible 'R' outcomes, possible 'B' outcomes, etc.

You could show the possible outcomes in Example 11 in a two-way table:

A two-way table that shows all possible outcomes is called a **sample space diagram.**

		Spinner			
		R	**B**	**G**	**Y**
Coin	**Heads (H)**	(R,H)	(B,H)	(G,H)	(Y,H)
	Tails (T)	(R,T)	(B,T)	(G,T)	(Y,T)

If there are a large number of outcomes, it is far simpler to list them in a sample space diagram.

Exercise 13D

1 A mother gives birth to triplets. List all of the possible outcomes in terms of boys and girls.

2 Three coins are tossed simultaneously.
   (a) List all of the possible outcomes systematically.
   (b) Show the outcomes in a sample space diagram.
   (c) Which method is better for recording the outcomes?

Think of 10p, 5p and 20p coins.

3 An ordinary six-sided dice is tossed together with a four-sided dice (tetrahedral dice).
   Draw a sample space diagram to record all of the possible outcomes.

A four-sided dice has scores 1, 2, 3, 4.

4 Two spinners are spun at the same time and the two scores added. List all the possible outcomes. What is the probability that the final answer is negative?

13.5 Independent events

When the outcome of one event *does not* affect the outcome of another event, the events are **independent** .

For example:
- rolling two dice together
- rolling a dice once and then rolling it again
- rolling a dice and tossing a coin.

In situations like these the outcome of one event has no influence on the outcome of the other.

Example 12

A six-sided dice and coin are thrown at the same time. One possible outcome is (3, Head).

(a) List all the other possible outcomes.

(b) What is the probability P(even number, Tail)?

> You already know how to list outcomes systematically.

		Dice					
		1	**2**	**3**	**4**	**5**	**6**
Coin	Heads (H)	1,H	2,H	3,H	4,H	5,H	6,H
	Tails (T)	1,T	2,T	3,T	4,T	5,T	6,T

> Construct a sample space diagram for the two events.

> The events with (even, T) are highlighted in red. There are 3 successful outcomes.

There are $6 \times 2 = 12$ outcomes all together.

$$P(even,T) = \frac{number\ of\ successful\ outcomes}{total\ number\ of\ outcomes}$$

$$= \frac{3}{12}$$

$$= \frac{1}{4}$$

For two independent events A and B:

$$\begin{array}{c}\text{Total number}\\\text{of outcomes}\end{array} = \begin{array}{c}\text{number of}\\\text{outcomes of}\\\text{event A}\end{array} \times \begin{array}{c}\text{number of}\\\text{outcomes of}\\\text{event B}\end{array}$$

Example 13

Two ordinary six-sided dice, one red and one blue, are thrown at the same time and their scores are added together.

Complete the sample space diagram to show all the possible outcomes.

Use a sample space diagram to answer these questions.

(a) How many possible outcomes are there?

(b) Which score is the most likely?

(c) What is the probability of this score?

(d) What is the probability of getting a score more than 8?

(e) What is the probability of the score being an odd number?

(f) What is the probability of getting a double 3?

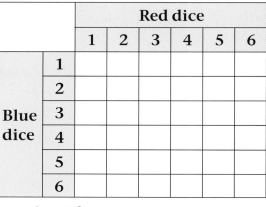

		Red dice					
		1	2	3	4	5	6
Blue dice	1						
	2						
	3						
	4						
	5						
	6						

		Red dice					
		1	2	3	4	5	6
Blue dice	1	2	3	4	5	6	7
	2	3	4	5	6	7	8
	3	4	5	6	7	8	9
	4	5	6	7	8	9	10
	5	6	7	8	9	10	11
	6	7	8	9	10	11	12

$4 + 5 = 9$

You *add* the scores from both dice.

(a) There are 36 possible outcomes.

(b) The score of 7 is the most likely. It occurs 6 times in the table.

(c) $P(7) = \frac{6}{36} = \frac{1}{6}$

(d) $P(\text{score} > 8) = \frac{10}{36} = \frac{5}{18}$

(e) $P(\text{odd}) = \frac{18}{36} = \frac{1}{2}$

(f) $P(3, 3) = \frac{1}{36}$

Count the values > 8 in the table.

Count the odd numbers in the table.

The total number of possible outcomes = (total number of outcomes on the red dice) × (total number of outcomes on the blue dice)
= 6 × 6
= 36

The probability of just one outcome = $\frac{1}{36}$.

Exercise 13E

1 A coin and a three-sided spinner with equal sectors coloured red, blue and green are used at the same time.

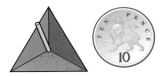

   (a) List all the possible outcomes systematically.

   (b) What is the probability of P(Red, Tail)?

2 A pizza can have any two of the following toppings: tomato, mushroom, ham, pineapple.

   (a) How many different ways are there to top a pizza?

3 A fruit machine has two windows. In each window only one of three different fruits can appear. The fruits are apple, strawberry and banana. List all the possible outcomes.

   What is the probability of getting:

   (a) no bananas

   (b) at least one apple

   (c) two identical fruits?

4 A four-sided dice and a spinner with three equal sectors coloured red, white and blue are used at the same time.

   (a) Construct a sample space diagram to show all the possible outcomes.

   (b) What is the probability of getting red *or* blue and an even number?   P(R or B, even).

5 An ordinary six-sided dice is thrown. It is then thrown a second time. The two scores are added together. Construct a sample space diagram and work out:

   (a) the probability of throwing two 3s

   (b) the probability that the sum is less than 5

   (c) the probability that the score on the first throw is double the score on the second throw.

Examination style questions

1 Danny has two fair spinners.

Spinner A has four equal sections, two are red, one is yellow and one is green.
Spinner B has six equal sections, three are red, one is yellow and two are green.

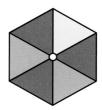

Spinner A   Spinner B

Danny spins each spinner once.

(a) Which colour is Spinner A most likely to land on?

(b) Which spinner is more likely to land on yellow, Spinner A or Spinner B?
Give a reason for your answer.

(c) What is the probability that Spinner A lands on green?

(d) The probabilities of three events have been marked on the probability scale below.

R: Spinner B lands on red
Y: Spinner B lands on yellow
G: Spinner B lands on green

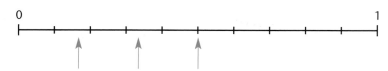

Copy the probability scale and label each arrow with a letter to show which event it represents.

(5 marks)
AQA, Spec B, 1F, June 2002

2 The probabilities of the following events have been marked on the probability scale below.

A: The next person to pass you will be less than 80 years old.
B: Tomorrow will be Sunday.
C: A fair three-sided spinner, coloured red, blue and green, will land on red.

Copy the probability scale and label each arrow with a
letter to show which event it represents.

(2 marks)

AQA, Spec B, 1F, March 2002

3 A bag contains three numbered discs, (2), (4) and (7).

A disc is drawn from the bag at random and a fair coin is
thrown.

If the coin lands heads, the score is 1 more than the number
on the disc.
If the coin lands tails, the score is 1 less than the number on
the disc.

(a) Copy the table and show the score for each possible.

Coin

Disc	Heads Disc + 1	Tails Disc − 1
(2)	3	
(4)		
(7)		6

(b) What is the probability of getting a score of 5?

(5 marks)

AQA, Spec B, 1I, March 2002

4 In a raffle 200 tickets are sold. There is one prize.
Mr Key buys 10 tickets. Mrs Key buys 6 tickets.
Their children, Robert and Rachel, buy 2 tickets each.

(a) Which member of the family has the best chance of
winning the prize? Give a reason for your answer.

(b) What is the probability that Mrs Key wins the prize?

(c) What is the probability that *none* of the family wins
the prize?

(7 marks)

AQA, Spec A, I, June 2003

5 A spinner has a red sector (R) and a yellow sector (Y). The arrow is spun 1000 times.

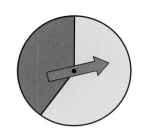

(a) The results for the first 20 spins are shown below.

R R Y Y Y R Y Y Y R Y Y Y Y Y R Y R Y Y Y

Work out the relative frequency of red after 20 spins.

(b) The table shows the relative frequency of red after different numbers of spins.
How many times was a red obtained after 200 spins?

Number of spins	Relative frequency of a red
50	0.42
100	0.36
200	0.34
500	0.3
1000	0.32

(4 marks)
AQA, Spec A, I, November 2004

Summary of key points

Probability (grades F to C)

$$\text{Probability} = \frac{\text{number of successful outcomes}}{\text{total number of possible outcomes}}$$

Relative frequency (grade D/C)

$$\text{Relative frequency} = \frac{\text{number of successful trials}}{\text{total number of trials}}$$

This is also called experimental or estimated probability

Mutually exclusive events (grade E/D)

If there are n mutually exclusive events all equally likely then the probability of each event is $\frac{1}{n}$.

If there are n mutually exclusive events and m successful outcomes then the probability of a successful outcome is $\frac{m}{n}$.

For any two events, say A and B, that are mutually exclusive and exhaustive then,

$$P(A \text{ or } B) = P(A) + P(B) = 1$$

For an event A the probability of the event A *not* happening is given by

$$P(\text{not } A) = 1 - P(A)$$

Listing outcomes (grades F to C)

When you list outcomes, work systematically to make sure you don't miss any out. For example, write a list or draw a sample space diagram.

Independent events (grade F/E)

For two independent events A and B

$$\text{total number of outcomes} = \text{number of outcomes of event A} \times \text{number of outcomes of event B}$$

14 Multiples, factors, powers and roots

Number 5

This chapter will show you how to:

✔ recognise odd, even and prime numbers
✔ write down multiples and factors of numbers
✔ recognise square and cube numbers and find square roots and cube roots
✔ revise and use the laws of indices
✔ find prime factors, the highest common factor (HCF) and the lowest common multiple (LCM)

14.1 Multiples, factors and prime numbers

Key words:
multiple
factor
prime

All whole numbers are either odd or even.

Even numbers divide exactly by 2.

The even numbers are 2, 4, 6, 8, 10, ...

Odd numbers do not divide exactly by 2.

The odd numbers are 1, 3, 5, 7, 9, ...

A **multiple** of a number is a number in its 'times table'.

The 'times table' gives the multiples of a number.

For example, the multiples of 4 are 4, 8, 12, 16, 20, ...

A **factor** of a number is a whole number that divides into it exactly. The factors of any number always include 1 and the number itself.

For example, the factors of 10 are 1, 2, 5 and 10.

The factors of a number come in pairs (except for square numbers).

You will find out more about square numbers in Section 14.2.

For example, for 10 you can link the factors like this:

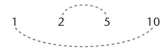

Each pair of factors multiplies to give 10.
$2 \times 5 = 10$
$1 \times 10 = 10$

A **prime** number has only *two* factors. The factors are always 1 and the number itself.

For example, 7 is prime, its only factors are 1 and 7.

The only *even* prime number is 2.
1 is *not* a prime number.

The first few prime numbers are, 2, 3, 5, 7, 11, ...

7 cannot be divided by any other whole numbers.

1 only divides by itself so it has only *one* factor.

Example 1

(a) Write down all the multiples of 4 which are less than 30.
(b) Write down all the multiples of 6 which are less than 30.
(c) Which multiples are common to both lists?

(a) 4, 8, 12, 16, 20, 24, 28
(b) 6, 12, 18, 24
(c) 12 and 24 are in both lists.

You need to remember these facts.

Do not include 30 in (b).

12 and 24 are multiples of both 4 and 6. They are called *common multiples*. You will meet these again in Section 14.5.

Example 2

Here is a list of numbers:

| 1 | 4 | 5 | 7 | 12 | 17 | 20 | 21 | 28 |

Write down: (a) the odd numbers
(b) the multiples of 3
(c) the factors of 24
(d) the factors of 28
(e) the prime numbers.

(a) The odd numbers are 1, 5, 7, 17 and 21.
(b) The multiples of 3 are 12 and 21.
(c) The factors of 24 are 1, 4 and 12.
(d) The factors of 28 are 1, 4, 7 and 28.
(e) The prime numbers are 5, 7 and 17.

1 and 4 are factors of both 24 and 28. They are called *common factors*. You will meet these again in Section 14.5.

Exercise 14A

1 From this card:

(a) list the even numbers.

(b) list the odd numbers.

(c) list the prime numbers.

[1]   [2]   [5]   [7]   [8]

[9] [11] [12] [13] [27]

[33] [51] [69] [81] [86]

(d) what is the largest prime number?

(e) what is the smallest odd number?

(f) which numbers are even and prime?

(g) which numbers are odd and prime?

2 Write down the first six multiples of:

(a) 3   (b) 9   (c) 11   (d) 15   (e) 21

3 Find all the factors of:

(a) 14   (b) 20   (c) 36   (d) 44

4 Which of these numbers are:

(a) multiples of 2   (b) multiples of 3

(c) multiples of 5   (d) multiples of 9

(e) multiples of both 2 and 3

(f) multiples of 2, 3 and 5?

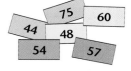

5 Find the missing factors.

(a) The factors of 12 are 1, ☐, 3, 4, 6, ☐.

(b) The factors of 36 are 1, 2, 3, ☐, 6, 9, ☐, 18, 36.

(c) The factors of 18 are 1, 2, ☐, 6, ☐, 18.

Make factor pairs.

6 (a) Which of these numbers are multiples of

(i) 4   (ii) 7   (iii) 13   (iv) 23?

(b) Which of these numbers are multiples of 7 and 13?

(c) Which of these numbers are multiples of 4, 7 and 13?

92   52
91   104
138 203
88   299
84

7 (a) List all the factors of 10.

(b) List all the factors of 25.

(c) Which factors are common to both lists?

8 Find the smallest number greater than 50 that is:
  (a) a multiple of 2   **(b)** a multiple of 3
  (c) a multiple of 5   **(d)** a multiple of 9.

9 **(a)** List all the factors of 24.
  (b) List all the factors of 16.
  (c) List all the factors of 30.
  (d) Which numbers are factors of 24 and 16?
  (e) Which numbers are factors of 16 and 30?
  (f) Are there any numbers that are factors of 24, 16 and 30?

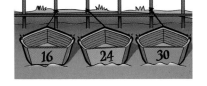

10 Find the biggest number less than 100 that is:
  (a) a multiple of 4   **(b)** a multiple of 6
  (c) a multiple of 7   **(d)** a multiple of 8.

11 How many even prime numbers are there?

14.2 Squares, cubes and roots

Square numbers and cube numbers

A **square** number is what you get when you multiply a whole number by itself.

For example, $4 \times 4 = 16$, $7 \times 7 = 49$, $15 \times 15 = 225$

16, 49 and 225 are square numbers.

They are called square numbers because you can arrange them in a square pattern of dots.

Here is the pattern for $4 \times 4 = 16$:

You write 4×4 as 4^2.

The ^{2} is called a **power** (or **index**).

You say '4 squared' or 'the square of 4' or '4 to the power of 2'.

Square numbers have an odd number of factors.

For example, the factors of 16 are

The factor pairs are 1 and 16, 2 and 8.
4 multiplies by itself
$(4 \times 4 = 16)$.

A **cube** number is what you get when you multiply a whole number by itself, then by itself again.

For example, $2 \times 2 \times 2 = 8$, $3 \times 3 \times 3 = 27$,
$10 \times 10 \times 10 = 1000$

8, **27** and **1000** are cube numbers.

They are called cube numbers because you can arrange them in a cube pattern.

Here is the pattern for $2 \times 2 \times 2 = 8$:

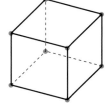

You write $2 \times 2 \times 2$ as 2^3.

The ^{3} is a power (or index).

You say '2 cubed' or 'the cube of 2' or '2 to the power of 3'.

You can cube or square any number. You may need to use a calculator.

A calculator has a key for squaring numbers $\boxed{x^2}$.

Some calculators have a key for cubing numbers $\boxed{x^3}$.

On others you need to use the power key: $\boxed{y^x}$ or $\boxed{x^y}$.

To calculate 2^3 you input $\boxed{2}$ $\boxed{y^x}$ $\boxed{3}$. Answer: 8.

> You met 'squared' and 'cubed' before, in Chapter 7.

> You will need a calculator to square or cube numbers that are not whole numbers.

> Check how to square and cube numbers on your calculator.

Example 3

Work out these cubes and squares:

(a) 8^2 (b) 15^2 (c) 4^3

(d) 30^3 (e) $(8.3)^2$ (f) $(3.4)^3$

(a) $8^2 = 8 \times 8 = 64$

(b) $15^2 = 15 \times 15 = 225$

(c) $4^3 = 4 \times 4 \times 4 = 64$

(d) $30^3 = 30 \times 30 \times 30 = 27\,000$

(e) $(8.3)^2 = 68.89$ (using a calculator)

(f) $(3.4)^3 = 39.304$ (using a calculator)

> You need to learn all the squares of numbers up to $15^2 = 225$. You may need them on a non-calculator paper.

Example 4

Write down:

(a) all the square numbers between 40 and 90

(b) all the cube numbers less than 100.

(a) $6 \times 6 = 36$ *(too small)*

    $7 \times 7 = 49$

    $8 \times 8 = 64$

    $9 \times 9 = 81$

    $10 \times 10 = 100$ *(too big)*

    The square numbers between 40 and 90 are 49, 64 and 81.

(b) $1 \times 1 \times 1 = 1$

    $2 \times 2 \times 2 = 8$

    $3 \times 3 \times 3 = 27$

    $4 \times 4 \times 4 = 64$

    $5 \times 5 \times 5 = 125$ *(too big)*

    The cube numbers less than 100 are 1, 8, 27 and 64.

> Start by trying any number.

> Stop when the answer is ≥ 100.

Exercise 14B

1 Work out these cubes and squares:

  (a) 5^2 **(b)** 11^2 **(c)** 3^3 **(d)** 15^2

  (e) 100^3 **(f)** 2^3 **(g)** 9^2 **(h)** 5^3

2 Work out these squares and cubes:

  (a) 17^2 **(b)** 200^2 **(c)** 12^3 **(d)** 25^2

  (e) $(4.9)^3$ **(f)** $(0.2)^2$ **(g)** $(2.5)^3$ **(h)** $(1.25)^2$

3 Work out the following:

  (a) $3^3 + 5^2$ **(b)** $6^2 - 4^2$ **(c)** $10^3 - 5^3$

4 Which is larger:

  (a) 4^3 or 5^2 **(b)** 2^3 or 3^2 **(c)** 6^3 or 7^2?

5 A square picture has a side of length 8.5 cm. What is its area? You may use a calculator.

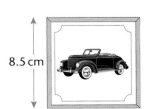

8.5 cm

Square roots and cube roots

In the last section you saw that $4 \times 4 = 16$,
which you write $4^2 = 16$,
and that 16 is a square number.

The *inverse* is the reverse, or opposite, process. For example, the inverse of 'add' is 'subtract'.

The inverse of squaring is finding the **square root** .

The square root of 16 is 4, because $4^2 = 16$.
You write $\sqrt{16} = 4$.
Also $\qquad (-4) \times (-4) = 16$ or $(-4)^2 = 16$
So -4 is also a square root of 16 or $\sqrt{16} = -4$.

For more on negative numbers see Chapter 1.

All positive numbers have *two* square roots.

negative $\times$ negative = positive
positive $\times$ positive = positive

For example $\sqrt{100} = 10 \qquad$ and $\quad \sqrt{100} = -10$
$\sqrt{36} = 6 \qquad$ and $\quad \sqrt{36} = -6$
You write $\quad \sqrt{100} = \pm 10 \quad$ and $\quad \sqrt{36} = \pm 6$

Squaring always gives a positive number, so (at this level) it is impossible to find the square root of a negative number.

The inverse of cubing is finding the **cube root** .

$\pm$ is shorthand for the positive and the negative value of a number.

The cube root of 8 is 2, because $2^3 = 8$.
You write $\sqrt[3]{8} = 2$.

$2^3 = 2 \times 2 \times 2 = 8$
8 is a cube number.

When you cube a positive number the answer is positive.
For example $2 \times 2 \times 2 = 8$.

When you cube a negative number the answer is negative.
For example $-2 \times -2 \times -2 = -8$.

$(-2)^3 = (-2) \times (-2) \times (-2)$
$= (+4) \times (-2)$
$= (-8)$
positive $\times$ negative = negative

Positive numbers have a positive cube root.
Negative numbers have a negative cube root.

$\sqrt[3]{8} = 2$
$\sqrt[3]{-8} = -2$

Any *positive* number has two square roots
(for example $\sqrt{36} = \pm 6$).

The $\sqrt{}$ key only gives the positive square root.

Any number (positive or negative) has a cube root.

Most calculators have a key for square roots $\boxed{\sqrt{}}$.

... and a key for cube roots 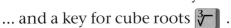 .

On some calculators the cube root is a 'second function'.

Square roots and cube roots are not always whole numbers.
You may need to round the calculator answer.
Remember to round to a sensible degree of accuracy.

Give answers to the same accuracy as the numbers in the question.

Example 5

Without using a calculator, work out:
(a) $\sqrt{49}$ **(b)** $\sqrt[3]{125}$

(a) $7 \times 7 = 49$, so $\sqrt{49} = 7$

(b) $3^3 = 3 \times 3 \times 3 = 27$ (too small)

    $5^3 = 5 \times 5 \times 5 = 25 \times 5 = 125$

    So $\sqrt[3]{125} = 5$

> You need to learn squares $2^2 = 4$, $3^2 = 9$, etc up to $15^2 = 225$.

> Try some values.

Example 6

Use your calculator to work out:
(a) $\sqrt{49}$ **(b)** $\sqrt{256}$ **(c)** $\sqrt{0.81}$
(d) $\sqrt{0.4624}$ **(e)** $\sqrt{264\,196}$

(a) $\sqrt{49} = 7$ (b) $\sqrt{256} = 16$ (c) $\sqrt{0.81} = 0.9$

(d) $\sqrt{0.4624} = 0.68$ (e) $\sqrt{264\,196} = 514$

> Give the positive square root unless you are asked for both.

Example 7

Use your calculator to work out:
(a) $\sqrt{67}$ **(b)** $\sqrt{0.4755}$ **(c)** $\sqrt{218.75}$
(d) $\sqrt[3]{70}$ **(e)** $\sqrt[3]{2344}$ **(f)** $\sqrt[3]{0.038}$

Give your answers correct to 3 s.f.

(a) $\sqrt{67} = 8.1853\ldots = 8.19$

(b) $\sqrt{0.4755} = 0.6895\ldots = 0.690$

(c) $\sqrt{218.75} = 14.790\ldots = 14.8$

(d) $\sqrt[3]{70} = 4.121\ldots = 4.12$

(e) $\sqrt[3]{2344} = 13.283\ldots = 13.3$

(f) $\sqrt[3]{0.038} = 0.3361\ldots = 0.336$

> Look back at Chapter 2, Section 2.1 for help with significant figures.

Exercise 14C

1 Without using a calculator, work out:
     (a) $\sqrt{100}$ **(b)** $\sqrt{25}$ **(c)** $\sqrt{64}$ **(d)** $\sqrt{81}$
     (e) $\sqrt[3]{8}$ **(f)** $\sqrt[3]{27}$ **(g)** $\sqrt[3]{64}$ **(h)** $\sqrt[3]{1000}$

> Remember to give the positive square root unless asked for both.

2 Use your calculator to work out:

    (a) $\sqrt{8}$ (b) $\sqrt{15}$ (c) $\sqrt{84}$ (d) $\sqrt{3.5}$

    (e) $\sqrt[3]{21}$ (f) $\sqrt[3]{285}$ (g) $\sqrt[3]{9.5}$ (h) $\sqrt[3]{15.7}$

> Give your answers to 1 d.p.

3 Use your calculator to work out:

    (a) $\sqrt{2}$ (b) $\sqrt[3]{9}$ (c) $\sqrt{14}$ (d) $\sqrt{110}$

    (e) $\sqrt[3]{500}$ (f) $\sqrt[3]{4008}$ (g) $\sqrt{8.3}$ (h) $\sqrt[3]{1.3}$

> Give your answers to 3 s.f.

It's a cube root!

14.3 Calculating powers

You can calculate any number raised to any power.

For example, $3^4 = 3 \times 3 \times 3 \times 3 = 81$

$$2^7 = 2 \times 2 \times 2 \times 2 \times 2 \times 2 \times 2 = 128$$

> In Chapter 7 you used index notation to simplify numerical or algebraic expressions. For example, $5 \times 5 \times 5 \times 5 = 5^4$ or $x \times x \times x = x^3$.

Calculations like this can lead to very large answers. When they involve decimals a calculator is essential.

> $2 \times 2 \times ...$ 7 times

The power key on a calculator will work out powers of any number.

It usually looks like x^y or y^x.

> On some calculators it is a 'second function' key.

To work out $(2.3)^6$ you press

      or

> Check how to calculate powers on your calculator.

which gives an answer of $148.035\,889 = 148.04$ (2 d.p.)

Example 8

Use your calculator to work out:

(a) 4^7 (b) 2^{12} (c) $(5.6)^4$ (d) $(0.8)^5$

(a) $4^7 = 16\,384$

(b) $2^{12} = 4096$

(c) $(5.6)^4 = 983.4496 = 983.4$ (1 d.p.)

(d) $(0.8)^5 = 0.327\,68 = 0.33$ (1 s.f.)

> Give your answers to the same accuracy as the numbers in the question.

Example 9

Find the value of x in each of the following:

(a) $2^x = 64$ **(b)** $12^x = 20\ 736$ **(c)** $3^x = 2187$

> **(a)** $2^4 = 16$ *(too small)*
>
> $2^5 = 32$ *(too small)*
>
> $2^6 = 64$
>
> $x = 6$
>
> **(b)** If $12^x = 20\ 736$ then $x = 4$
>
> **(c)** If $3^x = 2187$ then $x = 7$

Use the power button and try a value of x. Use trial and improvement to find the correct answer.

Exercise 14D

1 Use your calculator to work out:

(a) 2^5 **(b)** 5^7 **(c)** 15^4 **(d)** 9^6

(e) $(2.5)^4$ **(f)** $(0.2)^5$ **(g)** $(0.4)^4$ **(h)** $(0.1)^6$

Give your answers to 3 d.p.

2 Find the value of x in the following:

(a) $2^x = 256$ **(b)** $7^x = 16\ 807$ **(c)** $16^x = 4096$

(d) $9^x = 4\ 782\ 969$ **(e)** $2^x = 32\ 768$

Use trial and improvement and a calculator.

14.4 The laws of indices

Key words:
indices

You use the laws of **indices** (powers) to multiply or divide powers of the same number or variable.

A variable is a letter which stands for an unknown quantity.

The laws of indices

To multiply powers of the same number or variable you *add* the indices $x^n \times x^m = x^{n+m}$

You used the Laws of Indices for variables in Chapter 7.

To divide powers of the same number or variable you *subtract* the indices $x^n \div x^m = x^{n-m}$

These laws only work for powers of the *same* number (or variable).

Example 10

Write these as a single power:

(a) 4×4^6 **(b)** $7^8 \times 7^5$ **(c)** $12^9 \div 12^3$ **(d)** $8^{14} \div 8^4$

(e) $\dfrac{5^6 \times 5^9}{5^7}$ **(f)** $x^6 \times x^5$ **(g)** $y^2 \div y^5$

(a) $4 \times 4^6 = 4^1 \times 4^6 = 4^{1+6} = 4^7$ —————————————————— $4 = 4^1$

(b) $7^8 \times 7^5 = 7^{8+5} = 7^{13}$

(c) $12^9 \div 12^3 = 12^{9-3} = 12^6$ (d) $8^{14} \div 8^4 = 8^{14-4} = 8^{10}$

(e) $\dfrac{5^6 \times 5^9}{5^7} = \dfrac{5^{15}}{5^7} = 5^{15-7} = 5^8$ (f) $x^6 \times x^5 = x^{6+5} = x^{11}$

(g) $y^2 \div y^5 = y^{2-5} = y^{-3}$

From part (g) in the example, $y^2 \div y^5 = y^{-3}$.
Writing it out in full:

$$y^2 \div y^5 = \frac{y \times y}{y \times y \times y \times y \times y}$$

$$= \frac{1}{y^3}$$

$$\text{So } y^{-3} = \frac{1}{y^3}$$

This rule works for all negative powers:

$$x^{-n} = \frac{1}{x^n}$$

Any number (or variable) raised to the power of 1 is equal to the number (or variable) itself.

For example, $5^1 = 5$, $12^1 = 12$, $x^1 = x$, $y^1 = y$.

Exercise 14E

1 Write each of these as a single power:

(a) $3^6 \times 3^2$ (b) $5^4 \times 5^6$

(c) $4^2 \times 4^5 \times 4^2$ (d) $9^4 \times 9^8 \times 9^2$

(e) $6^4 \times 6^2 \times 6^6$ (f) $7^3 \times 7 \times 7^3 \times 5^4 \times 5^2$

2 Use the laws of indices to write each of these as a single power.

(a) $7^4 \div 7^2$ (b) $9^5 \div 9^2$ (c) $11^3 \div 11^2$

(d) $\dfrac{8^6}{8^1}$ (e) $\dfrac{5^9}{5^4}$ (f) $10^6 \div 10^3$

3 Use the laws of indices to write each of these as a single power.

(a) $8^2 \times 8^4$ (b) $10^6 \div 10^2$ (c) $y^2 \times y^3 \times y^2$

(d) $10^5 \times 10^2 \times 10^3$ (e) $3^5 \div 3^4$ (f) $m^3 \div m^2$

(g) $t^2 \times t^3 \times t^2$ (h) $3^4 \times 3^3 \div 3^6$ (i) $9^4 \div 9 \times 9^2$

4 Match each question from box A with an answer from box B.

A	B
$7^3 \times 7^5$	7^6
$3^4 \times 3^3$	7^3
$7^{10} \div 7^7$	3^8
$7^2 \times 7^3 \times 7^2$	3^6
$3^5 \div 3^3$	7^5
$7^{10} \div 7^5 \times 7$	7^8
$7^4 \times 7^2 \div 7^1$	7^9
$3^3 \times 3^4 \times 3$	3^2
$7^2 \times 7^3 \times 7^4$	3^7
$3^5 \times 3^3 \div 3^2$	7^7

5 Simplify the following by writing them as a single power.

(a) $\dfrac{4^3 \times 4^5}{4^6}$ (b) $\dfrac{7^6}{7^2 \times 7^3}$

(c) $\dfrac{2^5 \times 2^2 \times 2^4}{2^3 \times 2^2}$ (d) $\dfrac{8^7 \times 8^{10}}{8^{15}}$

(e) $\dfrac{6^5 \times 6^3 \times 6}{6^2 \times 6^4}$ (f) $\dfrac{2^5 \times 2^3}{2^4}$

(g) $\dfrac{3^3 \times 3^6 \times 3^2}{3^5}$ (h) $\dfrac{5^3}{5^4 \times 5^2}$

Calculate the value of your answers. You may use a calculator.

14.5 Using prime factors

Writing a number as the product of its prime factors

A factor of a number is a whole number that divides into it exactly.

If a factor is also a prime number, it is called a **prime factor** .

Key words:
prime factor
index form

See Section 14.1.

For example, the factors of 10 are 1, 2, 5 and 10.
2 and 5 are also prime numbers.
2 and 5 are prime factors of 10.

You can write any number as the product of its prime factors.
For example $10 = 2 \times 5$.

Product means multiply.

For larger numbers you need to work systematically to find
the prime factors.

Example 11

Write 84 as the product of its prime factors.

Divide 84 by the smallest
prime number: 2.

Divide by 2 as many times
as you can.

Then divide by the next
smallest prime number: 3.

Try dividing by 5 – doesn't
work. Try 7.

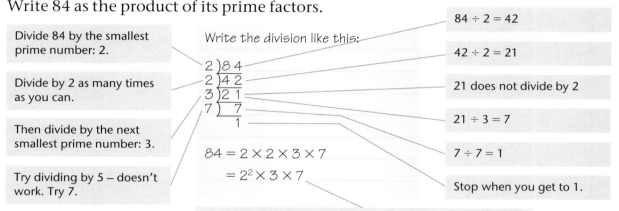

Write the division like this:

$$\begin{array}{r} 2\,)\overline{8\,4} \\ 2\,)\overline{4\,2} \\ 3\,)\overline{2\,1} \\ 7\,)\overline{7} \\ \overline{1} \end{array}$$

$84 = 2 \times 2 \times 3 \times 7$
$ = 2^2 \times 3 \times 7$

$84 \div 2 = 42$

$42 \div 2 = 21$

21 does not divide by 2

$21 \div 3 = 7$

$7 \div 7 = 1$

Stop when you get to 1.

The product of the numbers you divided by in **index form** .

The general method for writing a number as a product of its
prime factors is:

1 Start with the smallest prime.
   Divide as many times as you can.

2 Then try the next smallest prime.
   Divide as many times as you can.

3 Continue until you get to answer = 1.

4 The answer is the product of the primes you
   divided by.

Prime numbers 2, 3, 5, 7,
11, 13, 17, 19, 23, …

If you cannot divide by a
particular prime, try the
next one.

You can also write a number as the product of its prime
factors by using 'factor trees'. This is how it works.

1 Divide your number by *any* factor. Write the factor
   pair on 'branches' coming from the original number.

2 Keep dividing the factors until you only have
   prime numbers at the ends of the branches.

3 The answer is the product of the primes on the
   branches.

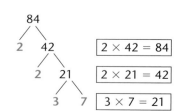

Example 12

Write 1980 as the product of its prime factors, giving your answer in index form.

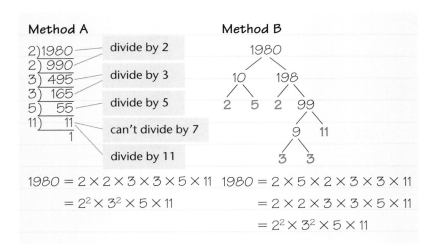

Method A

2)1980 —— divide by 2
2) 990
3) 495 —— divide by 3
3) 165
5) 55 —— divide by 5
11) 11 —— can't divide by 7
1
—— divide by 11

$1980 = 2 \times 2 \times 3 \times 3 \times 5 \times 11$
$= 2^2 \times 3^2 \times 5 \times 11$

Method B

1980
10   198
2 5 2 99
9 11
3 3

$1980 = 2 \times 5 \times 2 \times 3 \times 3 \times 11$
$= 2 \times 2 \times 3 \times 3 \times 5 \times 11$
$= 2^2 \times 3^2 \times 5 \times 11$

Exercise 14F

1 Write these numbers as the product of their prime factors.
Give your answers in index form.

(a) 20   (b) 12   (c) 45   (d) 24

(e) 100   (f) 72   (g) 210   (h) 108

Use the method you prefer.

2 Which of these numbers are *prime* factors of 64?

2   3   4   5   7   8   11   12

3 (a) Write 48 as the product of its prime factors.

(b) Write 66 as the product of its prime factors.

(c) Which prime factors are common to both 48 and 66?

Which prime factors are in both lists?

4 (a) Write 60, 72 and 120 as the product of their prime factors.

(b) What are their common prime factors?

Highest common factor (HCF)

The **highest common factor** (HCF) of two or more numbers is the largest number that divides into all of them exactly.

The HCF of two numbers is the highest factor they have in common.

To find the HCF:

1 Write each number as the product of its prime factors.

2 Pick out the prime factors common to all the numbers.

3 Multiply the common prime factors to give the HCF.

Use one of the methods from the previous section.

Example 13

Find the highest common factor of 54, 72 and 90.

You could use the factor tree method instead.

```
2)54        2)72        2)90
3)27        2)36        3)45
3) 9        2)18        3)15
3) 3        3) 9        5) 5
   1        3) 3           1
             3) 3
                1
```

Write each number as a product of its prime factors.

$54 = 2 \times 3 \times 3 \times 3$ $72 = 2 \times 2 \times 2 \times 3 \times 3$ $90 = 2 \times 3 \times 3 \times 5$

The factors common to all three are 2, 3, 3.

So the HCF of 54, 72 and 90 is $2 \times 3 \times 3 = 18$.

$54 \div 18 = 3$,
$72 \div 18 = 4$,
$90 \div 18 = 5$.

You can tell this from the factors left once you have taken out the common factors.

Pick out the factors common to all three.

Multiply them to get the HCF.

Exercise 14G

1 (a) Write 21 and 28 as the product of their prime factors.
   (b) Find the highest common factor (HCF) of 21 and 28.

2 Find the HCF of:
   (a) 15 and 25   (b) 32 and 56   (c) 27 and 45
   (d) 16 and 20   (e) 25 and 45   (f) 24 and 36.

3 Find the HCF of:
   (a) 4, 6 and 12   (b) 10, 30 and 35
   (c) 12, 16 and 18   (d) 27, 36 and 72
   (e) 40, 48 and 64   (f) 75, 100 and 150.

4 What is the highest number that will divide exactly into 60 *and* 140?

Is this the HCF?

5 Find two numbers less than 60 that have a common factor of:

(a) 5   (b) 16   (c) 28.

6 The quickest way to simplify a fraction is to divide both numbers by their HCF.

For example, $\overset{\div 12}{\overset{\frown}{\underset{\underset{\div 12}{\smile}}{\frac{48}{60}}}} = \frac{4}{5}$. 12 is the HCF of 48 and 60.

By finding the HCF of the two numbers, simplify:

(a) $\frac{20}{70}$ (b) $\frac{27}{36}$.

Lowest common multiple (LCM)

Key words:
lowest common multiple

The **lowest common multiple** (LCM) of two or more numbers is the smallest number that is a multiple of all of them.

To find the LCM:
1 Write each number as the product of its prime factors.
2 Pick out the *highest power* of *each* of the prime factors in the lists.
3 Multiply them to give the LCM.

Example 14

Find the lowest common multiple of 54, 72 and 90.

Using the prime factors of 54, 72 and 90 from Example 13.

$54 = 2 \times 3 \times 3 \times 3 \quad\quad = 2 \times 3^3$

$72 = 2 \times 2 \times 2 \times 3 \times 3 = 2^3 \times 3^2$

$90 = 2 \times 3 \times 3 \times 5 \quad\quad = 2 \times 3^2 \times 5$

The highest power of each of the prime factors is shown in **red**.

The LCM of 54, 72 and 90 is $2^3 \times 3^3 \times 5 = 1080$.

$1080 = 20 \times 54$
$1080 = 15 \times 72$
$1080 = 12 \times 90$

1080 is the smallest number in all of the 54, 72 and 90 'times tables'.

Exercise 14H

1 Find the LCM (lowest common multiple) of:

    (a) 10 and 8   **(b)** 8 and 9   **(c)** 18 and 12

    (d) 12 and 16   **(e)** 20 and 90   **(f)** 45 and 60

    (g) 45 and 54   **(h)** 12 and 66   **(i)** 33 and 88.

2 Find the LCM of:

    (a) 12, 16 and 24   **(b)** 15, 30 and 40

    (c) 15, 25 and 50   **(d)** 72, 48 and 36

    (e) 24, 40 and 120   **(f)** 30, 36 and 40

    (g) 18, 30 and 75   **(h)** 45, 90 and 105

    (i) 42, 56 and 90.

UAM **3** Toby the dog barks every 4 seconds.
Mini the cat mews every 6 seconds.
Jake starts the timer when they bark and mew at the same time. How many seconds does the timer show when they next bark and mew at the same time?

UAM **4** Dara is waiting for the Paddington bus and Ron is waiting for the Euston bus.
They decide to wait until both buses come together before they get on.
The Paddington bus runs every 10 minutes.
The Euston bus runs every 8 minutes.
The buses start running at the same time in the morning.
What is the longest time they will have to wait?

The quickest way to add fractions is to find the LCM of the denominators and use this as a common denominator.

For example, $\frac{3}{12} + \frac{1}{16}$

LCM of 12 and 16 is 48: $\frac{3}{12} + \frac{1}{16} = \frac{3 \times 4}{48} + \frac{1 \times 3}{48}$

$$= \frac{12}{48} + \frac{3}{48} = \frac{15}{48} = \frac{5}{16}$$

5 By finding the LCM of the denominators, work out:

    (a) $\frac{1}{6} + \frac{7}{9}$ **(b)** $\frac{5}{12} + \frac{4}{15}$ **(c)** $\frac{29}{60} - \frac{11}{24}$

    (d) $\frac{11}{50} - \frac{7}{60}$ **(e)** $\frac{19}{36} - \frac{7}{48}$ **(f)** $\frac{13}{48} + \frac{21}{40}$

See Section 3.4 for help with addition and subtraction of fractions.

6 In a group of handbell ringers:
Ann rings a bell every 6 seconds, Beth rings a bell every
9 seconds, Cheryl rings a bell every 10 seconds.
They start by all ringing their bells at the same time.
How long is it before they next ring their bells at the
same time?

Examination style questions

1 A list of numbers is given below.

4   5   7   2   15   19

State which of these are
(a) even numbers   **(b)** odd numbers.   *(2 marks)*
AQA, Spec B, 3F, June 2003

2 (a) Work out the cube of 4.
(b) Work out 0.2^2
(c) A list of numbers is given below.

15   16   19   27   34   42   45

From this list, write down
(i) a cube number   **(ii)** a prime number.   *(4 marks)*
AQA, Spec B, 3I, November 2002

3 Work out the value of
(a) 5^3 **(b)** 10^4 *(2 marks)*
AQA, Spec B, 5I, June 2003

4 (a) Express 24 as a product of its prime factors.
(b) Find the Least Common Multiple (LCM) of 24
and 60.   *(4 marks)*
AQA, Spec B, 3I, November 2003

5 36 expressed as a product of prime factors is $2^2 \times 3^2$.
(a) Express 45 as a product of its prime factors.
Write your answers in index form.
(b) What is the Highest Common Factor (HCF) of 36 and 45?
(c) What is the Least Common Multiple (LCM) of 36 and 45?   *(5 marks)*
AQA, Spec B, 3I, March 2003

6 Tom, Sam and Matt are counting drum beats.
Tom hits a snare drum every 2 beats.
Sam hits a kettle drum every 5 beats.
Matt hits a bass drum every 8 beats.

Tom, Sam and Matt start by hitting their drums at the same time.
How many beats is it before Tom, Sam and Matt next hit their
drums at the same time? *(2 marks)*
AQA, Spec A, I, June 2004

7 **(a)** Write down two multiples of 4.
(b) Write down two multiples of 7.
(c) Write down a number which is a multiple of both 4 and 7. *(3 marks)*
AQA, Spec B, 5F June 2004

Summary of key points

***Odd, even, multiples and factors (grade G) and prime
numbers (grade E)***

Even numbers divide exactly by 2, odd numbers do not.

A multiple of a number is a number in its 'times table'.
The multiples of 4 are 4, 8, 12, 16, ...

A factor of a number is a whole number that divides
into it exactly. The factors of any number always
include 1 and the number itself and come in pairs
(except for square numbers). The factors of 10 are:
$\quad$ 1 $\quad$ 2 $\quad$ 5 $\quad$ 10

A prime number has only two factors, 1 and the
number itself.
For example, 7 is prime. Its only factors are 1 and 7.

Square numbers (grade G) and cube numbers (grade E)

A square number is what you get when you multiply
a whole number by itself.
$4 \times 4 = 16$, so 16 is a square number. You write $4^2 = 16$.

Square numbers have an odd number of factors.
For example, the factors of 16 are:

$\quad$ 1 $\quad$ 2 $\quad$ 4 $\quad$ 8 $\quad$ 16

A cube number is what you get when you multiply a
whole number by itself, then by itself again.
$2 \times 2 \times 2 = 8$, so 8 is a cube number.
You write $2^3 = 8$.

Square roots (grade F) and cube roots (grade E)

The inverse of squaring is finding the square root.
Since $4^2 = 16$, the square root of 16 is 4. You write $\sqrt{16} = 4$.
All positive numbers have two square roots.

For example, $\sqrt{16} = \pm 4$.

Normally you will only give the positive square root.

Since $2^3 = 8$, 2 is called the cube root of 8. You write $\sqrt[3]{8} = 2$.

Positive numbers have a positive cube root. $\sqrt[3]{8} = 2$

Negative numbers have a negative cube root. $\sqrt[3]{-8} = 2$

Calculator keys

A calculator has keys for:

squaring $\boxed{x^2}$ cubing $\boxed{x^3}$ square root $\boxed{\sqrt{}}$

cube root $\boxed{\sqrt[3]{}}$ finding any power $\boxed{y^x}$ or $\boxed{x^y}$

Laws of indices (grade C)

To multiply powers of the same number or variable you *add* the indices: $x^n \times x^m = x^{n+m}$

To divide powers of the same number or variable you *subtract* the indices: $x^n \div x^m = x^{n-m}$

$$x^{-n} = \frac{1}{x^n}$$

Writing a number as the product of its prime factors (grade C)

Divide the number by prime numbers until you get to answer = 1.

Your answer is the product of the numbers you have divided by.

```
2)84
2)42        84 = 2 × 2 × 3 × 7
3)21        or
7) 7        84 = 2² × 3 × 7 (in index form)
   1
```

Factor tree method (grade C)

1 Divide the number by any factor. Write the factor pair on branches from the original number.

2 Keep dividing the factors until you only have prime numbers at the ends of the branches.

3 Your answer is the product of the primes on the branches.

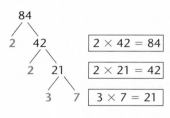

$$2 \times 42 = 84$$

$$2 \times 21 = 42$$

$$3 \times 7 = 21$$

$$84 = 2 \times 2 \times 3 \times 7$$
$$= 2^2 \times 3 \times 7$$

Highest Common Factor (HCF) (grade C)

The highest common factor (HCF) of two or more numbers is the largest number that divides into all of them exactly.

To find the HCF of two or more numbers:

1 Write each number as a product of its prime factors.

2 Pick out the prime factors common to all the numbers.

3 Multiply the common prime factors to give the HCF.

Lowest Common Multiple (LCM) (grade C)

The lowest common multiple (LCM) of two or more numbers is the smallest multiple that is a multiple of all of them.

To find the lowest common multiple of two or more numbers:

1 Write each number as a product of its prime factors.

2 Pick out the *highest power* of *each* of the prime factors in the lists.

3 Multiply them to give the LCM.

For example, to find the HCF and LCM of 90 and 105:

```
2)90              3)105
3)45              5) 35
3)15              7)  7
5) 5                 1
   1
```

$$90 = 2 \times 3 \times 3 \times 5 = 2 \times 3^2 \times 5 \qquad 105 = 3 \times 5 \times 7$$

HCF $= 3 \times 5 = 15$

LCM $= 2 \times 3^2 \times 5 \times 7 = 630$

15 Percentages and money

This chapter will show you how to:

✔ use percentages in calculations
✔ calculate percentage increase and decrease and find percentage change
✔ write one quantity as a percentage of another
✔ investigate money problems
✔ work out simple and compound interest and use repeated proportional change

15.1 Percentage

Using percentages in calculations

You will need to know:

- how to convert between fractions, decimals and percentages

Look back at Chapter 3 for help with percentages, decimals and fractions.

Percentage (%) means 'out of a hundred'.

$$78\% = \frac{78}{100} = 0.78$$

To convert a percentage to a decimal, divide by 100.

To convert a decimal to a percentage, multiply by 100:

$$0.25 = (0.25 \times 100)\% = 25\%$$

Multiplying by 100 is the inverse of dividing by 100.

To convert a fraction to a percentage:

1 convert the fraction to a decimal by dividing
2 multiply by 100.

$$\frac{1}{4} = 0.25 = (0.25 \times 100)\%$$
$$= 25\%$$

You need these skills for percentage calculations.

Percentages and fractions you need to remember:

$$100\% = 1 \qquad 50\% = \tfrac{1}{2} \qquad 25\% = \tfrac{1}{4} \qquad 75\% = \tfrac{3}{4}$$
$$10\% = \tfrac{1}{10} \qquad 33\tfrac{1}{3}\% = \tfrac{1}{3} \qquad 66\tfrac{2}{3}\% = \tfrac{2}{3}$$

Example 1

Work out 5% of £60.

'Of' means multiply.

Method A – Write 5% as a fraction.

$$5\% \text{ of } £60 = \frac{5}{100} \times 60$$

$$= \frac{300}{100}$$

$$= £3$$

Method B – Write 5% as a decimal.

$$5\% \text{ of } £60 = 0.05 \times 60$$

$$= £3$$

$5\% = \dfrac{5}{100} = 0.05$.

You can choose the method you prefer.

Example 2

In a village $37\frac{1}{2}\%$ of the population are under 17 years of age.

The village has a population of 1200. How many of them are under 17?

Number of people under 17 $= 37\frac{1}{2}\%$ of 1200

$$= \frac{37.5}{100} \times 1200$$

$$= 0.375 \times 1200$$

$$= 450$$

You do not really need this step, you can use your calculator on the line above.

To work out the percentage of a quantity:
write the percentage as a fraction or decimal and
multiply by the quantity.

Exercise 15A

1 Work out:
  (a) 10% of £350
  (b) 5% of 200 g
  (c) 25% of 64 litres
  (d) 1% of £16.

2 Work out:
  (a) 17% of 150 km
  (b) 45% of 630 kg
  (c) 88% of £1280
  (d) 11% of 32 mℓ.

3 Work out:
  (a) $2\frac{1}{2}\%$ of £160
  (b) 5.5% of 125 kg
  (c) 23.5% of 5800 g
  (d) $7\frac{1}{4}\%$ of £3560.

4 14% of a company's employees are left handed. There are 800 employees in the company. How many are left handed?

5 In a town, 4% of the houses need repairs. There are 4625 houses in the town. How many need repairs?

6 12% of children at a football match received a free ticket for the following game. If there were 1875 children, how many received free tickets?

7 Jim got 80% in his driving theory test. The test was out of 75. How many marks did he score?

8 Syeda earns £43 500 a year. She pays 24% of this in income tax. How much tax does she pay?

9 A survey in a village of 425 households found that 28% of them had satellite TV. How many households did not have satellite TV?

10 Glyn's journey to Manchester is 350 miles. 68% of this is on motorways. How many miles is on other roads?

Percentage increase and decrease

You sometimes need to increase (or decrease) a quantity by a given percentage. There are two ways to do this.

> **Key words:**
> percentage increase
> percentage decrease

Method A

Percentage increase	**Percentage decrease**
1 **Work out the actual** increase.	1 **Work out the actual** decrease.
2 **Add** it to the original amount.	2 **Subtract** it from the original amount.

Method B

Percentage increase	Percentage decrease
1 Add the % increase to 100%.	1 Subtract the % decrease from 100%.
2 Convert this % to a decimal.	2 Convert this % to a decimal.
3 Multiply the original amount by this decimal.	3 Multiply the original amount by this decimal.

Examples 3 and 4 use both methods.
You can choose the method you prefer.

Example 3

Liz earns £150 per week at the moment.
Next week she is due to receive a pay rise of 6%.
What will her new weekly pay be?

Percentage increase.

Method A	Method B
Pay rise = 6% of £150	Increase = 6%
$= \dfrac{6}{100} \times 150$	New % = 100% + 6%
	= 106%
= £9	= 1.06
Liz's new weekly pay	Liz's new weekly pay
= £150 + £9	= 1.06 × £150
= £159	= £159

Divide by 100 to convert a % to a decimal.

Example 4

A CD costs £14.50. With her student card, Alix gets 10% discount. What does she pay for the CD?

Percentage decrease.

A discount is a decrease in price.

Method A	Method B
Discount = 10% of £14.50	Discount = 10%
$= \dfrac{10}{100} \times £14.50$	New price = 100% − 10%
	= 90%
= £1.45	= 0.9
Alix pays £14.50 − £1.45	Alix pays 0.9 × £14.50
= £13.05	= £13.05

Divide by 100 to convert a % to a decimal.

Exercise 15B

1 Increase by 14%:

    (a) £420   **(b)** 600 g   **(c)** 84 litres

2 Decrease by 4%:

    (a) 500 miles   **(b)** £64   **(c)** 4250 kg

3 A pair of jeans cost £45. Chris gets a staff discount of 8%. What does Chris pay for the jeans?

4 Work out these percentage increases and decreases.

    (a) Increase £30 by 5%.

    (b) Increase 58 by 12%.

    (c) Decrease 3400 mℓ by 20%.

    (d) Decrease 480 by 32%.

    (e) Increase £135 by $7\frac{1}{2}$%.

    (f) Decrease 890 mℓ by 8.4%.

5 Firefighters were given a 4% pay rise this year. Before the pay rise their average wage was £420 per week. What is their average wage now?

6 A café wants to increase some of its prices by 5%. Work out the new cost of these items. Give your answers to the nearest penny.

Pasta	£2.60	Coffee	90p
Burger	£1.80	Tea	80p
Salad	95p		

7 There has been a 6% decrease in the number of reported thefts in Palton this year. There were 250 reported thefts last year. How many have there been this year?

8 A rail company announces that all rail fares will increase by $7\frac{1}{2}$%.
If I pay £28 for my ticket now, how much will it cost after the increase?

9 A new car costs £8450. After 2 years the value of the car will have decreased by 43%. How much will the car be worth then?

10 During one month the number of members of a gym increased by 65%. If there were 60 members at the beginning, how many were there at the end?

11 In a shop, sales of videos decreased by 11% this week. If they sold 430 last week, how many did they sell this week?

Remember to make your answer sensible.

Writing one quantity as a percentage of another

Test marks are usually written as a fraction.

For example, in a Maths test, a score of 17 out of 25 is written as $\frac{17}{25}$.

Here you are writing 17 as a percentage of 25.

You can write this score as a percentage:

$$\frac{17}{25} = \frac{17}{25} \times 100\% = 68\%$$

$\frac{17}{25} = 0.68$ or $\frac{68}{100}$

To write one quantity as a percentage of another:
1 write the two quantities as a fraction
2 multiply by 100 to convert to a percentage.

This is how you answer questions such as 'Write ___ as a percentage of ___'.

 ## Exercise 15C

1 Write these scores as percentages:

    (a) 6 out of 10   **(b)** 17 out of 20

    (c) 24 out of 40   **(d)** 30 out of 35.

2 Work out:

    (a) 3 as a percentage of 12

    (b) 8 as a percentage of 40

    (c) 22 as a percentage of 60

    (d) 65 as a percentage of 70.

3 In a maths test Jessie scored 16 out of 20. In her English test she scored 73%. In which subject did she do best?

Change both to percentages to compare them.

4 A sports centre has 1200 members, in the following groups.

Type	Men	Women	Girls	Boys
Number	602	392	118	88

What percentage of the members are:

(a) men   **(b)** girls

(c) children   **(d)** female?

Give your answers to 1 d.p.

Key words:
percentage change

Finding percentage change

You write one quantity as a percentage of another when you work out **percentage change** .

You use the formula:

$$\text{percentage change} = \frac{\textbf{actual change}}{\textbf{original amount}} \times \textbf{100\%}$$

'Change' means increase *or* decrease.

Always work out the actual change first.
Don't forget – the *original* amount goes on the bottom of the fraction.

Example 5

The original cost of a DVD player is £225.
In a sale it is reduced to £180.
What is the percentage decrease in the price?

Actual decrease in price = £225 − £180

= £45

Percentage decrease = $\dfrac{\text{actual decrease}}{\text{original amount}} \times 100\%$

= $\dfrac{45}{225} \times 100\%$

= 20%

The formula calculates the change as a percentage of the original amount.

Examples 5 and 6 are the type of question you can expect on your GCSE papers.

For a decrease, actual change = actual decrease.

Example 6

The average attendance at Newcastle United's home games last season was 55 000.
This year the average attendance is 58 575.
What percentage increase is this?

Actual increase in attendance = 58 575 − 55 000

$$= 3575$$

Percentage increase $= \dfrac{\text{actual increase}}{\text{original amount}} \times 100\%$

$$= \dfrac{3575}{55\,000} \times 100\%$$

$$= 6.5\%$$

For an increase,
$\dfrac{\text{actual}}{\text{change}} = \dfrac{\text{actual}}{\text{decrease.}}$

 ### Exercise 15D

1 In a sale the following price changes were made.
What was the percentage reduction for each item?

(a) coat: was £120, now £80

(b) shirt: was £24, now £18

(c) shoes: were £40, now £20

(d) trousers: were £30, now £18.

2 A car was for sale at £3600. The price was raised to £3780.
What was the percentage increase?

3 Ali had a pay rise from £240 per week to £255 per week.
What was his percentage increase in pay?

4 The number of tigers in an area of India has decreased from 108 to 96 in two years. What is the percentage decrease in tigers in this area?

5 The cost of a holiday will rise from £480 this year to £528 next year.
What is the percentage increase?

6 Peter's rent was increased from £280 per month to £315 per month.
What was the percentage increase?

7 A hospital reports that the number of sport injuries in people over 60 years is increasing. In 1995 there were 682 sport injuries but in 2004 there were 1088 in this age group.
What was the percentage increase?

8 There were 350 members of the Matis tribe living in the Amazon Basin.
The number fell to 198 one month after contact with diseases from the outside world.
What percentage decrease is this?

9 In a garage 15 cars were priced under £10 000. After a price rise the number under £10 000 had dropped to 9. What percentage reduction is this?

10 The population of a port in Chile increased from 183 000 people to 300 000 in just 20 years.
What percentage increase is this?

11 The number of sea lions seen at Bogoslof in Alaska had fallen from 5000 in 1973 to 120 in 2002.
What is the percentage reduction in sea lions seen in this area?

12 In 1970 the new survival suits gave 3 hours' protection in the Bering Sea. Modern survival suits will protect you for 24 hours. What is the percentage increase in time?

15.2 Working with money

Percentage profit and loss

A shopkeeper buys items from a wholesaler at **cost price** .

She sells the items at the **selling price** .

When you make money on the sale of an item you make a **profit** .

When you lose money on the sale of an item you make a **loss** .

You can use the percentage change formula to work out percentage profit or percentage loss.

$$\text{Percentage profit (or loss)} = \frac{\text{actual profit (or loss)}}{\text{cost price}} \times 100\%$$

where the actual profit (or loss) is the difference between the cost price and the selling price.

> **Key words:**
> cost price
> selling price
> profit
> loss

> If a garage buys a car for £1000 and sells it for £750, the garage makes a loss.

> The cost price is the *original* price of the item. It goes on the bottom of the fraction.

Example 7

A second hand car salesman buys a car for £2500 and
then sells it for £3200.
What is his percentage profit?

Actual profit = £3200 − £2500

= £700

Percentage profit = $\dfrac{\text{actual profit}}{\text{cost price}} \times 100\%$

= $\dfrac{700}{2500} \times 100\%$

= 28%

Cost price = £2500.
Selling price = £3200.

Example 8

Peter bought a new motorbike for £5450 but traded it in
for another one a year later.
He got a trade-in value of £4578 for his bike.
What percentage loss is this?

Actual loss = £5450 − £4578

= £872

Percentage loss = $\dfrac{\text{actual loss}}{\text{cost price}} \times 100\%$

= $\dfrac{872}{5450} \times 100\%$

= 16%

 ## Exercise 15E

1 A collector bought a stamp for £8 and sold it for £10.

(a) How much profit did he make?

(b) What was the percentage profit?

2 Tom's computer game cost £25 at Christmas. He sold
it for £15 in the summer.

(a) How much money did he lose?

(b) What was the percentage loss?

3 For each item find the percentage profit or loss.

	Item *a*	Item *b*	Item *c*	Item *d*	Item *e*	Item *f*
Cost price	£50	£34	£6	£73.50	£125	£3500
Selling price	£65	£24	£7.50	£69.50	£84	£3635

4 A bicycle shop paid £84 for a bike.
They sold it in a sale for £79.
What was the percentage loss?

5 Ahmed sold his CD player for £30. He had bought it
for £35.
What was his percentage loss?

6 Sarah restores skateboards. She bought one for £8 and
sold it for £15. What was her percentage profit?

7 Pet food normally costs £15.50 per sack.
If you buy two sacks the shop charges £13 per sack.
(a) What would two sacks cost without the discount?
(b) What percentage discount is this?

Discount is an amount
taken off the price. You
work out % discount in the
same way as % loss.

8 A shop buys a mobile phone for £22 and sells it for £35.
What is the percentage profit?

9 Floyd has a ticket for a rock concert. It cost him
£32.50 and he wants to sell it for £45. What will be his
percentage profit?

10 A video was sold for £14.99. Its cost price was £9.
What was the percentage profit?

VAT

Key words:
VAT

VAT stands for Value Added Tax.

**VAT at $17\frac{1}{2}$% is added to the cost of many items and
services.**

This means that you pay an extra $17\frac{1}{2}$% on top of the cost
of the item you buy.

You can use the percentage increase methods to calculate
VAT.

Usually the price of an
article includes VAT, but
sometimes it doesn't. You
need to look carefully to
see how much you will
have to pay.

Example 9

A digital camera is advertised for sale at £240 (excluding VAT). How much will you have to pay?

> 'Excluding' is the opposite of 'including'. You need to add VAT on to the advertised price.

Method A

$VAT = 17\frac{1}{2}\%$ of £240

$= \dfrac{17.5}{100} \times 240$

$= £42$

Cost of digital camera

$= £240 + £42$

$= £282$

Method B

Increase $= 17\frac{1}{2}\%$

New $\% = 100\% + 17\frac{1}{2}\%$

$= 117\frac{1}{2}\%$

$= 1.175$

Cost of digital camera

$= 1.175 \times £240$

$= £282$

> $117\frac{1}{2} \div 100 = 117.5 \div 100$
> $= 1.175$
> This is your multiplier to work out the % increase.

> The VAT makes quite a difference to the price!

 ## Exercise 15F

Take the VAT rate as $17\frac{1}{2}\%$ in this exercise.

1 Work out the VAT to be added to these prices. Give your answers to the nearest penny.

2 Work out the total cost of these:
   (a) a bag: £8 + VAT
   (b) a chair: £74 + VAT
   (c) a CD: £12 + VAT
   (d) a car: £2480 + VAT.

3 Two cars in a garage had these prices marked.
   Which one is the most expensive?

4 A computer costs £399 + VAT.
   What is the total cost?

5 Sarah's telephone bill is £56.84 + VAT.
   How much will she have to pay?

6 A meal for four cost £73.58 plus VAT.
   (a) How much VAT will be added to the bill?
   (b) What is the total cost?
   (c) The four people decide to share the bill equally. How much will each pay?

7 A meal for 3 people cost £55.50 + VAT. They decide to pay equal shares. How much will each person pay?

8 Kate's car needed some brake fluid, some antifreeze
and an adjustment to the fan-belt.
Labour costs were £60.50 and the cost of the parts used
was £30.23.
VAT was added to the total of these amounts.
How much was Kate's total bill? Give your answer to
the nearest penny.

Credit

Many items that you want to buy are expensive. You may
not have enough money to pay the full cost.

> When you buy **on credit** , you pay a **deposit**
> followed by a number of **regular payments** ,
> usually each month.

Key words:
on credit
deposit
regular payments
credit price
cash price

The cost of buying an item
on credit (the **credit price**)
is usually more than paying
cash for the full cost (the
cash price).

Example 10

Mr Round wants to buy a new washing machine. There
are two ways he can pay:

Option A: Pay the cash price of £470.
Option B: Pay a 10% deposit, then 24 monthly
payments of £19.30.

Mr Round chooses Option B.

(a) How much deposit does he pay?
(b) What is the total of his monthly payments?
(c) What is the total credit price?
(d) What is the extra cost of buying on credit rather than
paying the cash price?

(a) Deposit = 10% of £470 = $\dfrac{10}{100} \times 470$ = £47

The deposit is 10% of the
cash price.

(b) Total of monthly payments = 24 × £19.30

= £463.20

(c) Total credit price = deposit + monthly payments

= £47 + £463.20

= £510.20

Find the difference between
the credit price and the
cash price.

(d) Extra cost of buying on credit = £510.20 − £470

= £40.20

Exercise 15G

1 Sam is buying a new motorbike. He decides to buy it on credit. He pays a 10% deposit and 12 monthly payments of £275.

    (a) How much is the deposit?

    (b) How much does he pay in monthly instalments?

    (c) How much does Sam pay in total?

    (d) What is the difference between the cash price and the credit price of the motorbike?

2 You can buy a car in two ways.
    Option A: £6800 cash price.
    Option B: £2500 deposit and 36 monthly payments of £210.

    (a) What is the total cost of paying on credit (Option B)?

    (b) How much more does it cost to buy on credit?

3 A mobile phone costs £199 cash. The phone company also allows you to pay for the phone with a 20% deposit and £17 added to your monthly bill for a year. Which method is cheapest?
How much do you save using the cheapest method?

4 A sofa costs £384. The credit terms are 10% deposit and 15 monthly payments of £28.
What is the difference between the cash price and the credit price?

5 To pay for a surf board (cash price £124.50), Shona takes out a credit agreement. She pays a 15% deposit and 15 weekly payments of £9.
How much more is the credit price than the cash price?

6 A house costs £89 000. A company offers a mortgage of 5% deposit and monthly payments of £400 for 25 years.

    (a) How much is the deposit?

    (b) What is the total cost of the house?

Wages

If you work, you are probably paid by the hour. You get a **basic rate** of pay, which is a fixed amount per hour.

If you work more hours than your normal **working week** you get an **overtime** rate. This is more than your basic rate.

Key words:
basic rate
working week
overtime

Common rates of overtime pay are:
'Time and a quarter'
= $1.25 \times$ basic rate of pay
'Time and a half'
= $1.5 \times$ basic rate of pay.
'Double time'
= $2 \times$ basic rate of pay.

$$\text{Basic pay} = \begin{array}{c} \text{number of hours} \\ \text{in normal working} \\ \text{week} \end{array} \times \begin{array}{c} \text{basic rate} \\ \text{of pay} \end{array}$$

$$\text{Overtime pay} = \begin{array}{c} \text{number of hours} \\ \text{of overtime} \end{array} \times \begin{array}{c} \text{overtime} \\ \text{rate of pay} \end{array}$$

Example 11

Vijay is paid £6 per hour for a 38-hour working week.
His overtime is paid at 'time and a half'.
How much does he earn in a week when he works 50 hours?

Examples 11 and 12 are the type of question you could get in your GCSE exam.

Vijay works for 50 hours = 38 hours + **12** hours.

So he works his normal working week + **12** hours' overtime.

Basic pay = 38 × £6 = £228

Overtime rate of pay = 1.5 × £6 = £9

So overtime pay = 12 × £9 = £108

Vijay's total wage for the week = £228 + £108 = £336

'Time and a half' means
1.5 × the basic rate of pay.

Example 12

Helen is paid a basic rate of £7 per hour for a 40-hour working week.
Overtime is paid at the rate of 'time and a half'.
In a particular week she earns £353.50.
How many hours of overtime did she work?

Helen's basic pay = 40 × £7 = £280

Amount earned by working overtime

         = £353.50 − £280

         = £73.50   *continued* ▼

Work out her basic pay. Then you will know how much of the £353.50 came from overtime.

Rate of pay for overtime = 1.5 × £7

= £10.50 ──────── Work out the rate of pay for overtime.

Number of hours overtime = £73.50 ÷ £10.50

$$= \frac{73.50}{10.50}$$ ──────── Divide the overtime pay by the rate of pay for overtime.

= 7

Helen worked 7 hours' overtime.

Exercise 15H

1 What is the basic pay for a 38-hour week if the hourly rate is:
   (a) £6   **(b)** £8.45   **(c)** £9.30?

2 The hourly rate at McHill and Co. Ltd is £6.60.
   What is the hourly rate for overtime if they pay:
   (a) double time
   (b) time and a half
   (c) time and a quarter?

3 Jane is paid £7.20 per hour for a 40-hour week and then time and a half for any overtime. One week she works 47 hours. How much is she paid?

4 Three workers at 'Johnson's Builders' work 45 hours one week. Their normal working week is 38 hours. Overtime is paid at time and a half. How much will they each get if their hourly rate is:
   Scott, a joiner – £9.70
   Ali, a bricklayer – £8.30
   Liam, a labourer – £7.50?

5 At a call centre the pay is £6.80 per hour, Monday to Friday.
   The overtime rate of pay on Saturday is 'time and a quarter'.
   The overtime rate of pay on a Sunday is 'time and a half'.
   Here is Kim's timesheet.

Day	Mon	Tue	Wed	Thu	Fri	Sat	Sun
Hours	8	8	4	8	5	6	4

A timesheet shows the hours worked in a week.

Work out her total pay for the week.

6 Al is paid £8 per hour for a 40-hour week and double time for any overtime. One week he was paid £368. How many hours of overtime did he work?

7 A worker in a call centre is paid £323 for a 38-hour week. She is paid time and a half for working on a Sunday. What is she paid per hour on a Sunday?

8 A postman is paid £420 for a 40-hour week. When he works overtime he is paid time and a half.
How many hours' overtime has he worked in a week when he is paid £514.50?

Best buys

Cans of cola come in two sizes: a 330 mℓ can for 39p or a 500 mℓ can for 63p.
You can work out which is the best value for money in two ways.

> The 'best buy' is the best value for money.

- **Work out the cost of 1 mℓ of cola for each can.**
- **Work out how many mℓ you get for 1p for each can.**

> You are looking for the *cheapest* cost for 'best buy'.

> You are looking for the *most* mℓ for 'best buy'.

Example 13

Which is the best buy:
a 330 mℓ can for 39p or a 500 mℓ can for 63p?

Method A

330 mℓ can: 330 mℓ costs 39p

$$1 \, m\ell \text{ costs } \frac{39}{330}p = 0.118 \dots p$$

500 mℓ can: 500 mℓ costs 63p

$$1 \, m\ell \text{ costs } \frac{63}{500}p = 0.126p$$

1 mℓ costs less in the 330 mℓ can so it is the best buy.

> Find the cost of 1 mℓ.

> You must remember which method you are using so that you know whether to pick the biggest answer or the smallest one!

continued ▼

Method B

330 mℓ can: You get 330 mℓ for 39p

so you get $\dfrac{330}{39}$ mℓ for 1p

= 8.46 ... mℓ for 1p

500 mℓ can: You get 500 mℓ for 63p

Find how many mℓ you get for 1p.

so you get $\dfrac{500}{63}$ mℓ for 1p

= 7.93 ... mℓ for 1p

You get more mℓ for 1p in the 330 mℓ can so it is the best buy.

Exercise 15I

1 Find the number of grams for 1p.

(a) 10p for 40 g   (b) 18p for 324 g   (c) 500 g for 15p.

2 Find the cost of 1 gram.

(a) 25 g for £2   (b) 300 g for 75p

(c) £2.50 for 500 g.

Change the costs to pence.

3 A large box of cereal costs £2.78 for 500g.
A small box costs £1.45 for 250 g.

(a) How many grams do I get for 1p in each box?

(b) Which one is the best buy?

Change the costs to pence.

4 A mobile phone company has different tariffs.

Tariff	A	B	C
No. of calls	50	100	500
Total cost	£2.60	£5.15	£25.80

(a) Find the cost of 1 call for each tariff.

(b) Which tariff is the best buy?

UAM **5** I can buy a 545 g jar of jam for £1.35, or a 350 g jar for 90p. Which is the best buy?

UAM **6** A large block of chocolate costs £1.84 for 250 g. A small block costs 80p for 60 g. Which is the best buy?

Bills and services

Many people pay household bills and services (such as electricity and gas) **quarterly** . This means they pay four times a year.

Sometimes bills are made up of two parts:

- a fixed amount of money (a **standing charge**),
- an amount of money depending on the number of units of gas or electricity used.

> **Total amount = standing charge + cost of units used**

The number of units of gas or electricity you have used is worked out from the gas or electricity meter reading.

Key words:
quarterly
standing charge

4 times a year = every 3 months.

You pay the standing charge even if you haven't used any gas.

Example 14

Mr Sparks buys his gas from Cosygas.
The quarterly charges are
In March his meter reading was 12 027 units.
In June his meter reading was 14 967 units.

(a) How many units of gas did Mr Sparks use during the quarter from March to June?

(b) Calculate the cost **(i)** of the first 1200 units
                          (ii) of the remaining units.

(c) How much is Mr Sparks's gas bill for this quarter?

COSYGAS
Standing charge = £15.36
2p per unit for first 1200 units used
1.5p per unit for the remainder

You can use this method for any bill calculation where there is a standing charge.

(a) Units used = 14 967 − 12 027 = 2940

(b) (i) 2p per unit for 1200 units = 1200 × 2p = 2400p

    = £24

   (ii) Number of units used at 1.5p per unit = 2940 − 1200

        = 1740

     1.5p per unit for 1740 units = 1740 × 1.5p = 2610p

        = £26.10

(c) Gas bill = £15.36 + £24 + £26.10

      = £65.46

Convert these answers to £.

Standing charge + cost for 1200 units + cost for remainder.

Exercise 15J

1 Mrs Smith buys her electricity from *Top Power*. Their charges are:

TOP POWER
10.5p per unit for the first 100 units
6.5p per unit for the remainder

Mrs Smith uses 670 units.

(a) How much do the first 100 units cost?

(b) How much does the remainder cost?

(c) What is the total cost for 670 units?

2 Mr Patel also buys electricity from *Top Power*. His meter reading in December was 17 228. In January it was 18 003.

(a) How many units of electricity had he used?

(b) What is the total cost of his electricity for this month?

3 Work out the electricity bills for these houses. There is no standing charge.

House	Previous reading	Present reading	Unit price for first 100 units	Unit price for remaining units
(a) 32 Front St.	10 325	19 436	2.7p	1.5p
(b) 34 Front St.	34 219	35 106	3.4p	2.8p
(c) 36 Front St.	61 754	68 127	2.3p	1.75p

4 Morag's quarterly charge for gas from *Ngas* is:

- 3.2p per unit for the first 100 units
- 2.4p per unit for the remainder
- standing charge £12.75.

Her previous meter reading was 43 249. The new reading is 56 310.

(a) How many units of gas has she used?

(b) What is the cost for the gas used?

(c) What is her total bill for gas for this quarter?

5 Morag's next bill from *Ngas* was £183.95.
  (a) What was the cost for the gas (*without* the standing charge)?
  (b) What was the cost of the first 100 units used?
  (c) How much did the remaining units cost?
  (d) How many units did she use altogether?

> Work out how many units she used at the rate of 2.4p per unit then add on 100 units.

6 The Talkalot telephone company's quarterly charges are:

Talkalot ↘
Standing charge = £18.50
3p for cheap rate calls
7.5p for normal rate calls

  (a) Sam made 168 cheap rate calls. How much do these cost?
  (b) He made 98 normal rate calls. How much do these cost?
  (c) What is the total for the calls and standing charge?
  (d) 17.5% VAT is added to the bill. What is the total telephone bill?

Savings

> **Key words:**
> interest
> invest

If you put money into a bank savings account or a building society, the bank or building society pays you **interest** (extra money). The interest is worked out as a percentage of the money you **invest** (save).

If you borrow money from a bank or a building society, you have to pay interest on the money you borrow.

There are two ways of calculating interest: simple interest and compound interest.

> You will look at compound interest in Section 15.3.

Simple interest

> **Key words:**
> rate of interest
> principal
> per annum

The **rate of interest** is fixed.
The interest is a percentage of the **principal** (the money you invest).
You receive the same amount of interest each year.

Example 15

Jim invests £400 at 3% **per annum** simple interest.
How much interest will he receive over 5 years?

per annum (p.a.) means 'per year' or 'each year'.

Interest for 1 year = 3% of £400 = $\dfrac{3}{100} \times 400 = £12$

Interest received over 5 years = $5 \times £12 = £60$

You calculated:

$$\text{Simple interest} = \left(\dfrac{\text{rate}}{100} \times \text{principal} \right) \times \text{time}$$

For 5 years   Interest for 1 year   5 years

This gives you a formula for working out simple interest

$$I = \dfrac{P \times R \times T}{100} = \dfrac{PRT}{100}$$

$$\left(\dfrac{R}{100} \times P \right) \times T = \dfrac{R}{100} \times P \times T$$
$$= \dfrac{PRT}{100}$$

(put the letters in alphabetical order)

I = simple interest   **P = principal**
R = rate of interest (% *p.a.*)   **T = time (in years)**

Example 16

Carla invests £600 at a rate of 5.5% for 36 months.
How much simple interest will she receive?

$$I = \dfrac{P \times R \times T}{100} = \dfrac{600 \times 5.5 \times 3}{100} = £99$$

36 months = 3 years. *T* must be in years.

Carla receives £99 interest.

You can use the formula to calculate *P* or *R* or *T* if you are given the value of *I*.

Example 17

How much money do I need to invest over 6 years at a rate of interest of 4% to earn simple interest of £210?

$T = 6$, $R = 4$ and $I = 210$ P is unknown

$$I = \dfrac{P \times R \times T}{100}$$

continued ▼

$$210 = \frac{P \times 4 \times 6}{100}$$

$$210 = \frac{P \times 24}{100}$$

$$\frac{210 \times 100}{24} = P$$

Multiply both sides by 100 and divide by 24.
For help in solving equations like this, see Chapter 7.

$$P = 875$$

You need to invest £875.

Exercise 15K

1 Find the simple interest on
   (a) £400 invested for 2 years at a rate of 10% p.a.
   (b) £250 invested for 5 years at a rate of 4% p.a.
   (c) £875 invested for 48 months at a rate of 5.5% p.a.
   (d) £4350 invested for $2\frac{1}{2}$ years at a rate of 6.25% p.a.

2 Kim's grandmother gives her £150. Kim invests it at a rate of 8% for 3 years until she starts college.
   (a) How much simple interest will she earn on her money?
   (b) How much will she have in total at the end of the 3 years?

Add the interest earned to the principal (£150).

3 Michael has £2300 to invest. The Tiger Bank gives 7% interest for the first year and 5% for the following years.
The Panda Bank gives 6% p.a.
He wants to invest his money for 4 years. Which bank will pay him the most interest?

4 How much would I have to invest to receive £48 interest after 5 years at a rate of 8%?

5 How long will it take to make £96 interest if I invest £800 at a rate of 4% p.a?

6 I invested £300 for 5 years and received £52.50 simple interest. What was the rate of interest?

15.3 Repeated proportional change

Key words:
compound interest

Generally, when you invest money the interest you earn is calculated using **compound interest** .

If you borrow money, the interest you pay is calculated using compound interest.

- The interest you receive in year 1 is added to the principal.
- In year 2 your interest is calculated as a percentage of your principal *plus* your interest from year 1.
- In year 3 your interest is calculated as a percentage of your principal *plus* your interest from years 1 and 2.
- and so on ...

The rate of interest is fixed.

> **For compound interest:**
> - **The interest you receive each year is *not* the same.**
> - **The interest is calculated on the amount invested in the first place *plus* any interest already received.**

Example 18

Venetia puts £800 into a building society account.
The rate of interest is 5% p.a. compound interest.
How much does she have in her account after 2 years?

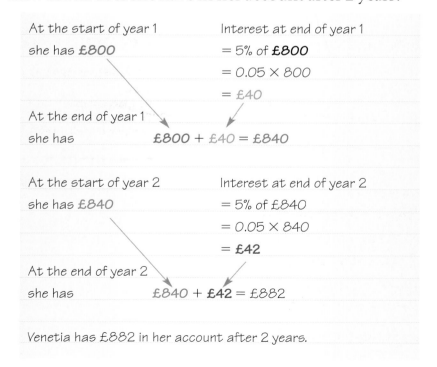

At the start of year 1
she has **£800**

Interest at end of year 1
= 5% of **£800**
= 0.05 × 800
= £40

5% = 0.05 as a decimal.

At the end of year 1
she has   **£800** + £40 = £840

Add the interest to the principal.

At the start of year 2
she has **£840**

Interest at end of year 2
= 5% of £840
= 0.05 × 840
= **£42**

At the end of year 2
she has   £840 + **£42** = £882

Venetia has £882 in her account after 2 years.

To calculate the amount of money you will have if you invest it at compound interest it is quickest to:

1 **Add the rate of interest on to 100%.**
2 **Convert this percentage to a decimal.**
3 **Multiply the original amount of money by this decimal as many times as the number of years the money is invested.**

Using this method for the last example:
Venetia invested £800 at a rate of interest of 5% pa. for 2 years.
● 100% + 5% = 105%
● 105% = 1.05
● £800 × 1.05 × 1.05 =
                      £882

This method gives you the final amount of money, not the interest.

In the problem above, 1.05 was the multiplier. You can use a multiplier where anything is increasing or decreasing. This is called repeated proportional change.

Example 19

There are 5000 whales of a certain species but scientists think that their numbers are reducing by 8% each year. Calculate an estimate of how many of these whales there will be in 3 years' time.

The number of whales reduces by 8% of the number *at the start of each year*. Each % calculation is done on a *different* number so you use the same method as for compound interest problems.

The whales are reducing by 8% each year.

100% − 8% = 92%

92% = 0.92 as a decimal

Number of whales in 3 years' time

$$= 5000 \times 0.92 \times 0.92 \times 0.92$$

$$= 3893.44$$

Estimated number of whales = 3900

The whales are *reducing* in number so you *subtract* the % from 100%.
0.92 is the multiplier.

3 years, so multiply by 0.92 3 times.

You cannot have 3893.44 whales!
3893 is acceptable.
3890 and 3900 are also acceptable as estimates.

Exercise 15L

1 Find the compound interest on:
   (a) £200 invested for 2 years at a rate of 10%
   (b) £750 invested for 2 years at a rate of 5%
   (c) £4500 invested for 2 years at a rate of 2%.

2 Amy invests £120 for 3 years at a rate of 4% compound interest.
   How much will she have in total at the end of the 3 years?

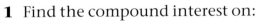

3 How much will Mike have at the end of 3 years if he invests £350 at a rate of 6% compound interest?

4 A new car cost £12 000. Each year the car depreciates by 8% of its value at the start of the year. What will the car be worth at the end of 3 years?

'Depreciate' means it reduces in price.

5 Paul and Sarah inherit £200 each. Sarah invests hers for 2 years at a rate of 6% compound interest. Paul invests his at a rate of 5% compound interest for 3 years.
Who will have the most money at the end of their investment?

6 The population of goldfinches is on the increase in parts of Britain. It is estimated that the present population of 2000 pairs is increasing at the rate of 4% each year. How many pairs will there be at the end of 3 years?

7 The seal population in Scotland is estimated to be declining at the rate of 15% each year due to pollution. In 2001 there were 3000 seals. How many seals would you expect there to be in 2004?

8 Johnson's Biscuits Ltd give staff an annual pay increase of 5% for every year they stay with the firm.
  (a) Kim earns £12 500 a year. How much will she earn in 2 years' time?
  (b) Ben earns £11 800 a year. How much will he earn in 3 years' time?

9 A baby octopus increases its body mass by 5% each day for the first week of its life. An octopus was born weighing 10 kg.
  (a) How much did it weigh at the end of the first day?
  (b) How much did it weigh at the end of the third day?
  (c) How much did it weigh at the end of one week?

10 Amina earns £21 000 a year. This increases by 4% each year. After how many years will she earn over £23 000 a year?

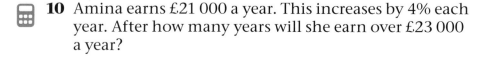

Revision exercise 15M

1 Find:

(a) 15% of £250   (b) $4\frac{1}{2}$% of 800 g.

2 Brooktown has a population of approximately 230 000 people. 7% of them are over 80 years of age. How many people are over 80?

3 An advertisement for "Chewy Gums" says you now get 15% more for the same price. If the original packs were 250 g, how much will the new packs contain?

4 A car manufacturer employed 1280 people. Due to a slump in car sales they reduced their workforce by 40%. How many people work there now?

5 Carl scored 21 out of 25 in his French test and 16 out of 20 in his geography test. Which was his best result?

6 What percentage of 300 g is 125 g?

7 In a survey in Durham the number of starlings had fallen from 18 000 in 1998 to 12 000 in 2002. What is the percentage decrease?

8 A school population increased from 650 in 2000 to 1010 in 2004. What was the percentage increase?

9 A bike was bought for £125 and sold for £145. What was the percentage profit?

10 A car bought for £12 500 new was sold for £8750. What was the percentage loss?

11 A TV was for sale at £220 plus VAT ($17\frac{1}{2}$%). What was the total cost of the TV?

12 Five friends went out for a meal which cost £52 plus VAT ($17\frac{1}{2}$%).

(a) Work out the full cost of the meal.

(b) How much did each of them pay if they shared the cost equally?

13 Liam is buying a washing machine (cash price £320) on credit. He pays 15% deposit and 12 monthly payments of £25.

(a) How much deposit does he pay?

(b) What is the total cost of the machine on credit?

UAM **14** To buy a car costing £6800 on credit you have to pay 20% deposit and 36 monthly payments of £210. How much more does the car cost if you buy it on credit?

UAM **15** Mamet is paid £7.20 per hour for a 38-hour week. When he works overtime he is paid at the rate of 'time and a half'.
One week he works 45 hours.
How much will he be paid?

UAM **16** Lee works 16 hours a week and earns £97.60. When he works overtime he is paid 'double time'. One week he earns £146.40.
How many hours of overtime did he work?

UAM **17** A 375 g box of icing sugar costs £1.25. A 750 g box costs £3.
Which size is the best buy?

UAM **18** A 50 mℓ tube of toothpaste costs 78p. A larger 120 mℓ tube costs £1.90.
Which one is the best buy?

UAM **19** Gas from *Hotburn* costs 3.5p per unit for the first 100 units and then 2.2p per unit for the rest. There is also a standing charge of £23.50.
How much is my total bill if I use 475 units of gas?

UAM **20** Miss Green's electricity meter reading was 14 358 in April and 15 842 in May. The cost of electricity is 5.3p per unit for the first 150 units and 3.4p per unit for the rest. What is Miss Green's total bill?

21 What is the simple interest earned on £375 invested for 6 years at a rate of 5.5%?

22 How many years would it take to earn £720 simple interest if I invested £2000 at a rate of 9%?

23 I invest £400 for 3 years at a rate of 4% compound interest. How much will I have in total at the end of this time?

24 The amount of water in a lake in Africa is estimated at 3 million litres. Due to drought, the amount of water is reducing by 15% each year. How much water will there be in the lake at the end of 3 years?

Examination style questions

1 In a school there are 300 students in year 11.
Of these 300 students, 60% have a part-time job on a Saturday.

  (a) How many have a part-time job on a Saturday?

  (b) Of these 300 students, $\frac{3}{4}$ decide to stay on at school into the sixth form. How many do **not** stay on at school in the sixth form?

  (c) Of the 300 students, 24 are in one tutor group. What percentage are in this tutor group?

(7 marks)
AQA, Spec B, 3F, June 2003

2 **(a)** Jake earns £4 an hour for a basic 35 hour week.
He earns £6 an hour for overtime.
One week he works the basic 35 hour week and 2 hours overtime.
How much does he earn altogether?

  (b) One morning, Jake works from 0815 to 1210.
How long does he work?
Give your answer in hours and minutes.

(6 marks)
AQA, Spec B, 3I, March 2003

3 A shop has a special offer.

  (a) Ian buys the following items:

A bottle of shampoo priced at	£1.75
A bar of soap priced at	£1.15
An aftershave spray priced at	£2.85

What is Ian's percentage saving using this special offer?

SPECIAL OFFER
This week only
BUY THREE ITEMS
AND
GET THE CHEAPEST
ONE FREE!!

  (b) The greatest possible percentage saving is when the three items are all the same price.
Calculate this percentage saving.

(5 marks)
AQA, Spec B, 3I, June 2003

4 Cobalt-60 is a radioactive substance that decays with time.
The mass of the cobalt reduces by 12% each year.
How many years will it take for 200 kg of cobalt-60 to decay to a mass of less than 120 kg?

(3 marks)
AQA, Spec B, 3I, June 2003

5 Mary buys a new kitchen for £5390.

    (a) Mary pays a 10% deposit.
        How much is the deposit?

    (b) After paying the deposit, Mary pays the rest of the cost in 12 equal
        instalments. How much is each instalment?   *(4 marks)*
                                        AQA, Spec B, 3F, March 2003

6 James invests £700 for 2 years at 10% per year compound interest.
How much interest does he earn?   *(2 marks)*
                                      AQA, Spec A, I, June 2003

7 This is part of Sari's electricity bill.
How much does she pay?

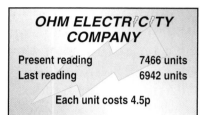

                                       (4 marks)
                                AQA, Spec A, F, June 2004

8 Find $17\frac{1}{2}\%$ of £174.80.   *(2 marks)*
                             AQA, Spec B, 5I, November 2003

9 Mineral water is sold in two sizes.
Which size is the better value for money?
You must show all your working.

                                    (2 marks)
                              AQA, Spec B, 3F,
                                  June 2003

Summary of key points

Work out a percentage of a quantity (grades G to E)

Write the percentage as a fraction or a decimal and
multiply by the quantity.
For example, 5% of £60 $= \frac{5}{100} \times 60 = £3$
or   5% of £60 $= 0.05 \times 60 = £3$.

***Increase or decrease a quantity by a given percentage
(grade D)***

Method A:
1 Work out the actual increase (or decrease)
2 Add to (or subtract from) the original amount.

Method B:

1 Add the % to 100% (or subtract the % from 100%)
2 Convert this % to a decimal
3 Multiply it by the original amount.

For example, to increase £20 by 15%:
115% of £20 = 1.15 × 20 = £23.

Write one quantity as a percentage of another
(grade E/D)

1 Write the two quantities as a fraction.
2 Multiply by 100 to convert to a percentage.
For example, 18 as a percentage of 45 = $\frac{18}{45}$ × 100 =
40%.

Calculate percentage change (grade C)

$$\text{Percentage change} = \frac{\text{actual change}}{\text{original amount}} \times 100\%$$

Calculate percentage profit or loss (grade C)

$$\text{Percentage profit (or loss)} = \frac{\text{actual profit (or loss)}}{\text{cost price}} \times 100\%$$

where the actual profit (or loss) is the difference
between the cost price (what you pay for an item) and
the selling price (what you sell the item for).

Value Added Tax (VAT) (grade D)

VAT at $17\frac{1}{2}$% is added to the cost of many items and services.

Buying on credit (grade E/D)

When you buy on credit you pay a deposit followed by a
number of regular (monthly) payments. It is usually more
expensive than paying the cash price for an item.

Wages (grade E/D)

$$\text{Basic pay} = \begin{array}{c}\text{number of}\\\text{hours in normal}\\\text{working week}\end{array} \times \begin{array}{c}\text{basic}\\\text{rate of}\\\text{pay}\end{array}$$

$$\text{Overtime pay} = \begin{array}{c}\text{number of hours}\\\text{of overtime}\end{array} \times \begin{array}{c}\text{overtime}\\\text{rate of pay}\end{array}$$

Best buys (grade D)

Either,   work out how much it costs for 1 unit (e.g. mℓ) of the item (choose the smallest answer)

or,   work out how many units (e.g. mℓ) you get for 1p (choose the biggest answer)

Bills for services (grades G to D)

Gas and electricity bills are sometimes made up of two parts:

Total amount = standing charge + cost of units of gas (or electricity) used

Simple interest (grade E)

Simple interest is calculated using the formula,

$$I = \frac{P \times R \times T}{100} = \frac{PRT}{100}$$

where I = simple interest, P = principal (original sum of money), R = rate of interest (% p.a.) and T = time (in years).

Compound interest and repeated proportional change (grade C)

The interest you receive each year is *not* the same. The interest is calculated on the amount invested in the first place *plus* any interest already received.

To work out how much money you will have:

1 Add the rate of interest on to 100%.
2 Convert this percentage to a decimal.
3 Multiply the original amount of money by this decimal as many times as the number of years the money is invested.

You can use this method to find how populations increase or decrease.

16 Sequences

Algebra 4

This chapter will show you how to:

✔ **describe how a sequence continues**
✔ **find the next term in a sequence of numbers or diagrams**
✔ **find and use rules for the *n*th term of a sequence**

16.1 Number patterns

Key words:
sequence
term

A number pattern or number **sequence** is a list of numbers. There is often a connection between the numbers in the list.

Each number in a number sequence is called a **term**.

Example 1

Here is a number sequence: 3, 5, 7, 9, 11, …
(a) Write down the 1st term.
(b) Write down the 4th term.
(c) Describe the rule for continuing the sequence.
(d) Write down the next three terms.

(a) 3, 5, 7, 9, 11, … The 1st term is 3.

(b) 3, 5, 7, 9, 11, … The 4th term is 9.

(c) The rule for this sequence is: add 2 to find the next term.

(d) 13, 15, 17.

> You add 2 to 3 to get 5, then add 2 to 5 to get 7, and so on.

> Because 11 + 2 = 13, 13 + 2 = 15, 15 + 2 = 17.

Example 2

For the sequence: 24, 20, 16, 12, …
(a) Describe the rule for continuing the sequence.
(b) Write down the 5th and 6th terms.

(a) Subtract 4 each time.

(b) 5th term = 8   6th term = 4.

> 24 − 4 = 20, 20 − 4 = 16, 16 − 4 = 12.

> Because 12 − 4 = 8 and 8 − 4 = 4.

Exercise 16A

1 Here is a number sequence: 4, 6, 8, 10, 12, ...
   (a) Write down the 1st term.
   (b) Write down the 4th term.
   (c) Describe the rule for continuing the sequence.
   (d) Write down the next three terms.

2 Here is a number sequence: 7, 11, 15, 19, 23, ...
   (a) Write down the 3rd term.
   (b) Write down the 5th term.
   (c) Describe the rule for continuing the sequence.
   (d) Write down the next three terms.

3 For each sequence:
   ● describe the rule for continuing the sequence
   ● write down the 5th and 6th terms.
   (a) 6, 9, 12, 15, ...   **(b)** 4, 9, 14, 19, ...
   (c) 22, 32, 42, 52, ...   **(d)** 6, 13, 20, 27, ...
   (e) −4, −2, 0, 2, ...   **(f)** 10, 7, 4, 1, ...

4 Here is a number sequence: 25, 21, 17, 13, ...
   (a) Write down the 2nd term.
   (b) Write down the 4th term.
   (c) Describe the rule for continuing the sequence.
   (d) Write down the next three terms.

5 Here is a number sequence: 40, 34, 28, 22, ...
   (a) Write down the 1st term.
   (b) Write down the 3rd term.
   (c) Describe the rule for continuing the sequence.
   (d) Write down the next three terms.

6 For each sequence:
   ● describe the rule for continuing the sequence
   ● write down the 5th and 6th terms.
   (a) 76, 66, 56, 46, ...   **(b)** 60, 54, 48, 42, ...
   (c) 32, 28, 24, 20, ...   **(d)** 87, 78, 69, 60, ...
   (e) 105, 90, 75, 60, ...   **(f)** 36, 29, 22, 15, ...

7 Write down the next two numbers in each sequence:

(a) 20, 15, 10, 5, ...   (b) 13, 10, 7, 4, ...

(c) 35, 25, 15, 5, ...   (d) 26, 19, 12, 5, ...

16.2 Using differences

Key words:
consecutive
difference

Terms next to each other are called **consecutive** terms.

In the sequence

4, 10, 16, 22, ...

4 and 10 are consecutive terms,
10 and 16 are consecutive terms.

To write the next terms in a sequence it helps to look at the differences between consecutive terms.

sequence: 4,   10,   16,   22,   ...

differences: +6   +6   +6

The difference between 4 and 10 is $10 - 4 = 6$.
or to get from 4 to 10 you add 6.

In this sequence the differences are all the same: +6.

In some sequences the differences are not all the same.

Example 3

For the sequence:

3, 5, 8, 12, ...

(a) Find the differences between consecutive terms.

(b) Describe the pattern of the differences.

(c) Write down the next two terms.

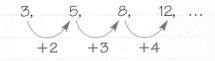

(a) sequence: 3,   5,   8,   12,   ...

difference: +2   +3   +4

(b) The difference goes up by 1 each time.

(c) 12 + 5 = 17

    17 + 6 = 23

The next two differences will be +5 and +6.

Exercise 16B

1 For the sequence:

   2, 4, 7, 11, ...

   (a) Find the differences between consecutive terms.

   (b) Describe the pattern of the differences.

   (c) Write down the next two terms.

2 For the sequence:

   6, 9, 13, 18, ...

   (a) Find the differences between consecutive terms.

   (b) Describe the pattern of the differences.

   (c) Write down the 5th and 6th terms.

3 For each sequence find the next two terms. Use the differences between consecutive terms to help you.

   (a) 10, 11, 13, 16, ...   **(b)** 4, 5, 8, 13, ...

   (c) 20, 25, 35, 50, ...   **(d)** 1, 4, 9, 16, ...

   (e) 1, 1, 2, 3, 5, 8, ...   **(f)** 4, 8, 16, 28, ...

   (g) 49, 36, 25, 16, ...   **(h)** 17, 12, 8, 5, ...

   (i) 55, 45, 36, 28, ...   **(j)** 13, 9, 6, 4, ...

4 For these sequences, find
   (i) the differences between consecutive terms
   (ii) the next two terms:

   (a) 3, 6, 12, 24, ...   **(b)** 1, 3, 9, 27, ...

   (c) 40, 20, 10, ...   **(d)** 10 000, 1000, 100, ...

16.3 Rules for sequences

Key words:
general rule
term number
general term
nth term

You can use the **general rule** for a sequence to work out any term in a sequence. You need to know the position of the term.

The **term number** is the position of the term.

Sequence:

1,	4,	7,	10, ...
1st	2nd	3rd	4th
term	term	term	term

Term number:	1	2	3	4

Example 4

The general rule of a sequence is:

$3 \times$ *the term number then add 1*

(a) Use the rule to find the first three terms.
(b) Use the rule to find the 15th term.

(a) 1st term: $3 \times 1 + 1 = 3 + 1 = 4$

2nd term: $3 \times 2 + 1 = 6 + 1 = 7$

3rd term: $3 \times 3 + 1 = 9 + 1 = 10$

(b) 15th term: $3 \times 15 + 1 = 45 + 1 = 46$

The term number of the 15th term is 15.

The general rule is also called the **general term** or the ***n*th term**.

Exam questions usually use nth term.

The letter n is used to stand for the term number.

Example 5

The nth term of a sequence is $2n + 5$.
(a) Write down the first four terms.
(b) What is the difference between consecutive terms?
(c) Which term has a value of 45?
(d) Explain why 36 cannot be a term in this sequence.

(a) 1st term: $2 \times 1 + 5 = 2 + 5 = 7$

2nd term: $2 \times 2 + 5 = 4 + 5 = 9$

3rd term: $2 \times 3 + 5 = 6 + 5 = 11$

4th term: $2 \times 4 + 5 = 8 + 5 = 13$

Substitute the term numbers for n in the formula.
For the 1st term $n = 1$
For the 2nd term $n = 2$
... and so on.

continued ▼

(b) sequence: 7, 9, 11, 13, ...

difference: +2 +2 +2

The difference between consecutive terms is +2.

(c) $2n + 5 = 45$

$2n = 40$

$n = 20$

The 20th term is 45.

Every term follows the rule $2n + 5$. Find which value of n satisfies the equation $2n + 5 = 45$.

(d) If 36 is in the sequence then there is a whole number n for which $2n + 5 = 36$.

If $2n + 5 = 36$

then $2n = 31$

and $n = 15\frac{1}{2}$

but $15\frac{1}{2}$ is not a whole number so 36 is not in the sequence.

Part **(d)** is an extension of the method used in part **(c)**. This is a UAM question where you have to think of a method to solve the problem.

Example 6

The nth term of a sequence is $n^2 - 1$.

(a) Write down the first four terms of the sequence.

(b) Write down the 12th term.

(a) 1st term: $1^2 - 1 = 1 - 1 = 0$

2nd term: $2^2 - 1 = 4 - 1 = 3$

3rd term: $3^2 - 1 = 9 - 1 = 8$

4th term: $4^2 - 1 = 16 - 1 = 15$

(b) 12th term: $12^2 - 1 = 144 - 1 = 143$

Remember $n^2 = n \times n$. So $2^2 = 2 \times 2 = 4$.

Exercise 16C

1 The general rule of a sequence is:

   $3 \times$ *the term number then add 2*

   (a) Use the rule to find the first three terms.

   (b) Use the rule to find the 15th term.

You can write this rule as: $3n + 2$.

2 The general rule of a sequence is:

    2 × the term number then add 7

  (a) Use the rule to find the first three terms.

  (b) Use the rule to find the 10th term.

3 The general rule of a sequence is:

    4 × the term number then subtract 1

  (a) Use the rule to find the first three terms.

  (b) Use the rule to find the 50th term.

4 The nth term of a sequence is $2n + 3$.

  (a) Write down the first four terms.

  (b) What is the difference between consecutive terms?

  (c) Write down the 20th term.

5 For each of the following sequences:

- find the first four terms
- write down the difference between consecutive terms
- find the 30th term.

  (a) nth term: $3n + 5$ **(b)** nth term: $2n - 1$

  (c) nth term: $4n - 3$ **(d)** nth term: $3n + 7$

  (e) nth term: $5n + 1$ **(f)** nth term: $6n - 2$

6 Look at your answers to question 5. What do you notice about the rule for the nth term and the difference between consecutive terms for each sequence?

7 The nth term of a sequence is $2n + 4$.

  (a) Work out the value of the 8th term.

  (b) Which term has a value of 46?

  (c) Explain why 35 is not a term in this sequence.

Use Example 5(d) to help you.

8 The nth term of a sequence is $3n - 1$.

  (a) Calculate the value of the 6th term.

  (b) Which term has a value of 59?

  (c) Explain why 90 is not a term in this sequence.

9 The *n*th term of a sequence is $5n + 7$.
    (a) Calculate the value of the 10th term.
    (b) Which term has a value of 82?

UAM
    (c) Explain why 110 is not a term in this sequence.

10 The *n*th term of a sequence is $n^2 + 1$.
    (a) Write down the first four terms of the sequence.
    (b) Write down the 12th term.

11 The *n*th term of a sequence is $n^2 + 4$.
    (a) Write down the first four terms of the sequence.
    (b) Write down the 9th term.

12 The *n*th term of a sequence is $n^2 - 3$.
    (a) Write down the first four terms of the sequence.
    (b) Write down the 13th term.

> You are expected to know the first 15 square numbers.

13 The *n*th term of a sequence is $n^2 + 3$.
    (a) Which term has a value of 28?

UAM
    (b) Explain why 50 is not a term in this sequence.

16.4 Finding the *n*th term

If the difference between consecutive terms is the same you can use it to find the rule for the *n*th term.

In the sequence: 7,   11,   15,   19, ...

         difference: +4   +4   +4

the terms go up in 4s.

The 4 × table also goes up in 4s.
This tells you that the *n*th term includes **4*n***.

$4n = 4 \times n$

To find the rest of the rule, compare the sequence to the 4 × table:

4 × table	4	8	12	16	...
Sequence	7	11	15	19	...

You have to **add 3** to each number in the 4 × table to get the numbers in the sequence.
So the rule for the *n*th term is: **4*n* + 3**.

$4 + 3 = 7, 8 + 3 = 11, ...$

Example 7

Find the nth term and the 50th term of the sequence:

2, 5, 8, 11, ...

Sequence: 2,   5,   8,   11, ...

Difference: +3   +3   +3

The difference is 3 so the nth term includes $3n$.

3 × table	3	6	9	12	...
Sequence	2	5	8	11	...

You have to subtract 1 from each number in the 3 times table to get the numbers in the sequence.

So the rule for the nth term is: $3n - 1$.

50th term $= 3 \times 50 - 1 = 150 - 1 = 149$.

Substitute $n = 50$ into the nth term.

Exercise 16D

Find the nth term and the 50th term of each of these sequences:

1 4, 7, 10, 13, ...
2 5, 7, 9, 11, ...
3 7, 11, 15, 19, ...
4 3, 8, 13, 18, ...
5 6, 9, 12, 15, ...
6 6, 13, 20, 27, ...
7 2, 12, 22, 32, ...
8 16, 25, 34, 43, ...
9 5, 6, 7, 8, ...
10 $-4, -1, 2, 5, ...$

16.5 Sequences of patterns

Sequences of diagrams can lead to number sequences.

The numbers of dots in these patterns make a number sequence.

Pattern 1   Pattern 2   Pattern 3

Number
sequence   1   4   7

Example 8

This sequence of patterns is made from square tiles.

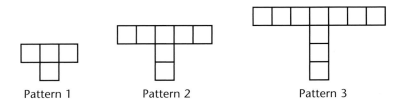

Pattern 1   Pattern 2   Pattern 3

(a) Draw pattern 4.
(b) How many tiles are needed for pattern 5?
(c) Describe the rule for continuing the sequence.
(d) How many tiles are needed for pattern *n*?
(e) Describe how the patterns relate to the *n*th term.
(f) Which pattern needs 37 tiles?

(a)

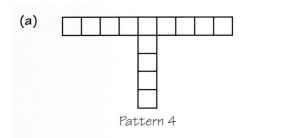

Pattern 4

You could draw pattern 5 to check, or you could work out that you add 1 tile to each arm, adding 3 in total.
The new tiles go on the end of each 'arm' of the pattern.

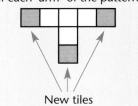

New tiles

(b) 16 tiles for pattern 5

(c) Add 3 tiles each time.

(d) The number sequence for the patterns:

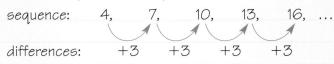

sequence: 4,   7,   10,   13,   16, ...

differences: +3   +3   +3   +3

The difference is 3 so the rule includes 3*n*.

Compare with the 3 × table:

3 × table	3	6	9	12	...
Sequence	4	7	10	13	...

The rule for the number of tiles needed in pattern *n* is: 3*n* + 1.

Pattern *n* is the *n*th term in the pattern sequence.

You have to **add 1** to each number in the 3 × table to get the numbers in the sequence.

continued ▼

(e)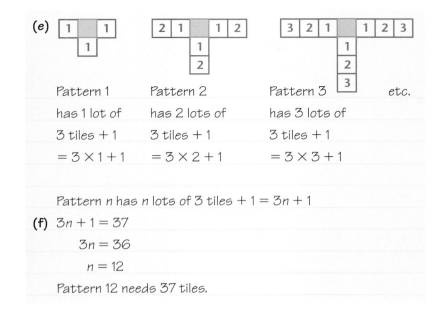

Pattern 1	Pattern 2	Pattern 3	etc.
has 1 lot of	has 2 lots of	has 3 lots of	
3 tiles + 1	3 tiles + 1	3 tiles + 1	
$= 3 \times 1 + 1$	$= 3 \times 2 + 1$	$= 3 \times 3 + 1$	

Pattern n has n lots of 3 tiles $+ 1 = 3n + 1$

(f) $3n + 1 = 37$

$\qquad 3n = 36$

$\qquad\quad n = 12$

Pattern 12 needs 37 tiles.

Use the same method as in Example 7.

Exercise 16E

1 Square tiles are used to make a sequence of patterns.

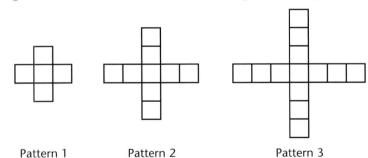

Pattern 1   Pattern 2   Pattern 3

(a) Draw pattern 4.
(b) How many tiles are needed for pattern 5?
(c) Describe the rule for continuing the sequence of the number of tiles needed.
(d) How many tiles are needed for pattern n?
(e) Which pattern needs 45 tiles?

2 Matchsticks are used to make triangles:

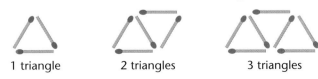

1 triangle   2 triangles   3 triangles

(a) Draw the pattern for 4 triangles.
(b) How many matchsticks are needed for 5 triangles?

(c) Describe the rule for continuing the sequence of the number of matchsticks needed.

(d) How many matchsticks are needed for *n* triangles?

(e) Copy and complete:

Pattern 1
1 lot of ____ matches + ____

Pattern 2
2 lots of ____ matches + ____

Pattern 3
3 lots of ____ matches + ____

Pattern n
n lots of ____ matches + ____

(f) How many triangles can you make with 51 matchsticks?

3 Matchsticks are used to make pentagons:

1 pentagon   2 pentagons   3 pentagons

(a) Draw the pattern for 4 pentagons.

(b) How many matchsticks are needed for 5 pentagons?

(c) Describe the rule for continuing the sequence of the number of matchsticks needed.

(d) How many matchsticks are needed for *n* pentagons?

(e) Describe how the patterns relate to the *n*th term.

(f) How many pentagons can you make with 85 matchsticks?

4 Matchsticks are used to make rectangles:

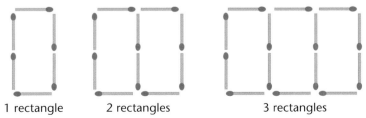

1 rectangle   2 rectangles   3 rectangles

(a) Draw the pattern for 4 rectangles.

(b) How many matchsticks are needed for 5 rectangles?

(c) How many matchsticks are needed for *n* rectangles?

(d) How many rectangles can you make with 66 matchsticks?

5 In a restaurant, tables are put together in a line to seat different numbers of people.

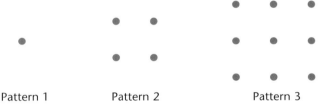

| 1 table | 2 tables | 3 tables |
| 4 people | 6 people | 8 people |

(a) Draw the pattern for 4 tables.

(b) How many people can sit at 5 tables in a line?

(c) How many people can sit at *n* tables in a line?

(d) How many tables do you need to seat 24 people?

6 Dots are used to make a sequence of patterns:

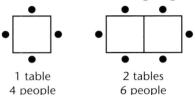
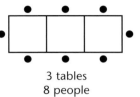

Pattern 1   Pattern 2   Pattern 3

(a) Draw pattern 4.

(b) How many dots are needed for pattern 5?

(c) Describe the rule for continuing the sequence of the number of dots.

(d) What is the name for the numbers in this sequence?

(e) How many dots are needed for pattern *n*?

(f) Which pattern will have 144 dots?

7 Dots are used to make a sequence of triangles:

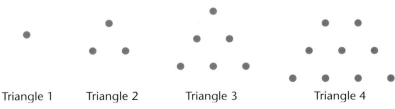

Triangle 1   Triangle 2   Triangle 3   Triangle 4

(a) Draw triangle 5.

(b) How many dots are needed for triangle 6?

(c) Describe the rule for continuing the sequence of triangles.

(d) The numbers in this sequence are called triangular numbers.
One rule for finding the nth triangular number is: $\frac{1}{2}n(n + 1)$.
What is the 10th triangular number?

Examination style questions

1 Patterns are made of sticks.

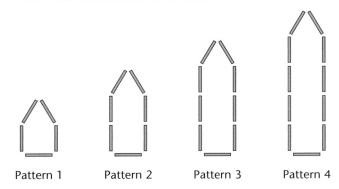

| Pattern 1 | Pattern 2 | Pattern 3 | Pattern 4 |

(a) Complete the table for pattern 4.

Pattern number	1	2	3	4
Number of sticks	5	7	9	

(b) Sketch pattern 5.

(c) Which pattern will have 25 sticks?

(d) Here is the rule for working out the number of sticks

> Multiply pattern number by 2 and add 3.

How many sticks will be in pattern 100?

(5 marks)
AQA, Spec B, 5F, June 2003

2 A sequence begins

1, 2, 5, ...

The rule for continuing this sequence is

> Multiply the last term by 3 and subtract 1

What is the next number in the sequence?

(1 mark)
AQA, Spec B, 5F, June 2003

3 **(a)** Here is a sequence of numbers.

30   25   20   15   10 ...   ...

Write down the next two numbers in the sequence.

(b) Here is another sequence of numbers.

160   80   ...   20   10   ...   2.5

Write down the two missing numbers in this sequence.   *(4 marks)*

AQA, Spec A, F, June 2005

4 Fill in the two missing numbers in this sequence.

31, 29, 25, 19, ..., 1, ...   *(2 marks)*

AQA, Spec A, I, November 2003

5 **(a)** A sequence of numbers is shown.

2   9   16   23 ...   ...

Write down the next two numbers in the sequence.

(b) Another sequence of numbers is shown.

2   6   12   20   ...

Write down the next number in the sequence.

(c) A different sequence begins

4   1   −2   −5

Write down the rule for this sequence.   *(4 marks)*

AQA, Spec A, F, June 2003

6 Patterns are made from shaded and unshaded squares.

Pattern 1   Pattern 2   Pattern 3   Pattern 4   Pattern 5

(a) Draw pattern 6.

(b) How many shaded squares will there be in pattern 20?

(c) How many unshaded squares will there be in pattern 20?   *(3 marks)*

AQA, Spec A, F, June 2005

7 **(a)** The nth term of a sequence is $4n + 1$

    (i) Write down the first three terms of the sequence.

    (ii) Is 122 a term in this sequence?
        Explain your answer.

(b) Tom builds fencing from pieces of wood as shown below.

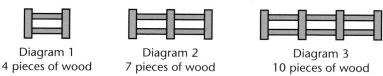

    Diagram 1   Diagram 2   Diagram 3
4 pieces of wood   7 pieces of wood   10 pieces of wood

How many pieces of wood will be in diagram n?   *(5 marks)*

AQA, Spec A, I, June 2003

8 A pattern using pentagons is made of sticks.

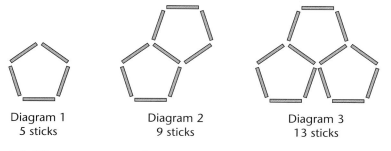

   Diagram 1   Diagram 2   Diagram 3
    5 sticks   9 sticks   13 sticks

(a) How many sticks are needed for diagram 5?

(b) Write down an expression for the number of sticks in diagram n?

(c) Which diagram uses 201 sticks?   *(7 marks)*

AQA, Spec A, I, Nov 2003

9 Patterns are made from shaded and unshaded squares.

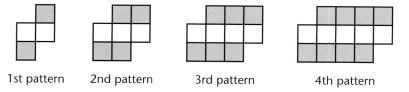

1st pattern   2nd pattern   3rd pattern   4th pattern

(a) How many shaded squares are there in the nth pattern?

(b) How many unshaded squares are there in the nth pattern?   *(2 marks)*

AQA, Spec A, I, June 2005

Summary of key points

Continuing sequences (grades G to E)

To write the next term in a sequence it helps to look at the differences between consecutive terms.

nth term (grade D)

The nth term is a general rule to work out any term in a sequence if you know its position.

For example, nth term $= 5n + 3$.

To work out the 40th term use $n = 40$.

40th term $= 5 \times 40 + 3 = 200 + 3 = 203$.

Finding the nth term of a sequence (grade C)

For example,

sequence: 2,   5,   8,   11,   ...

differences: $+3$ $+3$ $+3$

The difference between consecutive terms is $+3$, so the nth term includes $3n$.

Compare the sequence to the $3 \times$ table to find the rest of the rule.

3 × table	3	6	9	12	...
Sequence	2	5	8	11	...

You have to subtract 1 from each number in the $3 \times$ table to get the numbers in the sequence, so the rule for the nth term is: $3n - 1$.

17 Constructions and loci

This chapter will show you how to:

✔ use a straight edge and compasses to construct
  • a triangle given all three sides
  • the mid-point and perpendicular bisector of a line segment
  • the perpendicular from a point to a line
  • the perpendicular from a point on a line
  • the bisector of an angle
✔ construct angles of 60° and 90°
✔ understand, interpret and solve problems using simple loci (including scale drawing and bearings)

17.1 Constructions

Key words:
arc

You will need to know:
• how to use a pair of compasses
• how to measure accurately

Standard constructions use only a straight edge (ruler) and a pair of compasses to draw accurate diagrams.

Your drawings must be accurate.

NO ARCS, NO MARKS!

When you use compasses you must leave the construction **arcs** as evidence that you have used the correct method.

Arcs are parts of a curve drawn with compasses.

17.2 Construct a triangle given all three sides

Key words:
intersect

Example 1 shows the method for constructing a triangle given the lengths of all three sides.

It helps to draw and label a sketch first.

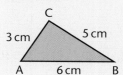

Example 1

Construct a triangle *ABC* where *AB* = 6 cm, *AC* = 3 cm and *BC* = 5 cm.

A ———————————————————— B

continued ▼

Use your ruler to draw the longest side 6 cm long. Label it *AB*.

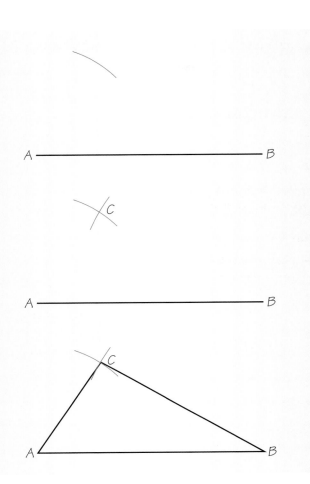

From the question,
$AB = 6$ cm.

Open your compasses to a radius of 3 cm. Put the point on A and draw an arc in the space above line AB.

$AC = 3$ cm.

$BC = 5$ cm.

Open your compasses to a radius of 5 cm. Put the point on B and draw an arc to intersect the first arc. Where the arcs intersect is point C.

Point C is 3 cm from A and 5 cm from B.

Join C to A and B to complete the triangle.

Remember: don't rub out the arcs!

Exercise 17A

Construct triangles ABC with these measurements:

1 $AB = 8$ cm, $AC = 6$ cm, $BC = 5$ cm.

2 $AB = 9$ cm, $AC = 8.5$ cm, $BC = 4$ cm.

3 $AB = 7.5$ cm, $AC = 10$ cm, $BC = 4.5$ cm.

4 $AB = 6$ cm, $AC = 8$ cm, $BC = 10$ cm.

5 Equilateral triangle ABC with sides 6 cm.

6 Equilateral triangle ABC with sides 9.5 cm.

In an equilateral triangle all sides are the same length.

17.3 Constructing perpendiculars

Construct the perpendicular bisector of a line segment

To **bisect** means to cut in half.

Key words:
bisect
line segment
mid-point

A straight line has infinite length so you will be looking at a finite part of it – a **line segment** .

line segment

A perpendicular bisector:

- cuts a line segment in half
- is perpendicular (at 90°) to the line segment.

Example 2

Construct the perpendicular bisector of the line segment *AB*.

A —————————————— B

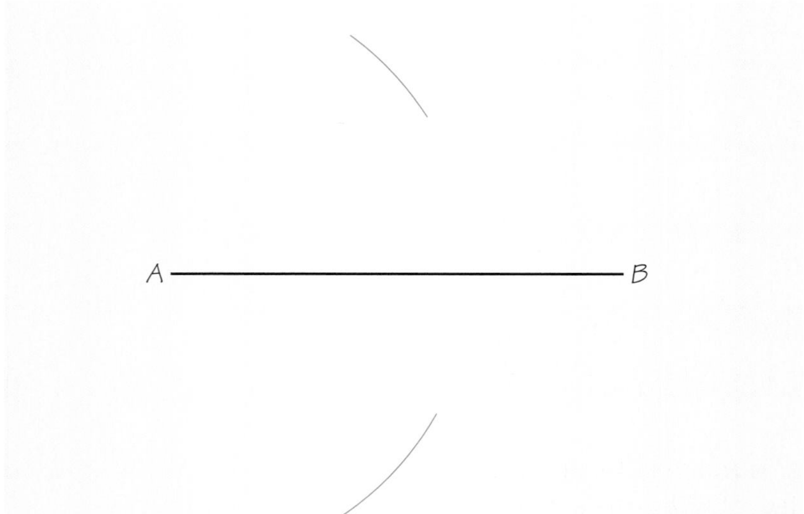

A —————————————— B

Open your compasses to a radius which is just over half the length of *AB*.

Put the compass point on *A* and draw one arc above *AB* and one below.

Keep the radius the same for both arcs.

continued ▼

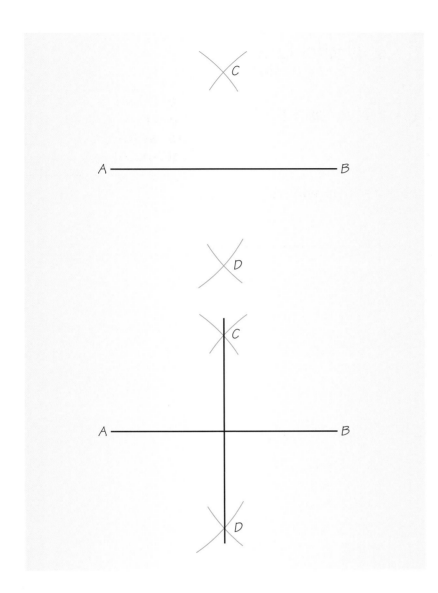

Keeping the radius the same, put your compass point on B and draw two more arcs to intersect the first pair.
Label these points C and D.

Join CD. This line is the perpendicular bisector of AB.

CD crosses AB at the **mid-point** of AB.

CD makes an angle of 90° with AB.

Exercise 17B

Draw six line segments of different lengths. Label the ends A and B. Construct the perpendicular bisector for each line segment.

Check by measuring that your perpendicular bisector passes through the mid-point of the line segment.

Check, using a protractor, that the angle between the two lines is 90°.

Construct the perpendicular from a point to a line

Example 3

Construct the line
which is perpendicular
to *AB* and passes
through point *P*.

The line from *P* to *AB* meets *AB* at 90°.

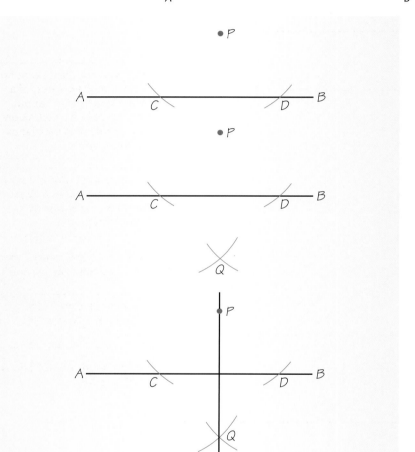

Open your compasses to a radius slightly longer than the distance of *P* from the line. Put the compass point on *P* and draw an arc to intersect *AB* twice. Label these points *C* and *D*.

From *C* and *D*, draw arcs of the same radius to intersect below *AB*. Label this point *Q*.

This need not be the same radius as the first arcs you drew from *P*.

Join *PQ*. This line is perpendicular from *P* to *AB*.

PQ is at 90° to *AB* and passes through *P*.

Exercise 17C

Draw six line segments *AB* with point *P*, similar to this:

• *P*

A ——————————————— B

For each, construct the perpendicular from the point *P* to
the line segment *AB*.

Construct the perpendicular from a point on a line

This time the point is **on** the line and you construct the perpendicular to the line.

You can use this construction to construct an angle of 90°. Another construction for an angle of 90° is shown in Example 7.

Example 4

Construct the line perpendicular to *AB* from point *P*, which lies on *AB*.

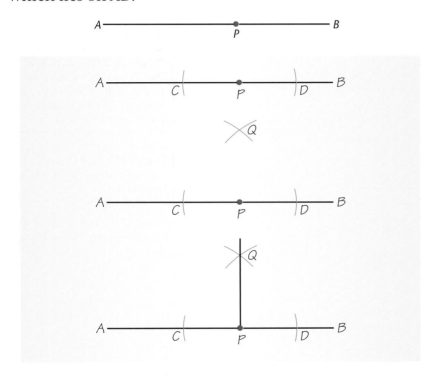

Put your compass point on *P* and draw two arcs, of a fairly small radius (about 1.5 cm), to intersect *AB* on either side of *P*. Label these points *C* and *D*.

From *C* and *D* draw arcs of the same radius to intersect in the space above *AB*. Label this point *Q*.

Make this radius larger than the previous one.

Join *PQ*. This line is the perpendicular from the point *P*.

PQ passes through *P* and is at 90° to *AB*.

Exercise 17D

Draw six line segments *AB*.
Mark a point *P* on each line segment.
Construct the perpendicular from point *P* for each.

17.4 Constructing angles

Construct the bisector of an angle

The bisector of an angle divides an angle into two equal parts.

Sometimes this is called an *angle bisector*.

Example 5

Construct the bisector of angle *A*.

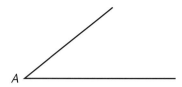

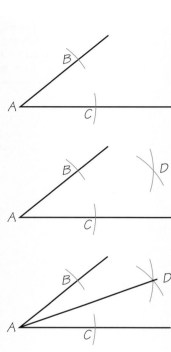

Open your compasses to about 2 cm. Put your compass point on *A* and mark off two small arcs, one on each arm of the angle. Label these points *B* and *C*.

From *B* and *C* draw two arcs of the same radius to intersect in the space between the arms of the angle. Label this point of intersection *D*.

Make this radius larger than the previous one.

Join *AD*. This line is the bisector of angle *A*.

AD divides angle *A* into two equal parts.

Exercise 17E

Draw six angles of different sizes. Construct the angle bisector for each of them.

Check the accuracy of your constructions by measuring the angles with a protractor.

Construct angles of 60° and 90°

Example 6

Construct an angle of 60° at point *P* on line segment *AB*.

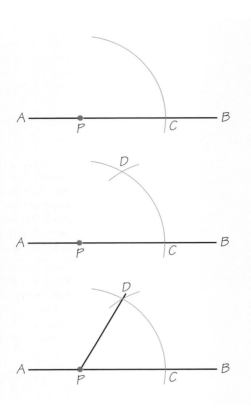

Open your compasses to a radius of about 3 cm.
Put the compass point on *P* and draw an arc which starts just below the line *AB* on one side of *P* and ends almost above *P*.
Label point *C*, where this arc intersects the line *AB*.

Keeping the radius the same, from point *C* draw an arc to intersect the first arc at point *D*.

Join *DP*. Angle *DPC* is 60°.

Can you see that △*DPC* is equilateral?

Example 4 showed one way of constructing an angle of 90°. Example 7 shows another.

Example 7

Construct an angle of 90° at point *P* on line segment *AB*.

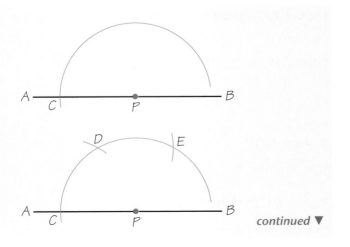

Open your compasses to a radius of about 3 cm.
Put the compass point on *P* and draw an arc which starts just below the line *AB* on one side of *P* and ends fairly close to the line on the other side of *P*.
Label point *C*, where this arc intersects *AB*.

Keeping the radius the same, from point *C* draw an arc to cut the first arc at point *D*.
Keeping the radius the same, draw an arc from point *D* to cut the first arc at point *E*.

Keep the radius the same for the first three arcs.

continued ▼

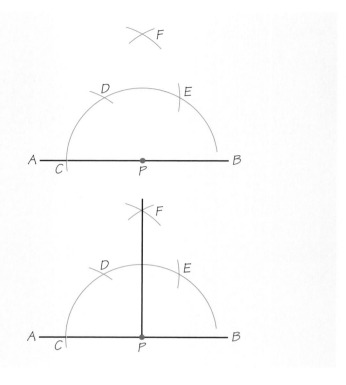

From points *D* and *E*, draw arcs of the same radius to intersect in the space above them. Label this point *F*.

This radius need not be the same as the first one.

Join *FP*. This line makes a 90° angle with line *AB*.

If you are asked to construct a 90° angle you can use the method you prefer.

∠FPB = ∠FPA = 90°

Notice that the first steps of Examples 6 and 7 are very similar.

Exercise 17F

Draw six line segments *AB*.
Mark point *P* on each line segment.
Construct a 60° angle at *P* for each line.

Repeat for angles of 90°.

17.5 Understanding and using loci

Key words:
locus (loci)
equidistant

The line *CD* is the perpendicular bisector of the line *AB*.

All the points on the line *CD* are exactly the same distance from *A* as they are from *B*. Some of them are shown on the diagram.

CD is an example of a **locus** (plural **loci**).

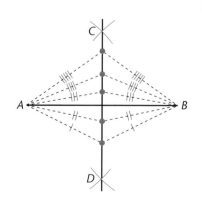

For more on the perpendicular bisector, see Example 2.

A locus is a set of points that obey a given rule.

For *CD* the rule is 'all points equidistant from *A* and *B*'. All the points on line *CD* obey this rule.

All the points to the left of *CD* are nearer to *A* than to *B*. All the points to the right of *CD* are nearer to *B* than to *A*.

Equidistant means 'the same distance from'. You will often see it in locus questions.

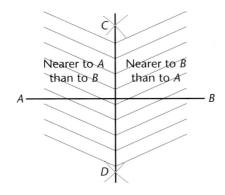

CD is the locus of points equidistant from *A* and *B*.

The locus of points equidistant from two fixed points is the perpendicular bisector of the two fixed points.

Example 8

What is the locus of points which are always 2 cm from a fixed point *P*?

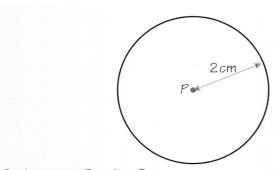

Circle, centre *P*, radius 2 cm.

All points on this circle obey this rule.

All points *inside* the circle are less than 2 cm from *P*. All points *outside* the circle are more than 2 cm from *P*.

The locus of points which are a fixed distance from a fixed point is a circle.

Example 9

What is the locus of points which are equidistant from the two lines *AB* and *AC*?

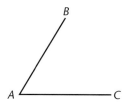

The locus of points which are equidistant from two fixed lines is the angle bisector of the two fixed lines.

Plot some points equidistant from *AB* and *AC*.
If you join the points you get a straight line – the angle bisector.

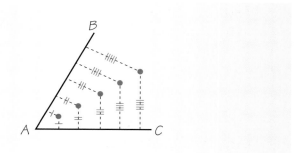

To construct the locus follow the steps in Example 5.

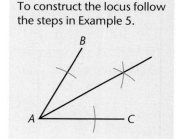

It can help to sketch a locus before you construct it.

Example 10

Construct the locus of points which are exactly 2 cm from the fixed line segment *AB*.

A ——————— B

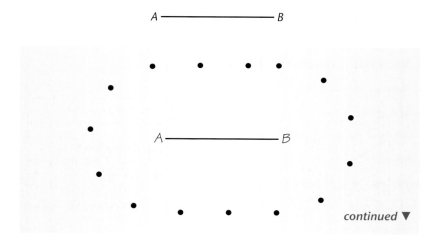

Using a ruler, mark points 2 cm from different points on *AB*.

continued ▼

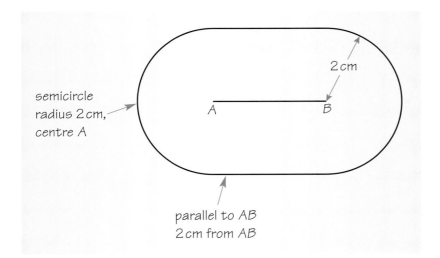

semicircle
radius 2 cm,
centre A

2 cm

A   B

parallel to AB
2 cm from AB

This shows the locus you
need to construct. The locus
will be a 'racetrack' shape
around the outside of the
line segment *AB*.

The locus of points which are a fixed distance from a line segment is a 'racetrack' shape. The shape has two lines parallel to *AB* and two semicircular ends.

For locus questions:
- **think about the points**
- **make a sketch**
- **construct the locus using standard constructions.**

Example 11

In $\triangle ABC$, $AB = 4$ cm, $AC = 6$ cm and $BC = 5$ cm. Shade the region inside the triangle where the points are

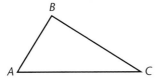

B

A   C

(a) less than 3 cm from *B*   **(b)** nearer to *C* than *A*.

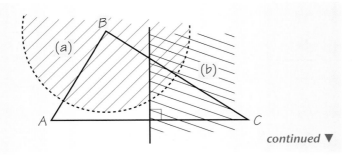

B

(a)

(b)

A   C

continued ▼

First make a sketch.

(a) Points less than 3 cm
from *B* are in a circle.
(b) Points nearer to *C* than
A are to the right of the
perpendicular bisector
of *AC*.

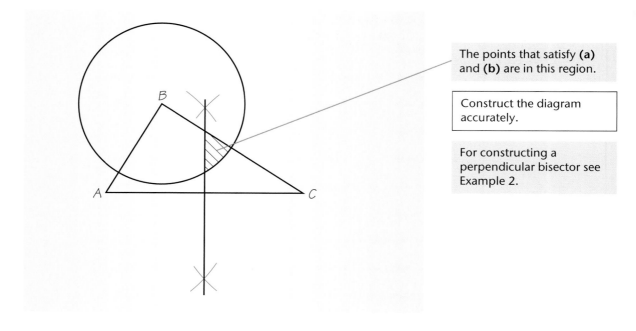

The points that satisfy **(a)** and **(b)** are in this region.

Construct the diagram accurately.

For constructing a perpendicular bisector see Example 2.

Exercise 17G

Use accurate constructions to answer these questions.

1 In △ABC, AB = 5 cm, AC = 7 cm and BC = 6 cm. Copy the triangle and shade the region where points are:

(a) less than 4 cm from C

(b) nearer to A than B.

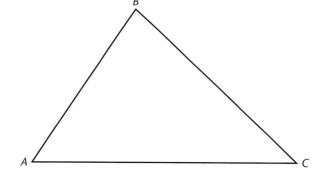

2 In △PQR, PQ = 6 cm, PR = 8 cm and QR = 4 cm. Copy the triangle and shade the region where points are:

(a) nearer to R than P

(b) closer to PR than QR.

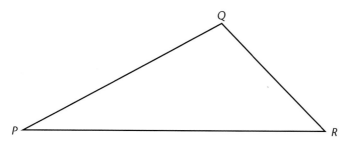

M **3** In $\triangle XYZ$, $XY = 6$ cm, $XZ = 8$ cm and $YZ = 10$ cm.
Copy the triangle and shade the region where points are:
(a) closer to YX than YZ
(b) more than 7 cm from Z.

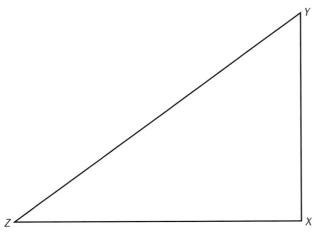

17.6 Using bearings and scale

Example 12

Two ships A and B are shown
in the diagram.

Ship A sails on a bearing of 070°.

Mark clearly all points where
ship A is within 10 km of ship B.

B
•

North
↑
|
•
A

Look back at bearings in
Section 6.2 if you need
some help.

Use a scale of 1 cm for 5 km.

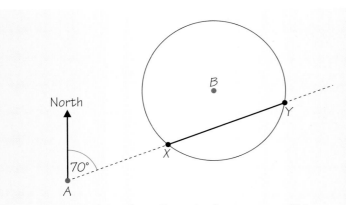

Draw in the line that A sails
on, using a protractor for
the angle.

The locus of points 10 km
from B is a circle, centre B.
On the diagram using the
scale 1 cm for 5 km, this is a
circle with radius 2 cm.

A is less than 10 cm from B on the line segment XY.

Example 13

The diagram shows a triangular field
ABC where $AB = 60$ m, $AC = 80$ m
and $BC = 70$ m. Some treasure is
buried in the field.

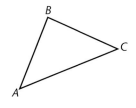

(a) It is between bearings of 040°
and 060° from A.

(b) It is closer to CA than to CB.

(c) It is less than 40 m from B.

Using a scale of 1 cm for 10 m, make accurate constructions
to find the area in which the treasure lies. Shade the region.

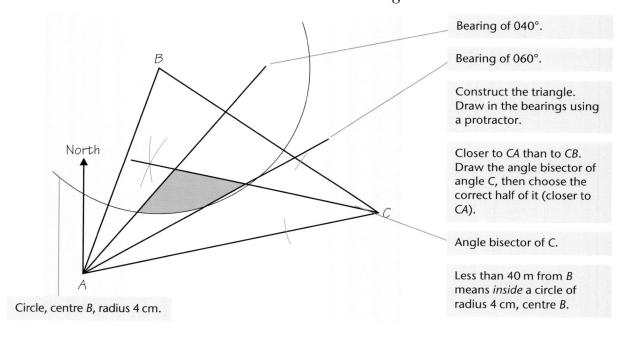

Bearing of 040°.

Bearing of 060°.

Construct the triangle.
Draw in the bearings using
a protractor.

Closer to CA than to CB.
Draw the angle bisector of
angle C, then choose the
correct half of it (closer to
CA).

Angle bisector of C.

Less than 40 m from B
means *inside* a circle of
radius 4 cm, centre B.

Circle, centre B, radius 4 cm.

Exercise 17H

1 Two ships A and B are shown
in the diagram.
Ship A sails on a bearing of
110° while ship B stays
stationary.
Mark clearly all points where
ship A is within 12 km of
ship B.

Use a scale of 1 cm for 4 km.

2 Main roads *AB*, *AC* and *BC* connect towns *A*, *B* and *C*.
AB = 16 miles, *AC* = 14 miles and *BC* = 12 miles.
A new leisure centre is to be built so that it is,

(a) closer to road *AC* than it is to road *AB*

(b) between 8 miles and 10 miles from *B*.

Construct the region where the leisure centre will be
built.

Use a scale of 1 cm for
2 miles.

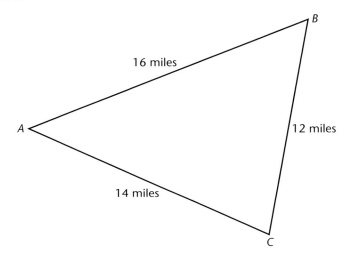

3 *A*, *B* and *C* are places on Treasure Island.
AB = 200 m, *AC* = 160 m and *BC* = 120 m.
The hidden treasure is,

(a) nearer to *C* than *B*

(b) closer to *AB* than to *AC*

(c) less than 70 m from *C*.

Using a scale of 1 cm for 20 m, construct the region
where the treasure is hidden.

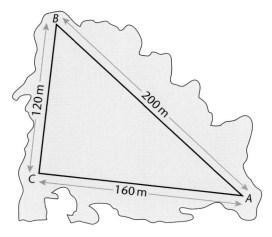

Examination style questions

1 Mutasem draws a rough sketch of a triangle with sides 300 m, 400 m and 500 m.

Using ruler and compasses only, make an accurate scale drawing of the triangle. Use a scale of 1 cm to represent 50 m.

You **must** show clearly all your construction arcs.   *(3 marks)*
                                        AQA, Spec B, 5I, June 2003

2 The map of an island is shown.

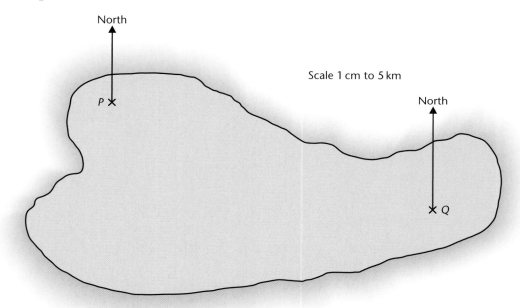

P and *Q* are the positions of two houses on the island.

(a) What is the bearing of *P* from *Q*?

(b) Calculate the actual distance from P to Q in kilometres.

(c) A house is 20 km from P on a bearing of 130°.
Trace the diagram and mark the position of the house with a **X**.

                                        (5 marks)
                                AQA, Spec A, I, November 2004

3 Copy the diagram and using ruler and compasses only, construct the bisector of angle *PQR*.

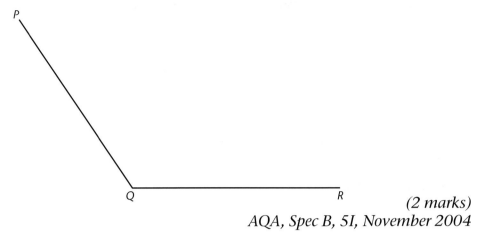

(2 marks)
AQA, Spec B, 5I, November 2004

4 In this question, you should use a ruler and compasses.

The diagram shows an equilateral triangle of side 10 cm.

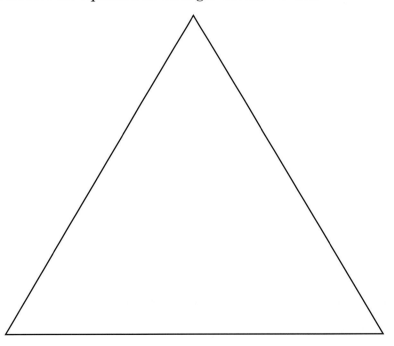

Copy the diagram and show all the points inside the triangle that are more than 5 cm from each vertex of the triangle.

You **must** show clearly all your construction arcs. *(3 marks)*
AQA, Spec B, 5I, November 2004

5 **(a)** The line *LM* is drawn below.

L ———————————————— M

Copy the line and use ruler and compasses to construct the perpendicular bisector of *LM*.
You must show clearly all your construction arcs.

(b) Complete the sentence.

The perpendicular bisector of *LM* is the locus of points which are _____

(3 marks)
AQA, Spec B, 5I, June 2004

6 The diagram shows a triangle, *ABC*.

(a) Copy the diagram and using a ruler and compasses only, construct the perpendicular bisector of *AB*.

(b) **(i)** Repeat this construction on another side of the triangle.

(ii) The point of intersection of the two bisectors is the centre of the circle which passes through *A*, *B* and *C*.

Draw this circle.

(5 marks)
AQA, Spec B, 5I, June 2003

7 *ABCD* is a square of side 8 cm.

Copy the diagram and show clearly the region inside the square that is both closer to the point *D* than to the point *A*, and closer to the side *CD* than the side *AD*.

(3 marks)
AQA, Spec A, I, June 2004

8 The map below shows three boats, *A*, *B* and *C*, on a lake.
Along one edge of the lake there is a straight path.

Treasure lies at the bottom of the lake.

The treasure is:
  between 150 m and 250 m from *B*,
  nearer to *A* than *C*,
  more than 100 m from the path.

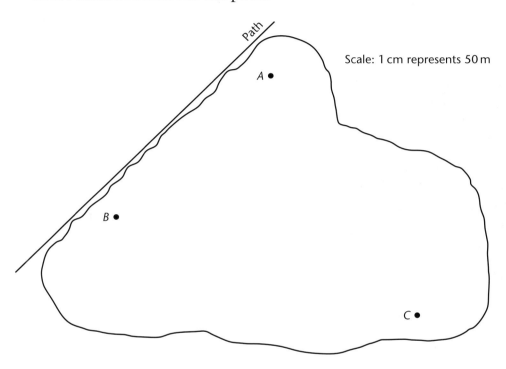

Scale: 1 cm represents 50 m

Trace the diagram. Using a ruler and compasses only, shade the region in which
the treasure lies.
You **must** show clearly all your construction arcs.   *(5 marks)*
                                                    AQA, Spec A, I, June 2003

Summary of key points

Constructions

Constructions must be drawn using *only* a straight edge (ruler) and pair of compasses.

Leave in the construction arcs as evidence that you have used the correct method.

Construction of a triangle given all three sides (grade E)

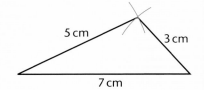

5 cm   3 cm

7 cm

Construction of perpendiculars (grade C)

Perpendicular bisector of a line segment

Perpendicular from a point to a line

Perpendicular from a point on a line (construction of a 90° angle)

   or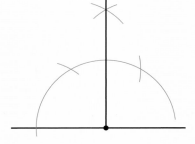

Construction of angles (grade C)

Construction of an angle of 60°

The bisector of an angle

Locus (grade C)

A locus is a set of points that obey a given rule.

The locus of points which are equidistant from two fixed points is the perpendicular bisector of the two fixed points.

The locus of points which are a fixed distance from a fixed point is a circle.

The locus of points which are equidistant from two fixed lines is the angle bisector of the two fixed lines.

The locus of points which are a fixed distance from a line segment is a 'racetrack' shape. The shape has two lines parallel to the line segment and two semi-circular ends.

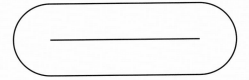

For locus questions:
- think about the points
- make a sketch
- construct the locus using standard constructions.

Some loci involve bearings and scale drawing (grade E/D)

You can measure bearings with a protractor.

This chapter will show you how to:
- ✔ write coordinates in all four quadrants
- ✔ find the mid-point of a line segment
- ✔ plot and draw straight-line graphs and draw quadratic curves
- ✔ work out coordinates of points of intersection when two graphs cross
- ✔ plot conversion graphs
- ✔ interpret and use distance–time graphs

18.1 Coordinates and line segments

Coordinates in all four quadrants

Coordinates describe the position of a point on a grid. In the diagram point A has coordinates (3, 4). The first value (3) gives the number of units left or right from the **origin** (O). The second value (4) gives the number of units up or down from the origin.

Key words:
coordinates
origin
x–y coordinate grid
quadrant

Values to the right and up from the origin are positive. Values to the left or down are negative.

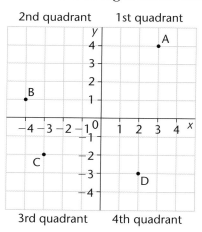

Coordinates are always given as (x, y).

The grid is called an **x–y coordinate grid**.

Left/right first ... x
Up/down second ... y

x comes before y in the alphabet.

The four quarters of the grid are called **quadrants**.

    $A(3, 4)$ is in the 1st quadrant.
    $B(-4, 1)$ is in the 2nd quadrant.
    $C(-3, -2)$ is in the 3rd quadrant.
    $D(2, -3)$ is in the 4th quadrant.

'Quad' means 4. Think of quad bikes, quadrilateral.

Example 1

Write down the coordinates of the points marked A, B, C, D and E on the x–y grid.

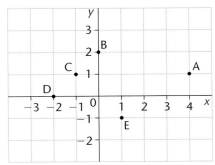

$A(4, 1)$ $B(0, 2)$ $C(-1, 1)$ $D(-2, 0)$ $E(1, -1)$

Write the x-value then the y-value in brackets, with a comma between.

Example 2

(a) Draw a coordinate grid with a horizontal axis (x-axis) from -4 to 4 and a vertical axis (y-axis) from -6 to 6.

(b) Plot the points on the grid:
$A(4, -3), B(-3, -3), C(-3, 1)$ and $D(4, 6)$.

(c) Join the points ABCD in order with straight lines. What have you drawn?

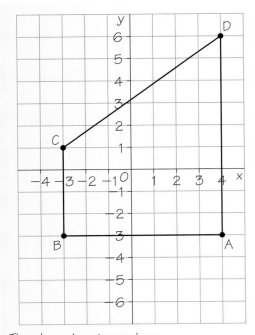

Use a ruler to join the points.

A trapezium is a four-sided shape with one pair of parallel sides.

The shape is a trapezium.

Coordinates in 3-D space

Just as a point on a plane (or flat surface) can be described by two coordinates, a point in 3-dimensional space can be described by three coordinates.

For example, the point, A, at one corner of the cuboid below, is at $(4, 3, 2)$.

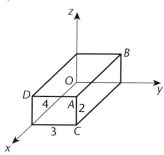

A is 4 units in the x-direction, 3 units in the y-direction and 2 units in the z-direction.

O is the origin and has the coordinates (0, 0, 0).

Example 3

Using the diagram of the cuboid above, write down the coordinates of the points B, C and D.

B is 0 units in the x-direction, 3 units in the y-direction and 2 units in the z-direction, so B is the point (0, 3, 2).

C is 4 units in the x-direction, 3 units in the y-direction and 0 units in the z-direction, so C is the point (4, 3, 0).

D is 4 units in the x-direction, 0 units in the y-direction and 2 units in the z-direction, so D is the point (4, 0, 2).

Finding the mid-point

> **Key words:**
> line segment
> mid-point

A straight line that joins two points is called a **line segment** .

A line continues forever. A line segment is a part of a line.

In Example 1, AB, BC, CD and DE are all line segments.

You calculate the **mid-point** of a line PQ by finding the means of the x- and y-values for the end points.

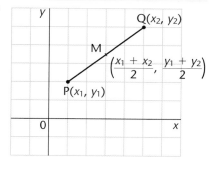

$$\text{Coordinates of mid-point } (x, y) = \left(\frac{x_1 + x_2}{2}, \frac{y_1 + y_2}{2} \right)$$

Example 4

Work out the coordinates of the mid-point of the line segment RS, where R has coordinates $(1, 7)$ and $S(4, 6)$.

$$(x, y) = \left(\frac{1+4}{2}, \frac{7+6}{2}\right)$$

$$= \left(\frac{5}{2}, \frac{13}{2}\right)$$

$$= (2\tfrac{1}{2}, 6\tfrac{1}{2})$$

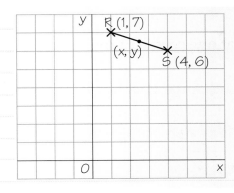

To find the mean of 2 values, add them and divide by 2.

Draw a sketch to check that your answer looks correct.

Example 5

Find the coordinates of the mid-point of the line joining $P(-5, 2)$ and $Q(-1, -6)$.

$$(x, y) = \left(\frac{-5 + -1}{2}, \frac{2 + -6}{2}\right)$$

$$= \left(\frac{-5 - 1}{2}, \frac{2 - 6}{2}\right)$$

$$= \left(\frac{-6}{2}, \frac{-4}{2}\right)$$

$$= (-3, -2)$$

You can use the same formula with coordinates that have negative values.

For rules for adding negative numbers, see Section 1.4.

Exercise 18A

1 Write down the coordinates of the points for P, Q, R, S and T.

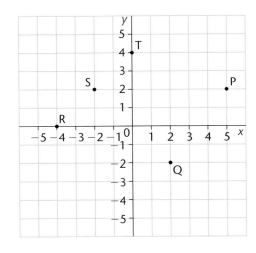

2 (a) On squared paper draw a coordinate grid with an x-axis from 0 to 10 and a y-axis from 0 to 10.

(b) Plot these points on the grid:

A(1, 4), B(4, 9), C(8, 9), D(8, 3) and E(5, 0).

Join them up in order.

What shape have you drawn?

3 For each line segment:

- write down the coordinates of the end points
- work out the coordinates of the mid-point.

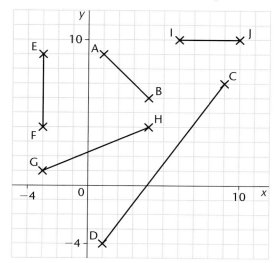

4

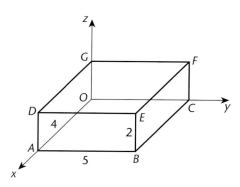

(a) Write down the coordinates of A, B, C, D, E, F and G.

(b) Write down the coordinates of the mid-points of
(i) OC **(ii)** CF **(iii)** BC **(iv)** BE.

(c) Write down the coordinates of the centre of the face
(i) OABC **(ii)** GDEF **(iii)** EBCF.

5 Without drawing these line segments, work out their mid-points.

   (a) *AB*: *A*(1, 1) and *B*(8, 1)

   (b) *CD*: *C*(7, 9) and *D*(7, 2)

   (b) *EF*: *E*(2, 3) and *F*(5, 9)

   (d) *GH*: *G*(−4, 5) and *H*(2, 5)

   (e) *IJ*: *I*(−2, 2) and *J*(3, −3)

18.2 Plotting straight-line graphs

Key words:
linear equation
linear graph

When points lie in a straight line on a grid, this means there is a connection between the *x*-value and the *y*-value of their coordinates.

> **In a straight-line graph there is a linear relationship between *x* and *y*.**

Plotting the points helps you to find the relationship between *x* and *y*.

You can describe this relationship using a **linear equation** or **linear graph** .

Lines parallel to the *x*-axis or *y*-axis

The graph shows four points in a straight line: *ABCD*.

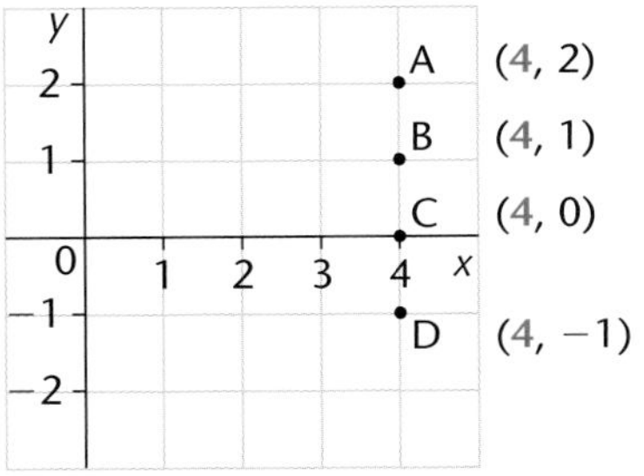

A	(4, 2)
B	(4, 1)
C	(4, 0)
D	(4, −1)

The points (4, 100) and (4, −100) also lie on this line. Any point with *x*-value 4 lies on the line.

Every point has exactly the same *x*-value. The relationship between this set of points is the equation of the line, $x = 4$.

> **A line parallel to the *y*-axis has equation $x = a$, where *a* is a number.**

The four points $E(2, -2)$, $F(1, -2)$, $G(0, -2)$, $H(-1, -2)$ all have the same y-value. The relationship is the equation of the line, $y = -2$.

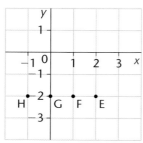

The equation of the x-axis is $y = 0$. The equation of the y-axis is $x = 0$.

A line parallel to the x-axis has equation $y = b$, where b is a number.

Example 6

Write down the equations of these lines.

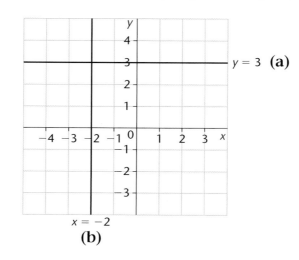

$y = 3$ **(a)**

$x = -2$
(b)

(a) All the points on the line have y-coordinate 3, so $y = 3$.

(b) All the points on the line have x-coordinate -2, so $x = -2$.

$(0, 3)$, $(-2, 3)$, $(7, 3)$,
$(-2, 4)$, $(-2, 0)$, $(-2, -1)$

Exercise 18B

1 Write down the equations of the lines

(**a**) parallel to the x-axis

(**b**) parallel to the y-axis.

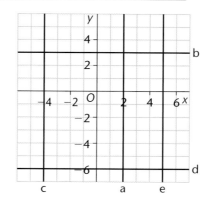

2 Draw a coordinate grid with both the *x*- and *y*-axes
from −5 to 5.
On your grid, draw and label the lines
(a) *x* = 4   (b) *y* = −1   (b) *x* = −5   (b) *y* = 4.

More straight-line graphs

For lines that are not parallel to the *x*-axis or the *y*-axis,
follow these steps to draw the graph:
1 Choose a minimum of three values for *x* (always
    include 0).
2 Draw a table of values and write in three *x*-values.
3 Substitute these values into the equation and work out
    the corresponding values for *y*.
4 Write the *y*-values in the table of values.
5 Draw a coordinate grid, making sure you draw it big
    enough so that all the points in your table will fit on it.
6 Plot the points from the table of values.
7 Draw a straight line through all the points.
8 Label the line with its equation.

> Pair up each *x*-value with its
> *y*-value to give coordinates
> (*x*, *y*).

Example 7

On a coordinate grid, using the same scale on both axes,
draw *x*- and *y*-axes between −5 and +5.

Draw the graph of $y = x + 1$ for values of *x* from −3 to +3.

x	−3	−2	−1	0	1	2	3
y	−2	−1	0	1	2	3	4

Substitute
x = 3 into
$y = x + 1$:
$y = 3 + 1 = 4.$

Plot the point (−3, −2).

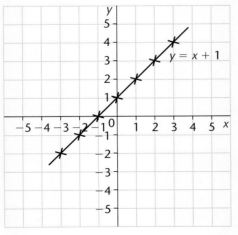

> Draw a table of values,
> choosing values of *x* from
> −3 to +3.
>
> Use the equation of the line
> to work out the values of *y*.
> Plot the *x*- and *y*-values as
> coordinate pairs on the
> grid. Join them with a
> straight line.
>
> Always extend your line
> just beyond the end points.
>
> Label your axes and label
> the line with the equation.
>
> You need a minimum of
> three points to draw a
> straight-line graph.
>
> Two points give the line −
> the third point is a 'check'.

Exercise 18C

1 In parts **(a)** to **(f)**, use the same scale on both axes and draw x- and y-axes between -5 and $+5$.

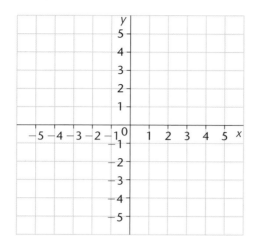

Draw these straight-line graphs. For each one:

- Make a table of values for at least three values of x (including 0).
- Substitute the x-values into the equation of the line to work out the corresponding y-values.
- Plot the points and draw a straight line through them.

    (a) $y = x - 3$ **(b)** $y = 2x + 3$

    (c) $y = 4 - x$ **(d)** $y = 3x - 1$

    (e) $y = 2 - 2x$ **(f)** $y = 1 - \frac{1}{2}x$

2 Using the same scale on both axes, draw x- and y-axes from -3 to $+3$.

Draw the graphs of $y = x$ and $y = -x$ on the same coordinate grid.

What do you notice about these graphs?

3 Using the same scale on both axes, draw x- and y-axes from -5 to $+5$.

Draw the graphs of $y = 2x - 1$ and $y = \frac{1}{2}x + 2$ on the same coordinate grid.

Write down the coordinates of the point where these graphs cross each other.

18.3 Equations of straight-line graphs

Calculating the gradient

Key words:
slope
steepness
gradient
coefficient

In Example 7 the **slope** of the line is from left to right in an upward direction. For every 1 unit moved to the right the line rises by 1 unit.

The ratio $\dfrac{\text{vertical distance}}{\text{horizontal distance}}$

gives a measure of the **steepness** of the slope of the line and is called the **gradient** of the line.

$$\textbf{Gradient} = \frac{\textbf{vertical distance}}{\textbf{horizontal distance}}$$

If the line slopes downward the gradient is negative.

In Example 7 the
gradient $= \dfrac{1}{1} = 1$

Positive gradient

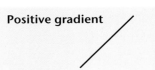

Negative gradient

Example 8

Find the gradient of the line joining the points
(a) $(1, 2)$ and $(5, 10)$ **(b)** $(-1, 6)$ and $(2, -3)$
(c) $(-4, -1)$ and $(0, -2)$.

Plot the points and join them with a line.

(a) Gradient $= \dfrac{\text{vertical distance}}{\text{horizontal distance}}$

$= \dfrac{8}{4}$

$= 2$

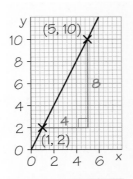

Draw a right-angled triangle as shown.

Use
gradient $=$

$\dfrac{\text{vertical distance}}{\text{horizontal distance}}$

Gradient 2 means for every 1 you go across, the line goes 2 up.

(b) Gradient $= \dfrac{\text{vertical}}{\text{horizontal}}$

$= -\dfrac{9}{3}$

$= -3$

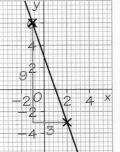

The line slopes downward, so the gradient is negative.

For every 1 you go across, the line goes down 3.

continued ▼

(c) Gradient = $\dfrac{\text{vertical}}{\text{horizontal}}$

$= -\dfrac{1}{4}$

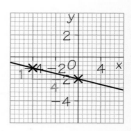

For every 1 you go across the line goes down $\frac{1}{4}$. Or, for every 4 you go across the line goes 1 down.

You can work out the gradient of a line from its graph.

Example 9

Find the gradient of these lines

(a) $y = 3x - 4$ **(b)** $y = -\frac{1}{2}x + 2$

(a) $y = 3x - 4$

x	0	1	2
y	−4	−1	2

Gradient is positive.

Gradient = $\dfrac{\text{vertical}}{\text{horizontal}}$

$= \dfrac{6}{2}$

$= 3$

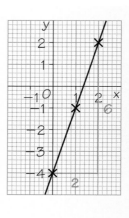

First draw the graph. Calculate and plot at least three (x, y) values.

To find the gradient choose *any* two points on the line and draw a right-angled triangle (as shown).

Remember, $3 = \frac{3}{1}$. For every 1 across, the line goes 3 up.

(b) $y = -\dfrac{1}{2}x + 2$

x	−2	0	2
y	3	2	1

Gradient is negative.

Gradient = $\dfrac{\text{vertical}}{\text{horizontal}}$

$= \dfrac{2}{4}$

$= -\dfrac{1}{2}$

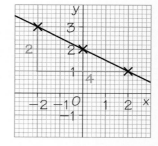

In Example 9:

Equation	Gradient
$y = 3x - 4$	3
$y = -\frac{1}{2}x + 2$	$-\frac{1}{2}$

This number in front of x is called the **coefficient** of x.

The gradient is the same as the number in front of the x in the equation of the line.

In an equation of the line in the form $y =$ ____ the coefficient of x is the gradient.

It must be $y =$ ____ not $2y$ or $3y$ or $-y$ or any other number of ys.

In Example 7 the equation of the line was $y = x + 1$ so the gradient was 1.

In algebra x means $1 \times x$.

Example 10

What are the gradients of the lines with these equations?

(a) $y = x + 8$ (b) $y = -3x + 3$

(c) $y = -\frac{1}{2}x - 5$ (d) $y = 2.6x + 0.4$

(a) gradient $= 1$

(b) gradient $= -3$

(c) gradient $= -\frac{1}{2}$

(b) gradient $= 2.6$

Gradients can be fractions and decimals.

$2.6 = \frac{2.6}{1}$. For every 1 across, the line goes 2.6 up.

If the equation of a line is *not* in the form $y =$ ____ you can rearrange the equation.

For rearranging equations see Chapter 7.

Example 11

Rearrange these equations into the form $y =$ ____.

(a) $y - x = 1$ (b) $-x = 1 - y$ (c) $y - x - 1 = 0$

(a) $y - x = 1$

$y - x + x = 1 + x$

$y = 1 + x$

$y = x + 1$

(b) $-x = 1 - y$

$-x + y = 1 - y + y$

$-x + y = 1$

$-x + x + y = 1 + x$

$y = 1 + x$

Do the same to both sides.

(c) $y - x - 1 = 0$

$y - x + x - 1 + 1 = 0 + x + 1$

$y = x + 1$

All three equations rearrange to $y = x + 1$. All have gradient 1.

Gradient of parallel lines

The graph shows the lines
$y = 2x$, $y = 2x + 2$ and
$y = 2x + 4$.

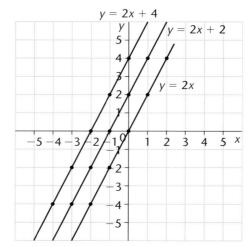

The graphs are parallel.
You can find the gradients from the equations.

Equation	Gradient
$y = 2x$	2
$y = 2x + 2$	2
$y = 2x + 4$	2

Straight lines that are parallel have the same gradient.

When the equations are in the form $y = \underline{\quad}$, lines are parallel if they have the same coefficient of x.

Exercise 18D

1 (a) Use the method of Example 9 to find the gradient of each of the straight lines you drew in Exercise 18C, question 1.

(b) For each, check that the gradient you calculate is the same as the coefficient of x in the equation of the line.

2 Write down the value of the gradient of each of these lines.

(a) $y = 3x + 2$ **(b)** $y = -x + 3$ **(c)** $y = 4x - 7$

(d) $y = -3x - 4$ **(e)** $y = \frac{2}{3}x + 8$ **(f)** $y = 0.8x - 0.3$

3 *Without plotting* these straight lines, identify the ones parallel to the line $y = x - 1$.

(a) $y = x + 1$ (b) $y = x - 7$

(c) $y = 2x - 1$ (d) $2y = x - 1$

(e) $y = x + \frac{1}{2}$ (f) $2y - 3 = 2x$

Remember: they need to be in the form $y = \underline{\quad}$.

4 *Without plotting* these straight lines, identify the ones parallel to the line $y = -3x - 2$.

(a) $y = 3x - 2$ (b) $y = -3x + 2$

(c) $y = -3x + 2$ (d) $y = -2 - 3x$

(e) $y = 2 + 3x$ (f) $x = -3y - 2$

The *y*-intercept

Another important feature of a straight line is where it crosses the *y*-axis. On the *y*-axis, the *x*-value is 0.

The point where a line crosses the *y*-axis is the **y-intercept** .

The graphs on page 471 all have gradient 2 but they cross the *y*-axis at different points.

Equation	*y*-intercept
$y = 2x$	0
$y = 2x + 2$	2
$y = 2x + 4$	4

When the equation of the line is $y = \underline{\quad}$ the *y*-intercept is the number term in the equation.

In the equation $y = x + 1$ the *y*-intercept is $+1$.

Straight lines that pass through the origin (0, 0) have *y*-intercept of 0 so their equations do not have a number term in them. For example, $y = 2x$.

Key words:
y-intercept

The graph crosses the *y*-axis at (0, 1). The *y*-intercept is 1.

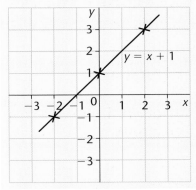

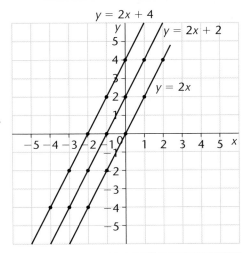

$y = 2x + 0 \rightarrow y = 2x$
$y = 2x$ is a straight line with gradient 2 passing through the origin.

Example 12

What are the y-intercept values of the lines with these
equations?

(a) $y = x + 8$ **(b)** $y = -3x + 3$

(c) $y = -\frac{1}{2}x - 5$ **(d)** $y = 2.6x + 0.4$

(e) $y = 7x$ **(f)** $y = 6 - 2x$

     (a) y-intercept = 8   **(b)** y-intercept = 3

     (c) y-intercept = −5   **(d)** y-intercept = 0.4

     (e) y-intercept = 0   **(f)** y-intercept = 6 ←

> Intercepts can be fractions
> and decimals.
>
> $y = ⑥ - 2x$
> The ↑ intercept is the
> number term.

If the equation of a line is *not* in the form $y = \_\_\_$
you can use algebra skills to rearrange the equation.

> See Chapter 7.

Example 13

Find the y-intercepts of these straight lines:

(a) $2y = x + 6$ **(b)** $3x - y = 5$ **(c)** $2x - 5y + 20 = 0$

     (a) $2y = x + 6$

$$\frac{2y}{2} = \frac{x}{2} + \frac{6}{2}$$

$$y = \tfrac{1}{2}x + 3$$

The y-intercept is +3

> All the equations need to
> be rearranged to the form
> $y = \_\_\_$.

     (b) $3x - y = 5$

$$3x - y + y = 5 + y$$

$$3x = 5 + y$$

$$3x - 5 = 5 - 5 + y$$

$$3x - 5 = y$$

The y-intercept is −5

> The y can be on the left of
> the equation or on the
> right.
> The y-intercept is the
> number term in the final
> equation.

     (c) $2x - 5y + 20 = 0$ or put $x = 0$:

$$2x - 5y + 5y + 20 = 0 + 5y \qquad 2 \times 0 - 5y + 20 = 0$$

$$2x + 20 = 5y \qquad\qquad -5y + 5y + 20 = 0 + 5y$$

$$\frac{2x}{5} + \frac{20}{5} = \frac{5y}{5} \qquad\qquad\qquad 20 = 5y$$

$$\frac{2x}{5} + 4 = y \qquad\qquad\qquad \frac{20}{5} = \frac{5y}{5}$$

$$\qquad\qquad\qquad\qquad\qquad 4 = y$$

The y-intercept is +4

> Remember the y-intercept
> is when $x = 0$.

Exercise 18E

1 What are the y-intercept values of the lines you drew in Exercise 18C, question 1?

2 Find the y-intercepts of these lines:

(a) $y = 3x + 2$ (b) $y = -x + 3$

(c) $y = 4x - 7$ (d) $y = -3x - 4$

(e) $y = \frac{2}{3}x + 8$ (f) $y = 0.8x - 0.3$

3 *Without plotting* the graphs of these lines, find their y-intercepts.

(a) $y = x + 1$ (b) $y + 7 = x$

(c) $y = x + \frac{1}{2}$ (d) $2y = x - 1$

(e) $3y = 2x - 1$ (f) $2y - 3 = 2x$

(g) $y + 2 = 3x$ (h) $3x + y = 2$

(i) $3y = 2 + 3x$ (j) $x = -3y - 2$

(k) $2y + 3x = 2$ (l) $4y + 2 + 3x = 0$

> You may need to rearrange some of the equations into the form $y = \underline{\quad}$.

The general equation of a straight line

When the equation of a straight line is in the form $y = \underline{\quad}$:

> It has to be $y = \underline{\quad}$ not $2y$ or $3y$ or $-y$ or any other number of ys.

- the gradient is the number in front of the x in the equation (the coefficient of x)
- the y-intercept is the number term in the equation.

You can rearrange an equation into the form $y = \underline{\quad}$ using algebra.

> **Equations of straight lines can always be written in the form**
>
> $$y = mx + c$$
>
> **where m represents the value of the gradient and c represents the value of the y-intercept.**

> This is an important result, you must remember it.

Example 14

Write the following equations in the form $y = mx + c$.

(a) $2y = 6x - 12$ **(b)** $5y = -10x + 20$ **(c)** $3x = y - 5$

(a) $2y = 6x - 12$

$y = \dfrac{6x}{2} - \dfrac{12}{2}$

$y = 3x - 6$

Rearrange each equation using algebra.

Divide both sides by 2.

gradient = 3
y-intercept = -6

(b) $5y = -10x + 20$

$y = -\dfrac{10x}{5} + \dfrac{20}{5}$

$y = -2x + 4$

Divide both sides by 5.

gradient = -2
y-intercept = $+4$

(c) $3x = y - 5$

$3x + 5 = y$

$y = 3x + 5$

Add 5 to both sides and rearrange the equation.

gradient = 3
y-intercept = $+5$

Exercise 18F

1 Write the following equations in the form $y = mx + c$. State the gradient and the y-intercept of each line.

Example 14 will help you.

(a) $3y = 9x + 18$ **(b)** $2y = -8x - 4$

(c) $8y = -24x + 8$ **(d)** $y - 7 = -2x$

(e) $2x + 6 = 2y$ **(f)** $-3x = 4 + y$

(g) $3x - y = 7$ **(h)** $2y - 10 = x$

(i) $15 - 5y = 10x$ **(j)** $6x + 2y - 7 = 0$

18.4 Curved graphs

Key words:
quadratic graphs

Graphs that have x^2 in their equation are curves.

They are called **quadratic graphs** .

All quadratic graphs are U-shaped. The U shape can be the right way up or upside down.

Example 15

Draw the graph of $y = x^2$ for values of x from -3 to $+3$.

x	−3	−2	−1	0	1	2	3
x^2	9	4	1	0	1	4	9
$y = x^2$	9	4	1	0	1	4	9

Draw a table of values. Include all whole-number values of x in the range.

Write the terms in the equation in the rows, so you can calculate one at a time.

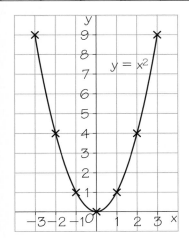

Plot the (x, y) points from your table of values and join them up with a *smooth* curve.

You will lose marks if you join the points with straight lines.

Example 16

Make a table of values and draw the graph of $y = x^2 - 4$ for values of x from -3 to $+3$.

x	−3	−2	−1	0	1	2	3
x^2	9	4	1	0	1	4	9
−4	−4	−4	−4	−4	−4	−4	−4
$y = x^2 - 4$	5	0	−3	−4	−3	0	5

x-values from -3 to $+3$.

Draw the table with a row for each term in the equation.
Add to get the y-values.

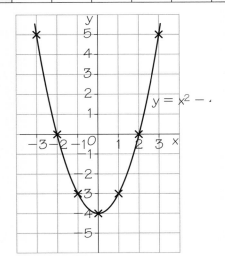

Look at the y-values in your table to see how far to draw the y-axis.

y-values are from -4 to $+5$.

Example 17

Complete the table of values and plot the graph of
$y = 2x^2 + 3$ for values of x from -2 to $+2$.

x	−2	−1	0	1	2
$2x^2$	8	2	0	2	8
+3	3	3	3	3	3
$y = 2x^2 + 3$	11	5	3	5	11

In an exam you will usually only have to calculate two of the *y*-values. The others will be given.

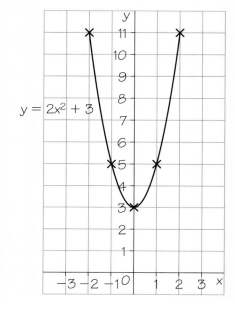

$2x^2 = 2 \times x^2$

Square first then multiply by 2.

In an exam the axes will be drawn for you.

Example 18

Complete the table of values and plot the graph of
$y = x^2 - 3x$ for values of x from -1 to $+4$.

x	−1	0	1	2	3	4
x^2	1	0	1	4	9	16
−3x	3	0	−3	−6	−9	−12
$y = x^2 - 3x$	4	0	−2	−2	0	4

For help with multiplying by a negative number, see Chapter 1.
Add to get the *y*-values.

y-values are from −2 to +4.

continued ▼

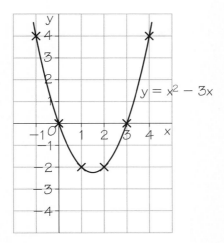

Exercise 18G

For each of these graphs:

- make a table of values
- draw the graph.

Use a scale of 1 square for 1 unit on each axis.

1 $y = x^2 + 2$ for values of x from -3 to $+3$

Use each value from -3 to $+3$ in your table.

2 $y = 2x^2 - 5$ for values of x from -2 to $+2$

3 $y = 3x^2 + 1$ for values of x from -2 to $+2$

4 $y = x^2 + x$ for values of x from -3 to $+2$

5 $y = x^2 - 5x$ I for values of x from 0 to $+5$

6 $y = x^2 + 3x$ for values of x from -4 to $+1$

7 $y = 2x^2 + 4x$ for values of x from -4 to $+2$

8 $y = 2x^2 - 6x$ for values of x from -1 to $+4$

Symmetry in quadratic graphs

The graphs in Examples 15, 16 and 17 are all symmetrical about the y-axis. The graph in Example 18 is symmetrical about the line $x = 1.5$

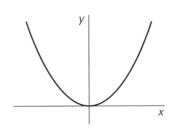

All quadratic graphs are symmetrical about a line parallel to the *y*-axis.

Quadratic graphs have x^2 in their equation.

A line parallel to the *y*-axis is a vertical line.

Example 19

Write a table of values and plot the graph of $y = x^2 + 2x - 3$ for values of *x* from -4 to $+2$.

State its line of symmetry.

x	−4	−3	−2	−1	0	1	2
x^2	16	9	4	1	0	1	4
+2x	−8	−6	−4	−2	0	2	4
−3	−3	−3	−3	−3	−3	−3	−3
$y = x^2 + 2x - 3$	5	0	−3	−4	−3	0	5

There are three terms in the equation. Give each a line of its own.

Add to find *y* for each *x*-value.

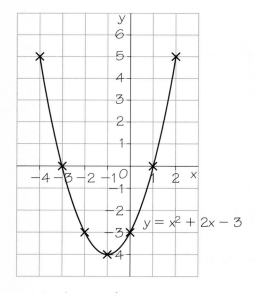

The line of symmetry is $x = -1$.

The line of symmetry passes through the lowest point on the curve $(-1, -4)$.
All points on this line have *x*-coordinate −1.

Example 20

Draw the graph of $y = 2x^2 - 6x + 7$ for values of *x* between -2 and 4.

What is the line of symmetry?

x	−2	−1	0	1	2	3	4
$2x^2$	8	2	0	2	8	18	32
$-6x$	12	6	0	−6	−12	−18	−24
$+7$	7	7	7	7	7	7	7
$y = 2x^2 - 6x + 7$	27	15	7	3	3	7	15

$2x^2 = 2 \times x^2$

$-6x = -6 \times x$

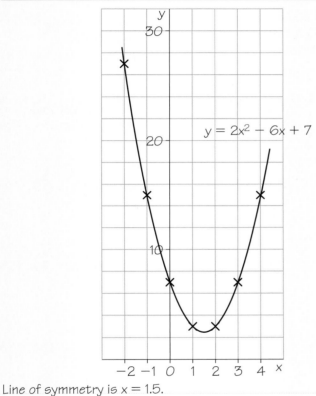

Line of symmetry is $x = 1.5$.

From the table, the y-values range rom 3 to 27. Draw the y-axis from 0 to 30, using the scale 1 square to 2 units.

Line of symmetry has x-value halfway between $x = 1$ and $x = 2$, so $x = 1.5$.

Exercise 18H

1 State the line of symmetry for each of the graphs you drew for Exercise 18G.

2 For each of these quadratic graphs:
- make a table of values
- draw the graph for the given range of values
- state the line of symmetry.

(a) $y = x^2 + x - 2$ for values of x from -3 to $+2$

(b) $y = x^2 + 4x - 1$ for values of x from -5 to $+1$

(c) $y = x^2 - 2x - 5$ for values of x from -2 to $+4$

Choose a sensible scale for each axis.

All the points on the graph must fit your squared paper.

(d) $y = x^2 + 3x + 1$ for values of x from -5 to $+2$
(e) $y = x^2 - 5x - 4$ for values of x from -1 to $+6$
(f) $y = 2x^2 - 2x + 3$ for values of x from -3 to $+4$
(g) $y = 2x^2 + 4x - 5$ for values of x from -4 to $+2$
(h) $y = 2x^2 - 4x - 9$ for values of x from -2 to $+4$

18.5 Graphs that intersect

Where do straight line graphs meet?

You can use graphs to help you find solutions to equations.

For example, to solve the equation $x + 1 = 0$ you can use the graph of $y = x + 1$.

$x + 1 = 0$ means that $y = 0$, so look to see where the line crosses the x-axis.

Key words:
point of intersection

On the x-axis $y = 0$.

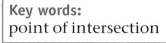

The line crosses the x-axis at $(-1, 0)$.

The solution to the equation $x + 1 = 0$ is $x = -1$.

You can use this graph to solve any equation of the form $x + 1 = \ldots$

You could also find this by rearranging using algebra.

For example to solve $x + 1 = 3$, find the **point of intersection** of the graph $x + 1 = 3$ with the line $y = 3$.

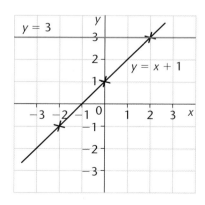

The point of intersection is the point where two graphs cross.

The solution to $x + 1 = 3$ is $x = 2$.

Check: $2 + 1 = 3$ ✓

Example 21

(a) Plot the graph of $y = 3x - 2$.

(b) It intersects the line $y = 3$ at a point P. What are the coordinates of P?

(c) What is the solution of the equation $3x - 2 = 3$?

(a)

x	−2	0	2
3x	−6	0	6
−2	−2	−2	−2
y	−8	−2	4

Make a table of values. Use x-values of −2, 0 and 2.

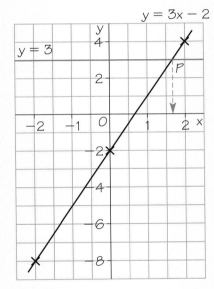

Plot the graph.

Draw in the line $y = 3$.

Draw your graph as accurately as possible.

In this example the point of intersection is not a whole number.

(b) The two straight lines intersect where $x = 1.7$ and $y = 3$,

i.e. at the point (1.7, 3).

(c) Using the graph, the solution of the equation $3x - 2 = 3$ is

$x = 1.7$.

The exact coordinates are $(1\frac{2}{3}, 3)$ but you are only expected to give your answer to an accuracy of 'half a small square'.

You can find the point of intersection of any two straight lines.

Example 22

Draw the graphs of the straight lines $y = 3x - 2$ and $y = x + 4$ on the same coordinate grid.

What are the coordinates of the point where these lines intersect?

$y = 3x - 2$

x	0	1	2
3x	0	3	6
−2	−2	−2	−2
y	−2	1	4

$y = x + 4$

x	0	1	2
x	0	1	2
+4	4	4	4
y	4	5	6

Draw a table of values for each graph.

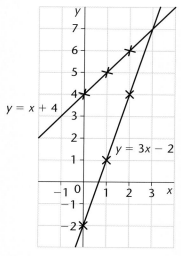

Extend the graphs so that the lines cross.
Make sure the y-axis extends far enough.

The graphs intersect at the point (3, 7).

Exercise 18I

1 Draw the graph $y = 2x + 5$. It intersects the line $y = 3$ at a point P.
What are the coordinates of P?

2 The line $y = 7$ intersects another line $y = 4x + 9$ at a point Q.
What are the coordinates of Q.
Where does the line $y = 4x + 9$ cross the x-axis?

Draw the graph.

3 For each part **(a)**, **(b)** and **(c)**, draw the two straight line graphs on the same coordinate grid.
Write down the coordinates of the point of intersection of the lines.
(a) $y = 2x + 1$ and $y = 10 - x$
(b) $y = 9 - 3x$ and $y = 2x - 1$
(c) $y = x - 2$ and $y = 3x + 1$

Solving quadratic equations graphically

Key words:
quadratic equations

A **quadratic equation** has an x^2 term. You can solve quadratic equations using quadratic graphs.

When a quadratic graph crosses the x-axis it does so in two places.

For the curve $y = x^2 - 4$ the graph crosses the x-axis at $x = -2$ and $x = 2$.

This is where $y = 0$.

So $x = -2$ and $x = 2$ are the solutions to the equation $x^2 - 4 = 0$.

$x^2 - 4 = 0$ is a quadratic equation because it has an x^2 term.

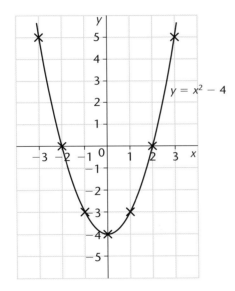

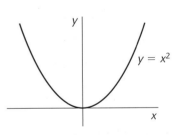

The graph of $y = x^2$ just touches the x-axis at the point $(0, 0)$, so the equation $x^2 = 0$ has only one solution, $x = 0$.

See Example 15.

Some quadratic graphs do not cross the x-axis at all. $y = 2x^2 + 3$ is one of these. This means there are no solutions to the equation $2x^2 + 3 = 0$.

A quadratic graph can intersect a line parallel to the x-axis.

This graph just touches the line $y = 3$ at the point $(0, 3)$. So the equation $2x^2 + 3 = 3$ has one solution, $x = 0$.

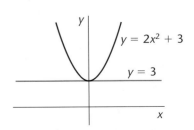

See Example 17.

Example 23

(a) Draw a pair of axes with the x-axis from -3 to $+3$ and the y-axis from -5 to $+10$. Draw the graph of $y = x^2 - 2$.

(b) What are the solutions to the equation $x^2 - 2 = 0$?

(c) Draw the straight line $y = 3$. Write down the coordinates of both points where the curve and line intersect.

(d) Write down the equation that is solved by the x-values in part **(c)**.

(a)

x	−3	−2	−1	0	1	2	3
x^2	9	4	1	0	1	4	9
−2	−2	−2	−2	−2	−2	−2	−2
$y = x^2 - 2$	7	2	−1	−2	−1	2	7

Draw a table of values for the graph.

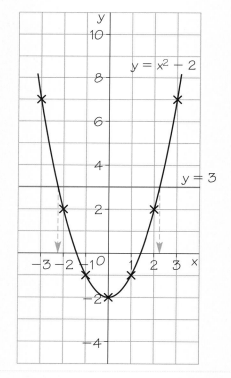

The solutions to $x^2 - 2 = 0$ are where the curve crosses the line $y = 0$ (x-axis).

Draw the line $y = 3$ on the graph. The two dashed lines show the x-values of the points of intersection.

(b) The curve crosses the x-axis ($y = 0$) at points $x = -1.4$ and $x = 1.4$.

The solutions to $x^2 - 2 = 0$ are $x = -1.4$ and $x = 1.4$.

continued ▼

(c) The coordinates of the points of intersection are $(-2.3, 3)$ and $(2.3, 3)$.

(d) The equation is $x^2 - 2 = 3$ which can be simplified to $x^2 = 5$.

You are solving
  'curve' = 'line'
in other words, $x^2 - 2 = 3$.

Example 24

(a) Using the graph of $y = x^2 + 2x - 3$, draw the graph of the straight line $y = 1$ on the same axes.

(b) Write down the x-values of the points where these graphs intersect.

(c) What equation is solved by these x-values?

(d) Use your graph to solve the equation $x^2 + 2x - 3 = 0$.

$y = x^2 + 2x - 3$ was plotted in Example 19.

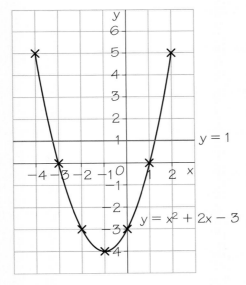

This is the graph from Example 19 with the line $y = 1$ added.

You need to draw the graph accurately to get accurate answers.

(a) The x-values of the points of intersection are $x = -3.2$ and $x = 1.2$.

(b) The equation solved is $x^2 + 2x - 3 = 1$ which simplifies to $x^2 + 2x - 4 = 0$.

(c) The solutions are $x = -3$ and $x = 1$.

You are solving
  'curve' = 'line'
in other words,
$x^2 + 2x - 3 = 1$.

To solve $x^2 + 2x - 3 = 0$ look to see where the curve crosses the line $y = 0$ (the x-axis).

Example 25

(a) Using the graph of $y = 2x^2 - 6x + 7$, draw the graph of the straight line $y = 12$ on the same axes.

From Example 20.

(b) What equation is solved by the x-values of the points of intersection?

(c) Use the graphs to find these x-values.

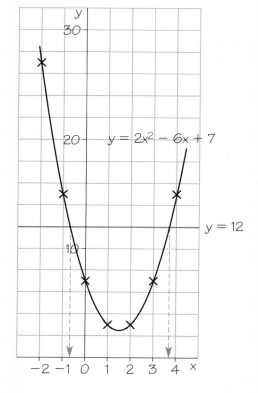

This is the graph from Example 20 with the line $y = 12$ added.

(b) The equation solved by the intersection is

$$2x^2 - 6x + 7 = 12$$

$$2x^2 - 6x - 5 = 0$$

(c) The solutions to $2x^2 - 6x - 5 = 0$ are the x-values of the points of intersection:

$$x = -0.8 \text{ and } x = 3.6$$

You are solving
  'curve' = 'line'
in other words,
$2x^2 - 6x + 7 = 12$.

Subtract 12 from both sides of the equation.

Draw dashed lines from the points of intersection to the x-axis. Read off the x-values.

You can solve a quadratic equation by seeing where its graph intersects a horizontal line.

Using $y = x^2 + 2x - 3$:

- solving $x^2 + 2x - 3 = 0$ means looking to see where the curve crosses the x-axis ($y = 0$).
  The solutions are $x = -3$ and $x = 1$.

You need to learn how to solve quadratic equations using graphs. This is always tested in the GCSE exam.

- Solving $x^2 + 2x - 3 = 1$ means looking to see where the curve crosses the line $y = 1$.
  The solutions are $x = -3.2$ and $x = 1.2$.

For the graph of $x^2 + 2x - 3$ see Example 24.

Exercise 18J

1 **(a)** Draw the graph of $y = x^2 + 4x + 2$ for values of x from -5 to $+1$.

Use a scale of 1 square for 1 unit on each axis.

   (b) On the same axes draw the graph of the line $y = 5$.

   (c) Write down the x-values of the points of intersection of the curve and the line $y = 5$.

   (d) What equation is solved by these x-values?

   (e) Use your graph to solve the equation $x^2 + 4x + 2 = 0$.

2 The graphs you drew in Exercise 18H question 2 are listed below.
For each, use the graph to solve the quadratic equations written alongside them.
Give your answers correct to 1 d.p.

Use the graphs you drew in Exercise 18H.

Graph	Solve these equations	
(a) $y = x^2 + x - 2$	$x^2 + x - 2 = 0$	$x^2 + x - 2 = 2$
(b) $y = x^2 + 4x - 1$	$x^2 + 4x - 1 = 0$	$x^2 + 4x - 1 = -3$
(c) $y = x^2 - 2x - 5$	$x^2 - 2x - 5 = 0$	$x^2 - 2x - 5 = -4$
(d) $y = x^2 + 3x + 1$	$x^2 + 3x + 1 = 0$	$x^2 + 3x + 1 = 8$
(e) $y = x^2 - 5x - 4$	$x^2 - 5x - 4 = 0$	$x^2 - 5x - 4 = -6$
(f) $y = 2x^2 - 2x + 3$	$2x^2 - 2x + 3 = 5$	$2x^2 - 2x + 3 = 20$
(g) $y = 2x^2 + 4x - 5$	$2x^2 + 4x - 5 = 0$	$2x^2 + 4x - 5 = 5$
(h) $y = 2x^2 - 4x - 9$	$2x^2 - 4x - 9 = 0$	$2x^2 - 4x - 9 = -8$

18.6 Conversion graphs

Key words:
conversion graph

You can use a **conversion graph** to convert one type of measurement into another, usually with different units.

Conversion graphs can be linear or curved.

Example 26

The table below shows the approximate conversion between degrees Celsius (°C) and degrees Fahrenheit (°F) using the equation $F = 2C + 30$.

> We used to measure temperature in °F, but now we use °C.

(a) Complete the table.

(b) Draw a coordinate grid with C-values (x-axis) from −40 to 100 and F-values (y-axis) from −50 to 250. Plot the points in the table and draw a straight line.

C	−40	−20	0	20	40	60	80	100
2C	−80		0		80		160	
+30	30		30		30		30	
F	−50		30		110		190	

(c) What is the temperature in °F when it is 10°C?

(d) If the temperature is 160°F, what is this in °C?

(a)

C	−40	−20	0	20	40	60	80	100
2C	−80	−40	0	40	80	120	160	200
+30	30	30	30	30	30	30	30	30
F	−50	−10	30	70	110	150	190	230

(b)

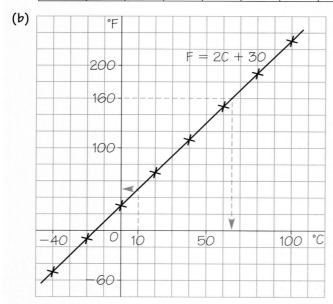

(c) 10°C = 50°F

(d) 160°F = 65°C

> Draw dashed lines on the graph at 10°C and 160°F to meet the conversion line. From these points draw another dashed line to meet the axes and read off the answers.

Exercise 18K

1 **(a)** Copy and complete this conversion table between km and miles using the conversion 5 miles = 8 km.

miles (x)	0	5	10	20	50	100
km (y)		8			80	

(b) Draw a conversion graph with x-values from 0 to 100 miles and y-values from 0 to 200 km.

(c) How many miles are equivalent to 75 km?

(d) How many km are equivalent to 75 miles?

2 **(a)** Copy and complete this conversion table between centimetres (to 1 d.p.) and inches.

cm	0				12.5			30
inches	0	1	2	3	5	6	10	12

5 inches = 12.5 cm
So 1 inch = ?

(b) Draw a conversion graph for cm and inches.

(c) How many cm are equivalent to 4 inches?

(d) How many inches are equivalent to 20 cm.

3 A stone is dropped down a well. The following table gives its distance from the top of the well after each second, using the approximate formula $d = 5t^2$.

Time in seconds (t)	0	1	2	3
Distance in m (d)	0	5	20	45

$d = 5t^2$ means
$d = 5 \times t^2$

(a) Draw a coordinate grid with x-values between 0 and 4 and y-values between 0 and 50. Plot the points and join them up with a smooth curve.

(b) How far did the stone drop in 1.5 seconds?

(c) How long did it take the stone to fall 35 m?

(d) If the well is 60 m deep, for how many seconds did the stone fall?

18.7 Distance–time graphs

Key words:
distance–time graph
speed
average speed

A **distance–time graph** shows information about a journey. You always plot 'time' on the horizontal axis (*x*-axis) and 'distance' on the vertical axis (*y*-axis).

You can use a distance–time graph to work out the **speed** for part of a journey using the relationship

$$\text{speed} = \frac{\text{distance}}{\text{time}}$$

If distance is measured in metres (m) and time in seconds then speed will be in m/s. Other units for speed are km/h and mph.

m/s can also be written ms^{-1}
km/h can be written kmh^{-1}
mph means miles per hour.

You can use a distance–time graph to help you work out an **average speed** for a journey using the relationship

Average speed is the same as the 'mean' speed.

$$\text{average speed} = \frac{\text{total distance}}{\text{total time}}$$

Total distance includes the return journey.

Example 27

The distance–time graph shows a railway journey from *A* to *D*.

(a) For how long did the train stop at station *B*?

(b) For how long did it stop at station *C*?

(c) What was the speed of the train between stations *A* and *B*?

(d) What was the average speed of the train over the whole journey from *A* to *D*?

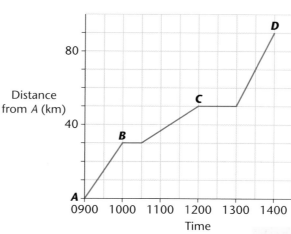

Distance from *A* (km)

Time

From the graph you can work out the times at each station and the distance travelled between each of the stations.

(a) The train arrived at station B at 1000 and departed at 1030. It stopped at B for 30 minutes.

(b) The train arrived at station C at 1200 and departed at 1300. It stopped at C for 1 hour.

(c) From A to B

distance = 30 km

time = 1 hr

$$\text{speed} = \frac{\text{distance}}{\text{Time}} = \frac{30}{1} = 30 \text{ km/h}$$

Read the times off the graph.

Distance in km
Time in hours
Speed in km/h

(d) Total distance from A to D = 90 km

Total time from A to D = 5 hours

$$\text{Average speed} = \frac{\text{total distance}}{\text{total time}}$$

$$= \frac{90}{5} = 18 \text{ km/h}$$

Total distance and total time for the journey (ignore the stops).

In Example 27 the train was travelling *away* from A. On the return journey from D to A the distance of the train from A is decreasing. The graph looks like this:

Graphs that represent an outward journey and a return journey always look like this.

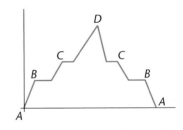

Example 28

Lisa drove to town in her car, a distance of 10 km in 15 minutes. She spent 45 minutes in town and then set off home again.

On the way home she stopped to get some petrol. The garage was 6 km from town and it took her 10 minutes to drive there.

Her stop for petrol took 5 minutes. She drove the rest of her journey home in 10 minutes.

(a) Draw a distance–time graph for Lisa's journey.

(b) What was her average speed on her journey into town?

Write all the distances and times in a table.

Lisa's journey		
	Distance	**Time**
To town	10 km	15 mins
In Town	0 km	45 mins
To garage	6 km	10 mins
At garage	0 km	5 mins
To home	4 km	10 mins

Total time
= 15 + 45 + 10 + 5 + 10
= 85 mins

Largest distance from home = 10 km

(a)

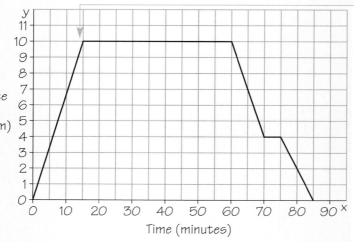

15 minutes, 10 km

Distance from home (km)

Time (minutes)

The long horizontal line is when she was in town. The short horizontal line is when she stopped for petrol.

The garage is 6 km *from town* which means 4 km from home.

(b) Average speed for journey into town

$$= \frac{\text{total distance}}{\text{total time}}$$

$$= \frac{10}{15} \text{ km per min}$$

$$= \frac{10}{15} \times 60 \text{ km/h}$$

$$= 40 \text{ km/h}$$

Distance in km and time in minutes will give speed in km per minute.

There are 60 minutes in 1 hr.
To convert km per min to km/h you multiply by 60.

Exercise 18L

1 Each morning Chloe walks to school and stops at the shop on the way. The distance–time graph shows her journey to school.

  (a) What time does she get to the shop?

  (b) How long is she in the shop?

  (c) How long does it take Chloe to walk to school?

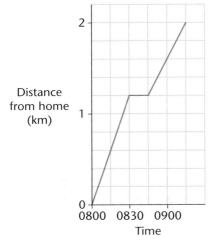

2 The distance–time graph shows Amina's cycle journey from home.
She stopped twice for breaks before returning home.

  (a) How long did Amina's cycle ride take?

  (b) On which part of her journey was she travelling at the greatest speed? How can you tell?

  (c) Calculate her average speed for the whole cycle ride.

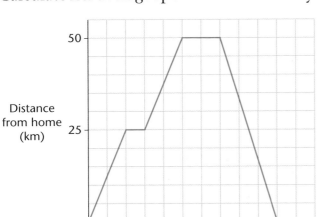

3 Dean walked to the village shop, a distance of 800 m. It took him 10 minutes to get there and he was in the shop for 5 minutes.
He then walked to the Post Office, another 200 m down the street. This took 3 minutes and he was in the Post Office for 6 minutes.
He then walked back home which took 16 minutes.
Draw a distance–time graph for his journey.

   (a) How fast did he walk to the shop?

   (b) What was his average speed for the whole journey?

> Before you start to draw the distance–time graph, write the times and the distances for each part, including the stops, in a table. Then you can see the total time and the total distance involved.
>
> Look at Example 28 if you need help.
>
> This will help you decide what scales to use for the axes.

4 Marcus is going to school. He leaves home at 0815 and walks 400 m to the bus stop in 6 minutes. He waits 5 minutes for the bus to arrive.
The bus journey is 4000 m and the bus arrives at the bus stop near school at 0838.
Marcus then walks another 100 m to school in 2 minutes.

   (a) Draw a distance–time graph for Marcus's journey.

   (b) What was the average speed of the bus journey? Give your answer in metres per minute then convert it to km/h.

> Use a sensible scale for the axes.
>
> Not the *whole* journey.

5 Alison went to town in her car. She drove the 12 miles into town in 25 minutes. She was in town for $1\frac{1}{4}$ hours. Then she left to drive home.
After 20 minutes of her journey home, 9 miles from town, she dropped off some library books at her aunt's house and stayed for a cup of tea for $\frac{1}{2}$ hour.
She then continued her journey home, taking another 10 minutes to get there.

   (a) Draw a distance–time graph for her journey.

   (b) What was her average speed on the way to town? Give your answer in mph.

Examination style questions

1 The parallelogram *ABCD* is drawn on a cm square grid.

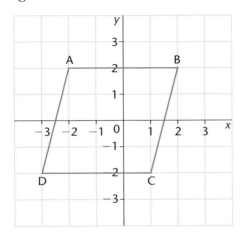

The coordinate of *A* is $(-2, 2)$.
Write down the coordinates of *B*, *C* and *D*

(2 marks)

AQA, Spec A, F, June 2003

2 Copy and complete the table of values for $y = 2x - 1$.

x	−1	0	1	2	3
y	−3		1		5

On a grid with *x*-values from -1 to $+3$ and *y*-values from -3 to $+5$
draw the graph of $y = 2x - 1$.

(3 marks)

AQA, Spec A, I, June 2003

3 Using *x*-values from -5 to $+5$ and *y*-values from -10 to $+12$ draw
the lines $y = -4$ and $y = 2x + 1$.
Write down the coordinates of the point where the lines $y = -4$
and $y = 2x + 1$ cross.

(5 marks)

AQA, Spec A, I, November 2003

4 Wayne cycles from Newcastle to Ashington, a distance of 20 miles. The diagram shows the distance–time graph of his journey.

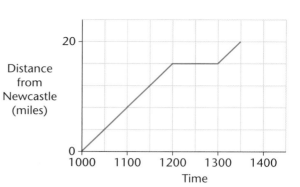

(a) Describe what is happening between 1200 and 1300.

(b) How far does Wayne travel in the first 2 hours of his journey?

(c) What is Wayne's average speed over the first 2 hours of his journey?

(d) What time does Wayne arrive in Ashington?

(5 marks)

AQA, Spec A, F, June 2003

5 Copy and complete the table of values for $y = 2x^2 - 4x - 1$.

x	-2	-1	0	1	2	3
y	15		-1			5

(a) Using a grid with y-values from -5 to $+15$, draw the graph of $y = 2x^2 - 4x - 1$ for values of x from -2 to $+3$.

(b) An approximate solution of the equation $2x^2 - 4x - 1 = 0$ is $x = 2.2$.
   (i) Explain how you can find this from the graph.
   (ii) Use the graph to write down another solution of this equation.

(6 marks)

AQA, Spec B, 5I, June 2004

Summary of key points

Plotting straight-line graphs (grades G to C)

Coordinates are always given as (x, y).

A line segment is a straight line joining any two points.

A straight-line graph is one where there is a linear relationship between x and y.

A line parallel to the y-axis has equation $x = a$, where a is a number. The x-axis has equation $y = 0$.

A line parallel to the x-axis has equation $y = b$, where b is a number. The y-axis has equation $x = 0$.

You can work out the gradient using:

$$\text{gradient} = \frac{\text{vertical distance}}{\text{horizontal distance}}$$

Straight lines that are parallel have the same gradient.

Equations of straight lines can always be written in the form $y = mx + c$, where m is the gradient and c is the y-intercept.

Mid-points of line segments (grade C)

$$\text{Coordinates of mid-point } (x, y) = \left(\frac{x_1 + x_2}{2}, \frac{y_1 + y_2}{2} \right)$$

where (x_1, y_1) and (x_2, y_2) are the coordinates of the end points.

Curved graphs (grade D/C)

Graphs that have x^2 in their equation are curves. They are called quadratic graphs.

All quadratic graphs are symmetrical about a line parallel to the y-axis.

You can use graphs to solve equations. The points of intersection of the graphs are the solutions to the equations.

Conversion graphs (grades F to E)

You can use a conversion graph to convert one type of measurement into another type of measurement, usually with different units, e.g. °F and °C. Conversion graphs can be linear or curved.

Distance–time graphs (grades E to C)

A distance–time graph shows information about a journey. You can use a distance–time graph to work out:

$$\text{average speed} = \frac{\text{total distance}}{\text{total time}}$$

19 Transformations

This chapter will show you how to:

✔ recognise and use the four types of transformation –
reflection, rotation, translation and enlargement
✔ find the centre and scale factor of an enlargement
✔ identify congruent shapes
✔ identify similar shapes
✔ use the congruence properties to identify congruent
triangles

19.1 Types of transformation

A **transformation** changes the position and/or the size
of an object. The transformed shape is called the **image**.

You need to know about four types of transformation:

- **reflection**
- **rotation**
- **translation**
- **enlargement**

Reflection, rotation and translation only change the
position of an object. The size and shape of the image and
the object are identical – they are **congruent**.

Enlargement changes the position of an object and its
size. The object and image shapes are **similar**.

> **Key words:**
> transformation
> image
> reflection
> rotation
> translation
> enlargement
> congruent
> similar

> 'Congruent' means
> identical.

> They are the same shape,
> but different sizes.

19.2 Reflection

> **Key words:**
> reflection
> mirror line

For a **reflection** you need a **mirror line**. You reflect
the object in the mirror line to produce an image. This
image is exactly the same size and shape as the object.

**When you reflect a shape in a mirror line, the
object and image are congruent.**

Points on the images are the same distance behind the mirror line as the corresponding points on the object are in front of the mirror line.

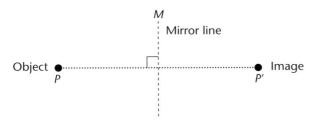

If you join a point *P* on the object to its corresponding point *P'* on the image, the line *PP'* crosses the mirror line *M* at right angles and

$$PM = P'M$$

P' is the reflection of *P*. You say '*P* dash'.

Every point on an object reflects to an image point behind the mirror line.

Example 1

The object *ABCD* is reflected in the mirror line shown. Draw the image of the object and label it *A'B'C'D'*.

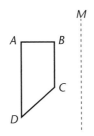

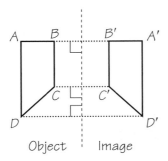

You could trace the object and the mirror line. Turn over the tracing paper and line up the mirror lines to see the reflection.

Reflect each corner (vertex) in the mirror line, then join them up. Draw a line from *C* at right angles to the mirror line. Continue this line to a point the same distance the other side and label it *C'*. Repeat this for points *A*, *B* and *D*.
Join the points in order with a straight line.

The object shape is labelled *A*, *B*, *C* and *D* in a clockwise direction. After reflection the image shape is labelled *A'*, *B'*, *C'* and *D'* in an anticlockwise direction.

Sometimes the mirror line passes *through* the object.
The reflections go in both directions.

Example 2

The triangle *ABC* is reflected in the mirror line shown.
Draw the reflection of this triangle and label it *A′B′C′*.

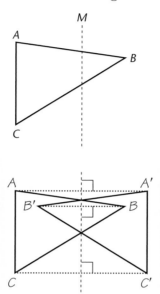

Start by finding the image
of *C* and *A*. For *B* the
reflection is from the right
side to the left. Lines *AB*
and *A′B′* cross exactly on
the mirror line. So do *BC*
and *B′C′*. This is always true
for lines that cross the
mirror line.

Mirror lines can be vertical, horizontal or diagonal.
You use the same method to draw the reflection

Example 3

Reflect the trapezium *ABCD* in the
diagonal mirror line.
Label the image *A′B′C′D′*.

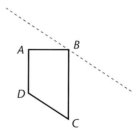

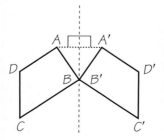

For diagonal mirror lines,
rotate the page until the
mirror line is vertical. Then
use the same method as
before.

Point *B* lies on the mirror,
so the image point *B′* also
lies on the mirror line. Any
point on the mirror line
reflects into itself.

When you reflect an object on squared paper and in a diagonal mirror line, you can count squares diagonally to find the position of an image point.

Do this for each point separately, then join up the points to produce the final image.

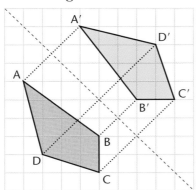

From A, count $1\frac{1}{2}$ squares diagonally to the mirror line. Count another $1\frac{1}{2}$ and mark the point A′. For B, count 1 square diagonally to the mirror line, then 1 more to B′. Do the same for C and D.

Sometimes you are given the object and image and asked to draw the mirror line.

> **To describe a reflection fully you need to give the equation of the mirror line.**

Example 4

The diagram shows an object *P* and its image *Q* after reflection.

(a) Draw in the mirror line as a dashed line.

(b) What is the equation of this mirror line?

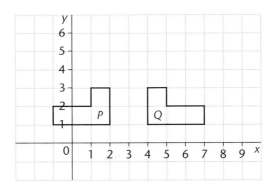

(a)

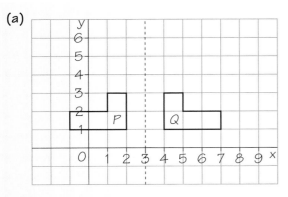

The mirror line must be the same distance from *P* as from *Q*.

The mirror line is parallel to the *y*-axis. The *x*-values of all points on it are 3. So the equation is $x = 3$. See Chapter 18 on equations of straight lines if you need help.

(b) The equation of the mirror line is $x = 3$.

Exercise 19A

You will need squared paper for each question.

1 Each diagram shows an object with its coloured image. Copy these diagrams onto squared paper and draw in the mirror line in each case.

(a)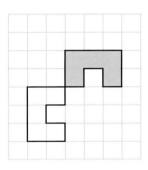

(b)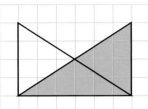

(c)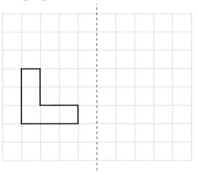

(d)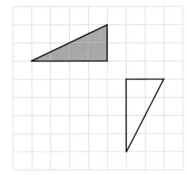

2 Copy these shapes and the dashed mirror line onto squared paper. Draw the reflected image in each case.

(a)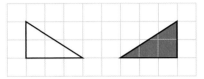

(b)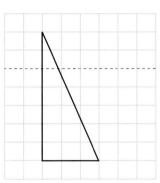

(c)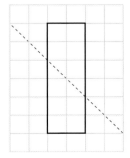

3 Copy the axes and triangle *PQR* onto squared paper.

(a) Reflect the triangle in the *y*-axis and label the image *P'Q'R'*.

> Use the *y*-axis as the mirror line.

(b) Reflect the triangle in the *x*-axis and label it *P"Q"R"*.

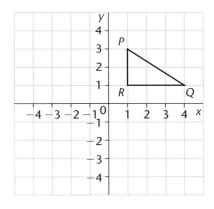

4 Copy the axes and the shape onto squared paper. Reflect the shape in the mirror line given by the equation $x = 1$.

> The line $x = 1$ is shown by a dashed line.

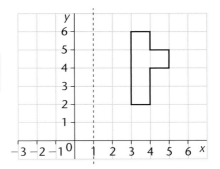

5 This five-sided polygon is to be reflected in the mirror line with equation $y = 1$.

(a) Copy the axes and the polygon onto squared paper.

(b) Draw in the mirror line as a dashed line.

(c) Draw the image after the reflection.

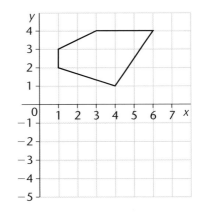

6 (a) Reflect triangle *P* in the line $y = x$. Call this *P'*.

(b) Reflect triangle *P'* in the line $y = x$. Call this *P"*.

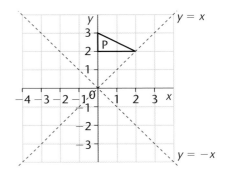

19.3 Rotation

Key words:
rotation
centre of rotation
quarter turn
half turn
three-quarter turn

A **rotation** turns an object either clockwise or anticlockwise through a given angle about a fixed point.

Always look carefully to see whether the rotation is clockwise or anticlockwise.

The fixed point is called the **centre of rotation**. It can be either inside or outside the object.

An object can be rotated through any angle (in degrees). Common rotations are **quarter turn** (90°), **half turn** (180°) and **three-quarter turn** (270°).

Example 5

Draw the image of this shape after it has been rotated through 90° anticlockwise about the centre of rotation at
(a) A   (b) B   (c) C.

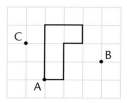

Tracing paper is very useful in all questions on rotation.

The image after each rotation is shown shaded.

Rotation about A
(a)

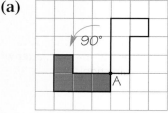

Rotation about B
(b)

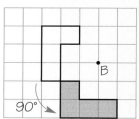

Trace the shape and the centre of rotation on to tracing paper. Hold the centre of rotation fixed with your pencil point. Turn the tracing paper through the required angle.

Rotation about C
(c)

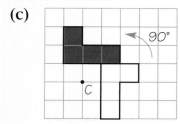

Draw the image you see on your diagram.

The image formed after rotation is the same size and shape as the original object.

An object and its image after rotation are congruent.

Example 6

Draw the image of the shape after rotation about $P(1, -1)$:

(a) a quarter of a turn anticlockwise

(b) rotation through 180° clockwise.

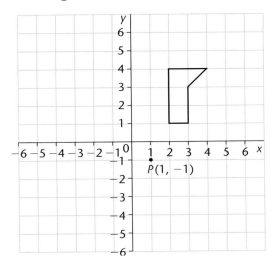

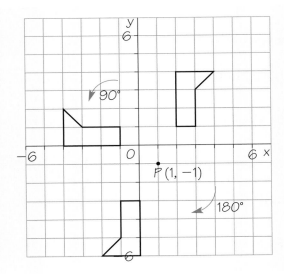

Use tracing paper to help you.

An image formed by a rotation through 180° clockwise is in the same position as an image formed by a rotation through 180° anticlockwise.
They are both 'half turns' and it doesn't matter which direction you turn.

To describe a rotation fully you need to give:

- **the centre of rotation**

- **the angle of turn**

- **the direction of turn**

Usually 90° or 180°.

Clockwise or anticlockwise.

If you want to describe fully the
rotation that takes shape *P* on to
shape *Q*, the easiest way is to use
tracing paper.

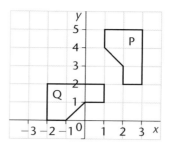

Place the tracing paper over the
whole of the diagram and draw
shape *P* on it.

Remember to ask for some
tracing paper when you
attempt any transformation
question on the exam
paper. It is particularly
useful for rotation questions.

The longest side of the diagram in
shape *P* is vertical. The corresponding side in shape *Q* is
horizontal. So the angle of rotation must be 90°
(or 270° if you rotate in the opposite direction).

Look for clues like this.

90° vertical

horizontal

1 Try a point somewhere near *P* and *Q*. Put your pencil
on the point and hold it fixed.

2 Turn through 90° (or 270°) and see if shape *P* lands
exactly on top of shape *Q*.

3 If it does not, try another centre of rotation. Keep
trying different centres of rotation until shape *P* fits
exactly on top of shape *Q*.

Once you have drawn the
diagram, you can try
several different centres of
rotation quite quickly.

4 Describe the rotation giving,
- the angle
- the direction
- the centre of rotation (as coordinates).

With experience, you will
find it a lot easier to identify
the correct centre of
rotation.

The rotation that takes *P* to *Q* is 90° anticlockwise about
point (2, 1).

Exercise 19B

1 Copy these shapes onto squared paper.

(a)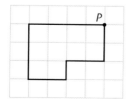

(b)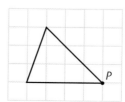

(c)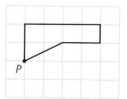

Rotate each shape
(a) a half turn clockwise.
(b) a quarter turn anticlockwise about the point *P*.

2 Copy this shape onto squared paper.

Draw the image of the shape after it has been rotated about the point P:

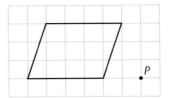

(a) 90° clockwise

(b) 180° anticlockwise

(c) three-quarters of a turn clockwise.

3 On squared paper draw x- and y-axes going from -6 to $+6$. Copy this shape onto your axes.

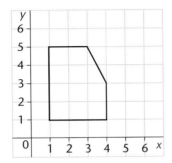

Draw the image of the shape after:

(a) a quarter turn clockwise about the origin $(0, 0)$. Label the image A.

(b) a quarter turn anticlockwise about the origin $(0, 0)$. Label the image B.

(c) 180° rotation anticlockwise about the origin $(0, 0)$. Label this image C.

4 Describe fully the transformation which maps shape P onto shape Q.

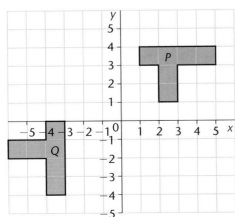

You need to give three pieces of information:
- the amount of turn,
- the direction of turn,
- the centre of rotation.

Use tracing paper to help you.

19.4 Translation

Key words:
translation
column vector

A **translation** slides a shape from one position to another.

In a translation every point on the shape moves the same distance in the same direction.

To describe a translation you need to give the distance and the direction of the movement.

Example 7

Translate this shape 4 squares to the right and 2 squares up.

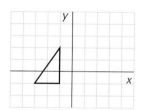

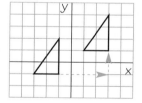

Choose any corner (vertex) of the shape and move this point 4 squares to the right and 2 squares up. Repeat this for the other vertices. Join up the vertices with straight lines.

An object and its image after a translation are congruent.

You can describe a translation by a **column vector** .

The translation in Example 7 has column vector $\begin{pmatrix} 4 \\ 2 \end{pmatrix}$.

Column vectors always have tall brackets round them.

The top number represents the movement in the *x*-direction and the bottom number represents the movement in the *y*-direction.

They are not fractions so do not draw a line between the numbers.

Notice that the translation to take the triangle back to its original position is $\begin{pmatrix} -4 \\ -2 \end{pmatrix}$.

Movements right and up are positive but left and down are negative.

Example 8

Translate this shape by the vector $\begin{pmatrix} -3 \\ -3 \end{pmatrix}$.

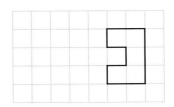

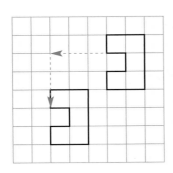

$\begin{pmatrix} -3 \\ -3 \end{pmatrix}$ means 3 squares *left* and 3 squares *down*.

Exercise 19C

1 Copy these shapes onto squared paper and translate them by the amounts shown.

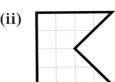 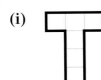

(i) **(ii)**

(a) $\begin{pmatrix} 3 \\ -2 \end{pmatrix}$ **(b)** $\begin{pmatrix} -4 \\ 2 \end{pmatrix}$ **(c)** $\begin{pmatrix} -5 \\ 0 \end{pmatrix}$ **(d)** $\begin{pmatrix} -1 \\ -6 \end{pmatrix}$ **(e)** $\begin{pmatrix} 0 \\ 3 \end{pmatrix}$

2 Copy this shape onto squared paper.
A translation of the shape moves the point P to the point P' on the image. Draw the complete image. What is the column vector that describes the translation?

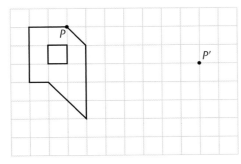

3 The triangle *A* is translated to new positions at *B*, *C*, *D* and *E*.
Describe each transformation by giving the column vector.

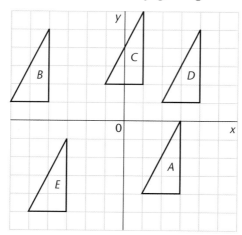

19.5 Enlargement

Key words:
enlargement
scale factor
multiplier
proportion
similar

An **enlargement** changes the size of an object but not its shape.

The number of times the shape is enlarged is called the **scale factor** or **multiplier** . This can be a positive whole number or a fraction.

> **In an enlargement, all the angles stay the same but all the lengths are changed in the same proportion . The image is similar to the object.**

For help with proportionality see Chapter 12 .

Similar shapes:
- have equal angles
- have lengths in the same proportion
- have perimeters in the same proportion.

Example 9

Enlarge the rectangle *ABCD* by a scale factor 2.

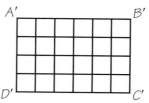

Every length on the object is multiplied by 2 (doubled) on the image.

$AB = 3$, so
$A'B' = 2 \times 3 = 6$
$BC = 2$, so
$B'C' = 2 \times 2 = 4$

Example 10

What is the scale factor of the enlargement
that takes shape *A* to shape *B*?

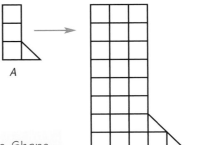

Shape *A* is only 1 square wide. Shape *B* is 3 squares wide. Shape
A is 3 squares long. Shape *B* is 9 squares long.

Shape *B* is 3 times longer and 3 times wider than shape *A*.

The scale factor is 3.

Compare the lengths of
corresponding sides.

The final position of an enlargement is determined by the
position of the centre of enlargement. In Examples 9 and
10 there is no centre of enlargement so the image can be
drawn anywhere.

If no centre of enlargement
is given you can then draw
it close to the original
shape.

When you enlarge from a centre of enlargement, the
distances from the centre to each point are multiplied by
the scale factor.

Example 11

Copy the triangle *ABC*. Enlarge the
triangle by scale factor 2 using the
point *O* as the centre of enlargement.

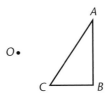

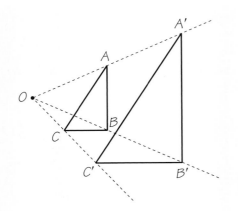

Multiply the distance *OA* by
2 to get *OA′*, and similarly
for the other vertices.
OA′ = **2** × *OA*
OB′ = **2** × *OB*
OC′ = **2** × *OC*

Always draw your diagram
as accurately as possible
using a pencil and ruler.
Leave the construction lines
on your diagram.

The centre of enlargement can be a point on the shape.

Example 12

Enlarge the triangle by scale factor 3
using point *P* as the centre of enlargement.

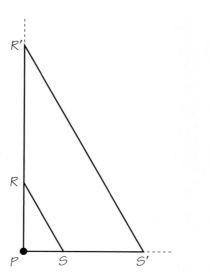

For an enlargement by a
scale factor **3**
$PR' = 3 \times PR$
$PS' = 3 \times PS$
The enlargement overlaps
the original shape.

**To describe an enlargement fully you need to give
the scale factor and the centre of enlargement.**

Exercise 19D

1 For each of the following shapes work out the scale
factor of the enlargement.

(a)

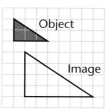

(b)

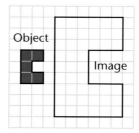

Use Example 10 to help
you.

(c)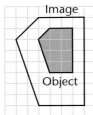

(d)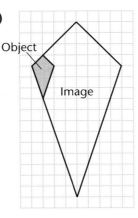

2 Copy each of the following shapes onto squared paper.
Enlarge each one by scale factor 2.

(a)   **(b)**   **(c)**

No centre of enlargement is given, so draw the enlargement close to the original shape.

3 Copy the following shapes onto squared paper.
Enlarge each one by scale factor 2 from
the centre of enlargement *C*.

(a) 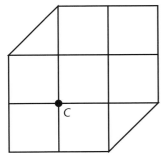   **(b)**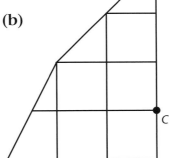

4 The vertices of triangle *K* are (1, 1), (1, 2)
and (4, 1).

Enlarge the triangle *K* by scale factor 3
with (0, 0) as the centre of enlargement.

What are the coordinates of the
image triangle *K′*?

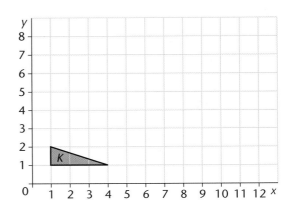

5 This right-angled
triangle is to be enlarged
by scale factor 4.
The centre of
enlargement is at
a point $P(1, 2)$.
Copy the diagram and
draw the enlargement.

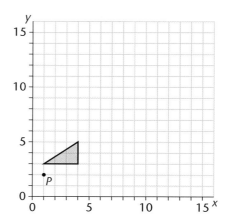

Use Example 11 to help
you.

Fractional scale factors

**A fractional scale factor of enlargement (<1) makes
the image smaller.**

Here, $ABCD$ has been
enlarged about centre O,
by scale factor $\frac{1}{3}$.

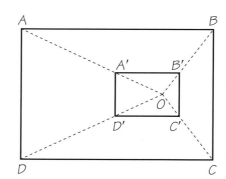

$OA' = \frac{1}{3}OA$
$OB' = \frac{1}{3}OB$
$OC' = \frac{1}{3}OC$
$OD' = \frac{1}{3}OD$

The enlargement is smaller
than the original.

To find the position of the centre of enlargement, join
vertices in the enlargement to the corresponding vertices
in the original and continue these lines until they meet at
a point.
This point is the centre of enlargement.

You draw in the
construction lines for the
enlargement.

Example 13

Triangle ABC has been enlarged
to produce the shaded triangle $A'B'C'$.

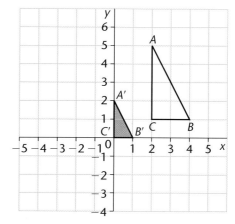

(a) What is the scale factor of the enlargement from ABC to $A'B'C'$?

(b) Mark on the grid the position of the centre of enlargement P, and give its coordinates.

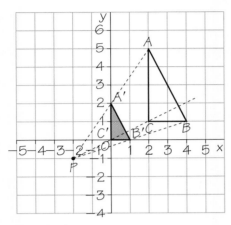

This enlargement has made a smaller image so the scale factor must be less than 1.

All lengths on triangle $A'B'C'$ are half the lengths on triangle ABC. So the scale factor is $\frac{1}{2}$.

Draw in the construction lines joining A to A', B to B' etc. These lines meet at point $P(-2, -1)$, the centre of enlargement.

(a) The scale factor is $\frac{1}{2}$.

(b) The centre of enlargement is at $P(-2, -1)$.

- A scale factor of 1 produces an image the same size as the object. The object and image are congruent.
- A scale factor > 1 produces an image larger than the object and the two shapes are similar.
- A scale factor < 1 produces an image that is smaller than the object and the two shapes are similar.

Exercise 19E

1 The rectangle X is enlarged to produce the image Y.

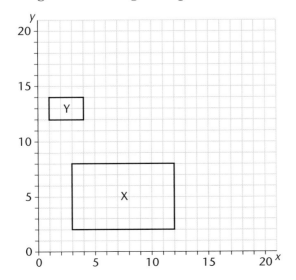

Use Example 13 to help you.

(a) What is the scale factor of this enlargement?

(b) Construct lines to show the position of the centre of enlargement.

(c) What are the coordinates of the centre of enlargement?

2

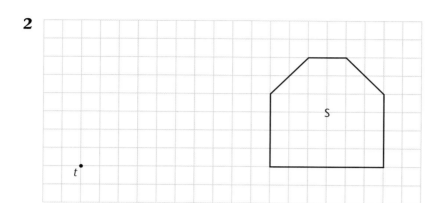

Copy the diagram and draw the image of shape S after an enlargement with scale factor $\frac{1}{2}$ and centre of enlargement at point t.

3

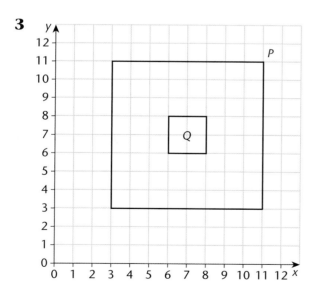

(a) Find the coordinates of the centre of enlargement when shape P has been enlarged to shape Q.

(b) What is the scale factor of this enlargement?

19.6 Congruence and similarity

Key words:
congruent
similar

Congruent shapes are exactly the same shape and exactly the same size.

All corresponding lengths and angles in the object and image are equal.

> You may need to turn shapes over before they fit exactly. Reflections in a mirror line are congruent.

Similar shapes have exactly the same shape but are not the same size.

All corresponding angles in the object and image are equal. All lengths are in the same ratio or proportion. The ratio is the same as the scale factor of the enlargement.

All circles are mathematically similar.
All squares are mathematically similar.

Example 14

(a) Which shapes are congruent to shape *A*?
(b) Which shapes are similar to shape *A*?

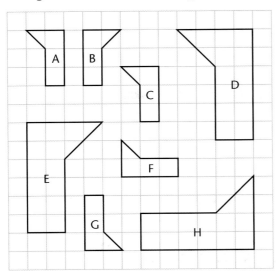

(a) *B, C, F* and *G* are all congruent to *A*.

(b) *D, E* and *H* are all similar to *A*.

> *D, E* and *H* are all enlargements, scale factor 2, of *A*.

Example 15

Which of these shapes are similar to *B*?
Explain your answer.

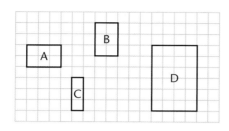

Shape A: shapes A and B are both of length 3 units and width
    2 units so they are identical.

Shape C: length the same as B
    width not the same as B   C and B are not similar.

Shape D: length of D = 2 × length of B
    width of D = 2 × width of B   D and B are similar.

A and B are not similar but they are congruent.

D is an enlargement of B, scale factor 2.

There are four different rules you can use to prove that
two triangles are congruent.

You need to remember these rules.

- **Three sides equal (Side, Side, Side – SSS)**

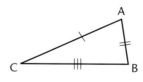

When two triangles have 3 equal sides they *must* be congruent.

- **Two sides equal and the included angle the same
  (Side, Angle, Side – SAS)**

This case is easy to spot.

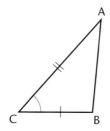
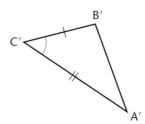

- Two angles the same and a corresponding side is equal
  (Angle, Side, Angle – ASA or Side, Angle, Angle – SAA)

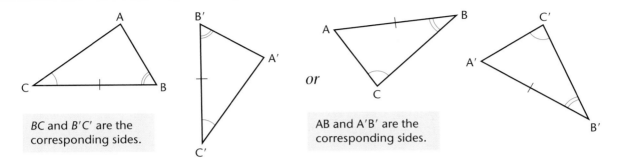

BC and B′C′ are the
corresponding sides.

or

AB and A′B′ are the
corresponding sides.

Corresponding side means 'the side in the same position in both triangles'.

- A right angle, a hypotenuse and a corresponding side
  are equal (Right angle, Hypotenuse, Side – RHS)

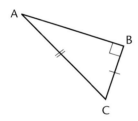

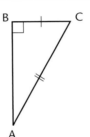

The hypotenuse is the side
opposite the right angle.

BC and B′C′ are the corresponding sides.
They are the shortest sides in each triangle.

Example 16

State whether any two of these triangles are congruent and
give your reasons.

 A  B  C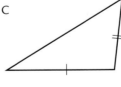

A and B are congruent, as they have 2 pairs of equal angles, and
corresponding sides (the side between the angles) equal.
Triangle C cannot be congruent with either A or B, as C has an
obtuse angle and A and B each have 3 acute angles.

There is one case where triangles are sometimes thought to be congruent, but they are not. Look at these diagrams:

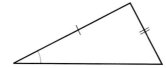

These triangles have two sides which are the same length and an angle which is the same size.

You might think that **SSA** is a reason for congruence but one look at these triangles should convince you that it is not!

Exercise 19F

1 Look at the shapes in this diagram.

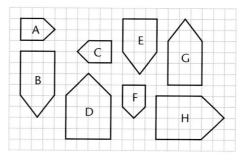

   (a) Which shapes are congruent to shape *A*?

   (b) Which shapes are similar to shape *A*?

2 Copy this shape onto squared paper. On the same squared paper draw one shape that is similar and one shape that is congruent to this shape.

> In similar shapes, all sides are enlarged by the same scale factor.

3 For each set of shapes, write down the letters of the shapes that are similar to each other.

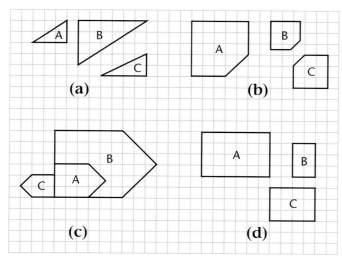

4 Look at the following pairs of triangles. State which pairs are congruent and give the appropriate reason.

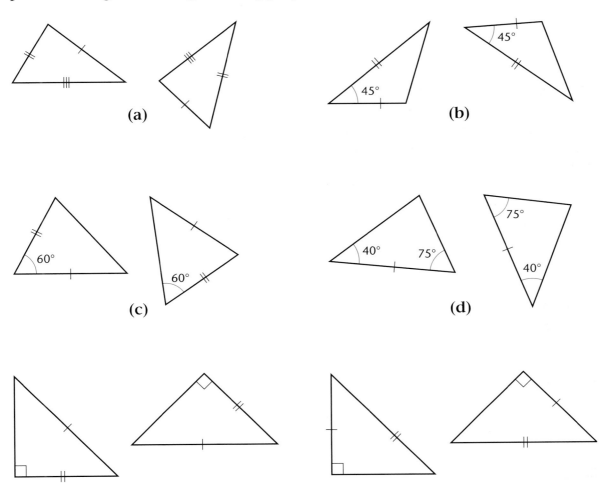

(a)

(b)

(c)

(d)

(e)

(f)

Examination style questions

1 Copy and reflect this shape using the dotted line as the mirror line.

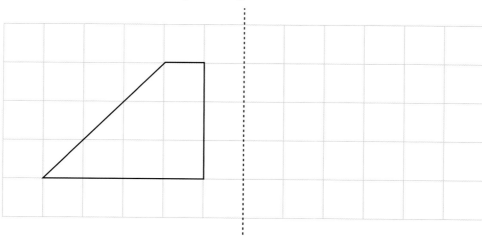

(2 marks)
AQA, Spec A, F, June 2003

2 Triangle *P* is an enlargement of triangle Q.

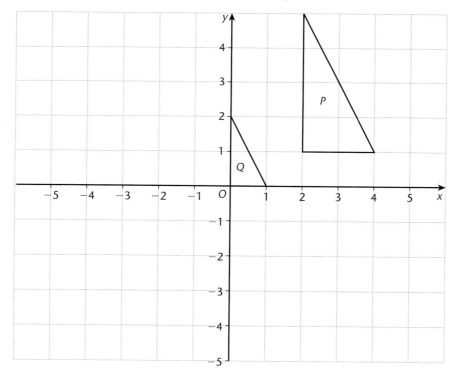

(a) What is the scale factor of the enlargement?

(b) What are the coordinates of the centre of enlargement?

(c) How many times bigger is the area of *P* than the area of Q? *(4 marks)*

AQA, Spec A, I, June 2003

3 Triangle *A* is drawn on the grid. Copy the grid and

  (a) reflect triangle *A* along the line $y = -1$.
    Label the triangle *B*.

  (b) rotate triangle *A* a quarter of a turn clockwise
    about the origin *O*. Label the triangle *C*.

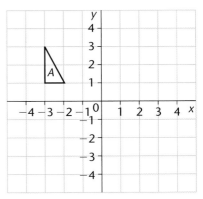

(5 marks)
AQA, Spec A, F, June 2004

4 Here are six rectangles on a grid.

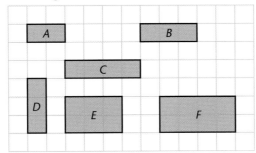

  (a) Which two rectangles are congruent?
  (b) Which two rectangles are similar?

(2 marks)
AQA, Spec B, 5F, June 2003

5 The grid shows several transformations of
the shaded triangle.

  (a) Write down the letter of the triangle
    (i) after the shaded triangle is reflected
      in the line $x = 3$
    (ii) after the shaded triangle is translated
      by 3 squares to the right and
      5 squares down
    (iii) after the shaded triangle is rotated
      90° clockwise about *O*.

  (b) Describe fully the single transformation
    which takes triangle *F* onto triangle *G*.

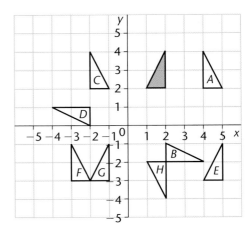

(5 marks)
AQA, Spec A, I, November 2003

6 The diagram shows a right-angled triangle A.

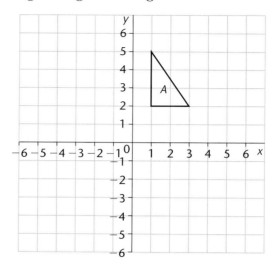

Copy the diagram and draw the new position of triangle A after a rotation of 90° clockwise about the point $(0, 0)$.

(3 marks)

AQA, Spec A, I, November 2003

Summary of key points

Reflection (grades G to C)

When you reflect a shape in a mirror line, the object and image are congruent.

If you join a point P on the object to its corresponding point P' on the image, the line PP' crosses the mirror line M at right angles and $PM = P'M'$.

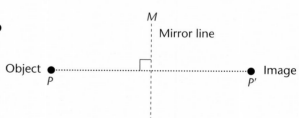

To describe a reflection fully you need to give the equation of the mirror line.

Rotation (grade D/C)

A rotation turns an object either clockwise or anticlockwise through a given angle about a fixed point – the centre of rotation.

An object can be turned through any angle. Common rotations are 90° (quarter turn), 180° (half turn) and 270° (three-quarter turn).

An object and its image after rotation are congruent.

To describe a rotation fully you need to give:
- the centre of rotation
- the angle of turn
- the direction of turn.

Translation (grade D/C)

A translation slides a shape from one position to another.

An object and its image after a translation are congruent.

To describe a translation you need to give the distance and the direction of movement.

You can describe a translation by a **column vector**. The top number represents the movement in the x-direction and the bottom number represents the movement in the y-direction.

Enlargement (grades F to C)

An enlargement changes the size of an object but not its shape. The number of times each length of a shape is enlarged is called the scale factor.

In an enlargement, all the angles stay the same but all lengths are changed in the same proportion. The image is similar to the object.

To describe an enlargement fully you need to give the scale factor and the centre of enlargement.

Congruency and similarity (grade G)

Congruent shapes are identical in size and shape. Reflection, rotation and translation transformations all produce images that are congruent to the object.

Similar shapes have exactly the same shape but are not the same size. Enlargement produces objects and images that are similar.

All squares are mathematically similar. All circles are mathematically similar.

Congruent triangles (grade C)

There are four different rules you can use to prove that two triangles are congruent:

- SSS (side, side, side)
- SAS (side, angle, side)
- ASA (angle, side, angle) and
- RHS (right angle, hypotenuse, side).

20 Averages

This chapter will show you how to:

✔ find the 'averages' – mean, median and mode for discrete data
✔ find the range and understand what is meant by the spread of the data
✔ find the mean, median and mode from frequency distributions for both discrete and continuous data
✔ compare data and offer sensible interpretations

20.1 Averages and spread

You will need to know:

● that discrete data is data that can be counted
● how to order numbers

Key words:
average
mean
mode
median
range
spread
sigma

'On average girls are better at mathematics than boys.'
'The average cost of a car, new or used, in the UK is £6500.'
'Joe is of average height.'
'Bill and Ali were pleased when they saw their exam results because they thought that they had both done better than average.'

In these statements 'average' describes something that typically happens.

In mathematics, an **average** is a value that represents a set of data.

For a set of discrete data you can find three 'averages' – the **mean**, the **mode** and the **median**. The **range** describes how the data values are **spread**.

The mean, mode, median and range describe the main features of the data. You can use these measures to compare data sets. For some data sets, one of these measures is more suitable than the others.

The mean

This list shows the amounts of pocket money received each week by nine friends.

Amy	£12	Beth	£7.20	Amina	£8.60
Debbie	£7.50	Emmie	£10	Fay	£7.50
Siobhan	£13.40	Hannah	£10	Imogen	£7.50

The mean of a set of data is the sum of all the values divided by the number of values.

The mean is often called the average value.

Mean =

$$\frac{12 + 7.20 + 8.60 + 7.50 + 10 + 7.50 + 13.40 + 10 + 7.50}{9}$$

Add up all the values. Divide by the number of values.

$$= \frac{83.70}{9}$$

$$= £9.30$$

This is the amount each person would have if the total amount was shared equally between them.

The mode

The mode of a set of data is the value that occurs most often.

Mode = £7.50

The median

The median of a set of data is the middle value when the data is arranged in order of size.

In ascending order the amounts are:

£7.20, £7.50, £7.50, £7.50, £8.60, £10, £10, £12, £13.40

First write the data in order.

The median is £8.60.

£8.60 is the middle value in the list.

There is an *odd* number of data values, so there is *only one middle value*. Notice that there are the same number of items either side of the median.

If you have an *even* number of data values then there are *two middle values* and the median is the number mid-way between them.

The data set:
1 5 7 9 10 12 14
15 15 17
has 10 values. The two middle values are 10 and 12. The median is
$$\frac{10 + 12}{2} = \frac{22}{2} = 11$$

The range

> **The range or spread of a set of data is the difference between the highest and the lowest values.**

Range = £13.40 − £7.20 = £6.20

Exercise 20A

1 Find the mean, mode, median and range for these sets of data.

(a) 4   4   7   10   3   7   6   6   3   1   4

(b) 34   62   24   80   76   61   80   54   51

(c) 27   21   15   16   31   19   32   23

(d) 0.5   0.25   0.75   0.25   1.25   1.5   2.5   0.25

Special notation

If there are n items of data in a list you could number them as $x_1, x_2, x_3, \ldots, x_n$

The special symbol Σ (pronounced **sigma**) means '*the sum of*'.

So, Σx means 'the sum of all the x values'.

or $\Sigma x = x_1 + x_2 + x_3 + \cdots + x_n$

n is the number of items of data in the set

Using this notation

$$\text{mean} = \frac{x_1 + x_2 + x_3 + \cdots + x_n}{n}$$

$$= \frac{\Sigma x}{n}$$

You usually write

$$\bar{x} = \frac{\Sigma x}{n}, \text{ where } \bar{x} \text{ represents the mean.}$$

Example 1

Find the mean, median, mode and range for this set of values.

17   6   11   12   12   16   2   2   7   12   8   9

$$\text{mean} = \frac{\Sigma x}{n}$$

$$= \frac{17 + 6 + 11 + 12 + 12 + 16 + 2 + 2 + 7 + 12 + 8 + 9}{12}$$

$$\bar{x} = \frac{114}{12} = 9.5$$

Writing the data in ascending order:

2, 2, 6, 7, 8, ⑨, ⑪, 12, 12, 12, 16, 17

$$\text{Median} = \frac{9 + 11}{2} = 10$$

Mode = 12

Range = 17 − 2 = 15

$\Sigma x = 17 + 6 + 11 + ...$
(the sum of all the data values)
$n = 12$ the number of pieces of data.

$n = 12$ (even) so there are two middle values. The median is halfway between them.

The value 12 appears three times in the list.

The difference between the largest and the smallest.

Example 2

(a) Work out the mean of these amounts of money:

£10   £15   £12   £9   £14   £12

When one more amount is added the new mean value is £11.

(b) What amount of money is added?

(a) $\bar{x} = \dfrac{\Sigma x}{n} = \dfrac{10 + 15 + 12 + 9 + 14 + 12}{6} = \dfrac{72}{6}$

mean = £12

(b) New mean value = £11

There are now 7 pieces of data.

So $11 = \dfrac{\Sigma x}{7}$

Rearranging: $\Sigma x = 7 \times 11 = 77$

Amount added = sum of 7 values − sum of 6 values

$= 77 − 72$

Amount added = £5

$11 = \dfrac{\Sigma x}{7}$

$7 \times 11 = 7 \times \dfrac{\Sigma x}{7}$

$7 \times 11 = \Sigma x$

From **(a)**, the sum of 6 values is 72.

Exercise 20B

1 Find the mean, median, mode and range of these sets of data.

(a) 6   3   9   12   1   9   8   8   6   1   3   6

(b) 20.1   20.7   21.4   22.7   29.6   22.6

(c) $\frac{1}{4}$ $\frac{1}{2}$ $\frac{3}{4}$ $\frac{1}{4}$ $\frac{5}{4}$ $\frac{3}{2}$ $\frac{5}{2}$ $\frac{1}{4}$

(d) 151   154   161   179   180   124   162   180   134

Use $\bar{x} = \dfrac{\Sigma x}{n}$ to find the mean.

2 In a driving theory test the following percentage marks were recorded:

75 61 52 82 64 71 90 46 55 57 64 63 67

Find the mean, median and mode for the marks.

3 In a ten-pin bowling game these scores were recorded:

7   8   4   1   10   8   3   6   5   9

Find the mean score (to the nearest whole number) and the mode.

4 The prices of new Ford Focus Zetecs in different dealerships were:

£9900   £10 200   £9625   £9865   £10 150   £9950

What is the mean price for a new Ford Focus Zetec?

5 (a) Work out the mean of these amounts of money:

£150   £75   £62   £87   £46   £102

When one more amount is added the new mean value is £83.

(b) What amount of money is added?

6 The mean height of 13 men in a theatre company is 179 cm. The mean height of 12 women in the company is 166 cm. What is the mean height for the whole company? Give your answer to the nearest cm.

7 In a mental test the following results were recorded:

21   25   18   27   22   23   19   16   21   27
24   24   16   18   23   24   25   20   28

Find (a) the mean score (b) the median score (c) the mode and (d) the range of scores.

8 Construct a list of six whole numbers with:
- range = 4
- mode = 6
- mean = 6

What is the median for this set of numbers?

9 This stem-and-leaf diagram shows the temperatures recorded in different cities around the world at noon yesterday.
Find the mean, mode, median and range of the temperatures.

> Look back at Chapter 11 if you need help with stem-and-leaf diagrams.

```
0 | 1 4 5 8 8 9
1 | 0 0 3 5 6 6 9 9
2 | 0 1 1 1 2 5 8
3 | 0 0 1
```

Key: 1|2 represents 12°C

10 A class of pupils had their heights measured to the nearest centimetre.
The results for the girls and the boys are shown separately in these stem-and-leaf diagrams.

```
Girls                    Boys
12 | 4 9                 12 | 8
13 | 2 7 7 7             13 | 0
14 | 2                   14 | 7
15 | 1 3 5 5 6           15 | 3 5 5 8 9
16 | 0 2 8               16 | 0 1 1 1
17 | 0                   17 | 5 7
```

Key: 14|7 represents 147 cm.

Find the mean, mode, median and range of these heights:
- **(a)** for the girls only
- **(b)** for the boys only
- **(c)** for the whole class.

Which is the best average to use?

> **Key words:**
> representative
> extreme value (outlier)

Sometimes an 'average' can give the wrong impression of a data set.

You need to decide which average is the most sensible one to use.

Example 3

A factory making components for a mobile phone has a managing director, a works' supervisor and 12 employees.

Their net weekly wages (in £) are:

135   135   135   135   150   150   150
164   164   178   193   193   276   957

Calculate the mean, mode and median wage.

$$\text{Mean} = \frac{135 + 135 + \dots + 276 + 957}{14}$$

$$\bar{x} = \frac{3115}{14}$$

$$\bar{x} = £222.50$$

$$\text{Mode} = £135$$

$$\text{Median} = \frac{150 + 164}{2} = £157$$

Which average makes the most sense?

'The average wage is £222.50'

This is the mean value.

All of the 12 employees earn less than this. Some of them earn nearly £90 less! The only reason the *mean* is so high is because the managing director's £957 is included in the total of £3115.

If you re-calculate the mean *without* the £957 you get

$$\text{Mean} = \frac{2158}{13}$$

$$\bar{x} = £166$$

This value gives a fairer picture of people's earnings.

which is much more **representative** of the data.

The value 957 is an **extreme value** and is best left out of the calculations.

An extreme value is either a lot larger or a lot smaller than the rest of the data. It is also called an **outlier**.

'The average wage is £135'

> This is the mode.

This is not representative of the data. No-one earns less than this and eight of the staff earn quite a lot more.

'The average wage is £157'

> This is the median.

All of the 12 employees wages are fairly close to this. As it is the median, it is in the middle of the list and is the most sensible answer to use in this case.

> You need to use your common sense to decide which is the best average to use in a situation.

Exercise 20C

1 The wickets taken by members of a village cricket team one season were as follows:

    39 38 38 35 34 33 29 28 26 9

(a) Find the mode, median and mean and the range for this set of data.

(b) Which 'average' would best represent the data? Explain your reasons.

2 The numbers of cars parked during the day in a city centre car park during a two-week period were

    52 45 61 67 48 70 12 56 41 57 53 62 70 9

(a) Find the mode, median and mean number of cars. Comment on your results.

(b) Which would be the best 'average' to use?

Retail price index

An index number compares one number (often a price) with another.

> The index number is a percentage of the base, but the percentage sign is left out.

The UK retail price index in Dec 2005 was 194.1. This compares with a base price of 100 in 1987. This means that average prices increased by 94.1% between 1987 and 2005.

Example 4

The index for gold was 190 in January 2006, compared with a base of 100 in January 2001.

(a) What was the percentage increase in gold prices in that period?

(b) Gold cost £170 per ounce in January 2001. What was the price in January 2006?

(a) $190 - 100 = 90$, percentage increase $= 90\%$

(b) £170 $\times \frac{190}{100} =$ £323

20.2 Frequency distributions

Key words:
frequency table
frequency

You will need to know:
- **how to tally results and record the frequency**

When you have a large amount of data, you can tally it into a **frequency table** .

The **frequency** is the number of times an answer or result occurs in the data.

Example 5

In a local soccer tournament the numbers of goals scored in each game were:

```
0 2 5 4 3 2 0 0 3 5 3 4 2 2 2 3 1 1 0 1
6 1 0 0 3 1 0 1 2 2 1 4 3 0 0 1 2 1 2 1
5 0 1 3 2 0 6 3 1 1 1 0 2 0 2 1 0 2 2 0
1 2 3 0 4 2 3 1 5 1 6 1 1 3 2 2 0 0 3 0
```

Put this data into a frequency table.

Find: **(a)** the mean **(b)** the median **(c)** the mode
  (d) the range of goals scored in the tournament.

Remember we are finding a representative number for the number of goals scored in each game.

Use x for the data. Use f for the frequency (the number of times each value occurs).

Number of goals scored (x)	Number of games (f)	Total number of goals (fx)
0	19	0
1	20	20
2	18	36
3	12	36 ←
4	4	16
5	4	20
6	3	18
Totals	$\Sigma f = 80$	$\Sigma fx = 146$

Count (or tally) the zeros in the data, then the 1s, the 2s, etc.

The column fx records the total number of goals. For example, 12 teams scored 3 goals, which is $12 \times 3 = 36$ goals in total.

Σ means 'the sum of'.

Σf is the total number of games played.

Σfx is the total number of goals scored.

(a) Mean $= \dfrac{\text{total goals scored}}{\text{total games played}}$

This is the same as the number of items of data.

$$\bar{x} = \frac{\Sigma fx}{\Sigma f} = \frac{146}{80} \qquad x = 1.825 \text{ goals per game}$$

$\bar{x}$ is the symbol for the mean.

There are 80 items of data. When they are arranged in order of size, the median will be midway between the 40th and 41st values.

In 80 pieces of data, the 40th and 41st pieces are the two middle values.

Number of goals scored (x)	Number of games (f)	Total number of goals (fx)
0	19	0
1	20	20
2	18	36

First 19 pieces of data.

20 + 19 = 39 pieces of data in first 2 rows.

40th and 41st pieces must be in here.

(b) The 40th and 41st values are both 2.

Median = 2 goals per game

(c) Mode = 1 goal per game

The mode is the number of goals with the highest frequency.

(d) Range = 6 − 0 = 6 goals

The range is the difference between the largest and the smallest number of goals.

For grouped data the mean is $\bar{x} = \dfrac{\Sigma fx}{\Sigma f}$

How to find the median

- Odd number of data items:

  Middle of 5 and 1 is $\dfrac{5+1}{2} = 3$

  so median value is 3rd item.

  1st 2nd 3rd 4th 5th

  median value

- Even number of data items:

  Middle of 4 and 1 is $\dfrac{4+1}{2} = 2.5$

  so median is mean of 2nd and 3rd items.

  1st 2nd 3rd 4th

  median value

So we can generalise by saying for *n* items in a data set:

1st 2nd 3rd 4th ... *n*th

the median value is the $\left(\dfrac{\boldsymbol{n+1}}{\boldsymbol{2}}\right)$**th item.**

Exercise 20D

1 A fair, six-sided dice is rolled 50 times. The scores are recorded in a frequency table:

$fx = f \times x$

Number on dice (x)	Frequency (f)	Total score (fx)
1	7	
2	9	
3	6	
4	11	
5	10	
6	7	
Totals	$\Sigma f =$	$\Sigma fx =$

(a) Copy and complete the table.

(b) Work out the mean, median and mode of the scores for this dice.

2 The police recorded the speeds of cars (to the nearest 10 mph) on a road in a built-up area.

Speed mph (x)	Frequency (f)	(fx)
10	0	
20	23	
30	48	
40	16	
50	2	
60	1	
Totals	$\Sigma f =$	$\Sigma fx =$

(a) Copy and complete the table.

(b) Find the mean, median and mode speeds for this road.

(c) What do you think is the likely speed limit for this road?

3 In a hockey tournament the following number of goals were scored in different games.

Goals scored	x	0	1	2	3	4	5	6	7
Frequency	f	12	14	12	9	4	6	0	3

Write the table vertically, as in Example 4.

Find the mean, median and mode of the goals scored in the tournament.

4 75 job applicants were given a mental arithmetic test. These are their scores out of 10.

```
5  7  8  2  1   9  3  8  7  4   2  8   4  9   8
2  4  9  5  5   6  6  5  8  9  10  6  10  7   3
8  3  4  2  5   7  3  7  7  8   9  7   8  4  10
6  4  3  8  2   9  9  2  3  8   5  6   6  8  10
5  6  6  2  8  10  5  5  6  7   1  8   7  4   6
```

(a) Put this data into a frequency table.
(b) Find the mean, median, mode and range of the scores.

20.3 Grouped frequency distributions for discrete data

When there are a large number of data items it is best to group the data.

This makes recording easier, but information about individual data items is not known.

The groups of data are called **class intervals** . The frequency table now becomes a **grouped frequency table** .

The class intervals do not have to be the same size, but they usually are.

The numbers at the start and end of each class interval are called the **class limits** .

You can group large amounts of data in class intervals in a grouped frequency table.

Key words:
class intervals
grouped frequency table
class limits
estimated mean
mid-interval value
median class interval
modal class interval

For the class interval 1–5, the class limits are 1 and 5.

The table shows the time taken to travel to work for 60 people. The times are recorded to the nearest minute.

Time (minutes)	Number of people (*f*)
1–5	6
6–10	12
11–15	8
16–20	11
21–25	14
26–30	4
31–35	3
36–40	2
Total	60

Class intervals must not overlap or have any gaps between them.

The class intervals are 1–5, 6–10, 11–15, etc. You could write them as $1 \leqslant x \leqslant 5$, $6 \leqslant x \leqslant 10$, etc.

Estimated mean

You can calculate an **estimated mean** from this grouped frequency table. This is not a guess, because it comes from a calculation. It is an estimate because you do not know the exact time taken by each person.

To calculate the estimated mean, you need a single number instead of a class interval in the 'Time' column.

The single number you use is the **mid-interval value** of each of the class intervals.

> **The mid-interval value is the mean of the two class limits.**

You can extend the grouped frequency table and write the mid-interval values. These mid-interval values replace the class intervals. They will be the '*x*-values' in your calculation.

To calculate the mean you would add up all the *exact* times and divide by 60.

Then you can calculate the estimated mean as you did in Example 3.

For the 11–15 class interval, the mid-interval value is
$$\frac{11 + 15}{2} = \frac{26}{2} = 13$$

Add the Total time (*fx*) column and calculate the totals.

Time (minutes)	Number of people (*f*)	Mid-interval value (*x*)	Total time (minutes) (*fx*)
1–5	6	3	18
6–10	12	8	96
11–15	8	13	104
16–20	11	18	198
21–25	14	23	322
26–30	4	28	112
31–35	3	33	99
36–40	2	38	76
	$\Sigma f = 60$		$\Sigma fx = 1025$

6 people take 3 minutes, total = $f \times x = 6 \times 3 = 18$

Total time taken.

Total number of people.

$$\text{Estimated mean} = \frac{\text{total time taken}}{\text{number of people}}$$

$$\bar{x} = \frac{\Sigma fx}{\Sigma f} = \frac{1025}{60}$$

$$\bar{x} = 17.1 \text{ minutes}$$

For grouped data you can calculate an estimated mean.

The median

With grouped data, you cannot give an exact answer for the median.

You need the *exact* data values for an exact median.

The class interval containing the median value is the median class interval .

There are 60 items of data. The median is mid-way between the 30th and the 31st items.

Time (minutes)	Number of people (f)
1–5	6
6–10	12
11–15	8
16–20	11

18
26
37

Both the 30th and the 31st items must be in the 16–20 minute interval.

The median class interval is 16–20 minutes.

The mode

With grouped data you cannot give an exact value for the mode.

> The class interval with the highest frequency is the **modal class interval** .

You can see from the table that this is the 21–25 minute interval.

Exercise 20E

1 The number of telephone calls made from a particular house over a 72-day period was recorded.

Calls made	Frequency (f)	Mid-interval value (x)	Total number of calls (fx)
0–2	12		
3–5	18		
6–8	31		
9–11	11		
Totals	$\Sigma f =$		$\Sigma fx =$

(a) Copy and complete the table.

(b) Work out an estimate of the mean.

(c) Find the mode and median class intervals.

2 A shoe shop recorded the sizes of shoes sold one day.

Shoe size	Frequency (f)	Mid-interval value (x)	Total number of shoes (fx)
3–4	5		
5–6	8		
7–8	12		
9–10	5		
11–12	2		
Totals	$\Sigma f =$		$\Sigma fx =$

(a) Copy and complete the table.

(b) Work out an estimate of the mean.

(c) Find the mode and median class intervals.

3 Some students were asked how many CDs they had. The results are shown in this grouped frequency table.

$1 \leqslant x < 25$ means x can take any value in the range 1–24. 25 *is not* included.

Number of CDs	Frequency (f)
$1 \leqslant x < 25$	12
$25 \leqslant x < 50$	0
$50 \leqslant x < 75$	11
$75 \leqslant x < 100$	9
$100 \leqslant x < 125$	15
$125 \leqslant x < 150$	33
$150 \leqslant x < 175$	28
$175 \leqslant x < 200$	41
$200 \leqslant x < 225$	16

(a) Copy the table and extend to include the mid-interval values.

(b) Use your table to find an estimate for the mean number of CDs.

(c) Work out the median and modal class intervals.

4 The number of pupils in a particular year 11 set using the school library was recorded over a term.

Number of visits	Frequency (f)
0–4	84
5–9	46
10–14	38
15–19	51
20–24	22
25–29	13

Using this data, work out an estimate for the mean number of visits to the library for this set.
How does the estimated mean compare with the median and modal class intervals?

Examination style questions

1 Josh counted the number of people living in each of ten houses.
His results were

    3   6   5   4   2   1   3   2   6   8

(a) Work out the range.
(b) Calculate the mean.   *(4 marks)*

AQA, Spec B, 1F, June 2002

2 Eleven pupils took part in a sponsored basketball match. The amount collected, in pounds, by each pupil is shown below.

    5   1   6   8   8   8   4   2   3   7   5

(a) Find the median of the amounts.
(b) Work out the range of these amounts.
(c) Write down the mode of these amounts.   *(4 marks)*

AQA, Spec B, 1F, November 2003

3 The stem-and-leaf diagram shows the test scores of some pupils.

(a) How many pupils scored less than 15?
(b) What was the median score?

An extra pupil takes the test and scores 29.

(c) Copy the stem-and-leaf diagram and add this score. Find the new median score.

```
0 | 9
1 | 0 2 2 6 7 8
2 | 0 3 3 5 6
3 | 2 7
4 | 3
```

Key: 1|6 means 16

(6 marks)

AQA, Spec B, 1F, March 2002

4 Phil counts the number of people in 50 cars that enter a car park.
His results are shown in the table.

Number of people	Frequency
1	25
2	17
3	6
4	2
More than 4	0

You can use the wording 'More than' when you don't know the maximum number possible.

Calculate the mean number of people per car.

(3 marks)

AQA, Spec B, 1F, November 2002

5 The following test scores were recorded by a year 11 class

```
38  25  11  87  66  54  53  39  23  14
43  35  52  64  27  70  23  14  58  38
19  61  26  51  59  63  73  24  38  63
27
```

(a) Construct a grouped frequency table for this data using the intervals 0–9, 10–19 etc.

(b) What is the modal class interval?

(c) Work out an estimate of the mean score for this set of data. *(6 marks)*

6 The lengths of 10 boxes are measured. The results are summarised in the table.

Length x (cm)	Frequency	Mid-point value
$1 < x \leqslant 3$	5	
$3 < x \leqslant 5$	2	
$5 < x \leqslant 7$	2	
$7 < x \leqslant 9$	1	

Copy the table and complete the mid-point value column.
Use it to calculate an estimate of the mean length. *(3 marks)*

AQA, Spec B, 1I, June 2002

Summary of key points

Averages and spread (grades G to C)

An average is a value that represents a set of data.

The mean of a set of data is the sum of all the values divided by the number of values.

For individual data the mean is given by $\bar{x} = \dfrac{\Sigma x}{n}$.

The mode of a set of data is the value that occurs most often (the most frequent).

The median of a set of data is the middle value when the data is arranged in order of size.

For a set of n data values:

The median is the middle value in the ordered list. If n is even, the median is halfway between the two middle values.

The range or spread of a set of data is the difference between the highest and the lowest values.

Frequency

The frequency is the number of times an answer or result occurs in the data.

Grouped data (grades E to C)

You can group large amounts of data in class intervals in a grouped frequency table.

For grouped data you can calculate an estimated mean, using frequency and mid-interval value (x).

The mid-interval value (x) is the mean of the two class limits.

For grouped data the mean is given by $\bar{x} = \dfrac{\Sigma fx}{\Sigma f}$

The class interval with the highest frequency is the modal class interval.

The class interval containing the median value is the median class interval.

This chapter will show you how to:
- ✔ understand Pythagoras' theorem
- ✔ calculate the longest side (the hypotenuse) of a right-angled triangle
- ✔ calculate a shorter side of a right-angled triangle
- ✔ apply Pythagoras' theorem to solving 'real' problems

21.1 Pythagoras' theorem

You will need to know:
- how to use your calculator
- how to square numbers
- how to find square roots

Key words:
Pythagoras' theorem
right-angled triangle
hypotenuse

Pythagoras' theorem only applies to **right-angled triangles**.

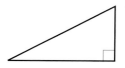

A right-angled triangle has one angle of 90°.

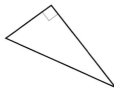 shows a right angle.

This right-angled triangle has sides $a = 4$ cm, $b = 3$ cm and $c = 5$ cm.

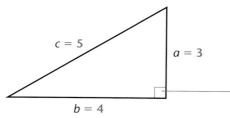

c is the hypotenuse.

This is the right angle.

In a right-angled triangle the longest side is called the **hypotenuse**.
The hypotenuse is always opposite the right angle.

You can construct a square on each side of the triangle.

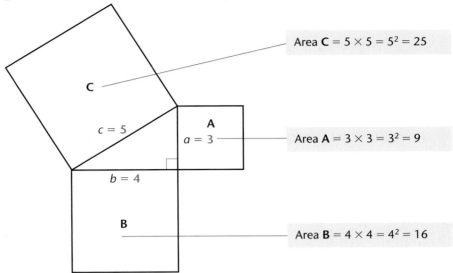

Area **C** $= 5 \times 5 = 5^2 = 25$

Area **A** $= 3 \times 3 = 3^2 = 9$

Area **B** $= 4 \times 4 = 4^2 = 16$

Area **A** + Area **B** $= 9 \text{ cm}^2 + 16 \text{ cm}^2 = 25 \text{ cm}^2 =$ Area **C**

$$\text{Area } \mathbf{A} + \text{Area } \mathbf{B} = \text{Area } \mathbf{C}$$
$$3^2 + 4^2 = 5^2$$

In other words $a^2 + b^2 = c^2$

Use squared paper to check this using different triangles.

This leads to Pythagoras' theorem:

> In any right-angled triangle the square of the hypotenuse (c^2) is equal to the sum of the squares on the other two sides ($a^2 + b^2$).

'Sum' means 'add'.

For a right-angled triangle with sides of lengths a, b and c, where c is the hypotenuse, Pythagoras' theorem states that $a^2 + b^2 = c^2$

Exercise 21A

1 In the following triangles, write the letter that represents the hypotenuse.

(a)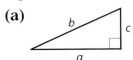

(b)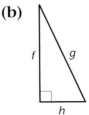

(c)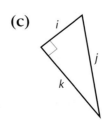

2 For the triangles above, write out the formula using Pythagoras' theorem.
Make the longest side (the hypotenuse) the subject of the formula.

> The subject appears on its own on one side of the equals sign.

21.2 Finding the hypotenuse

You can use Pythagoras' theorem to find the length of the hypotenuse.

Example 1

Work out the length of the hypotenuse (the side marked x) in this right-angled triangle.

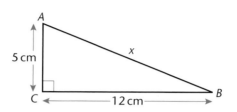

$a^2 + b^2 = c^2$

$c^2 = a^2 + b^2$

$x^2 = 5^2 + 12^2$

$x^2 = 25 + 144$

$x^2 = 169$

$x = \sqrt{169}$

$x = 13 \text{ cm}$

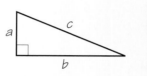

> Write out Pythagoras' theorem and then substitute in the values given.

Example 2

In the triangle PQR angle Q = 90°, QP = 4 cm and QR = 7 cm.
Calculate y, the length of PR, to 1 decimal place.

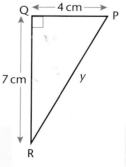

$c^2 = a^2 + b^2$

$y^2 = 4^2 + 7^2$

$y^2 = 16 + 49$

$y^2 = 65$

$y = \sqrt{65}$

$y = 8.06 \approx 8.1 \text{ cm (to 1 d.p.)}$

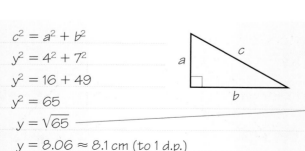

> PR is the hypotenuse, because it is opposite the right angle.

> Use your calculator to find the square root.

> Always put the units in your answer.

Example 3

In this isosceles triangle EF = EG, M is the mid-point of the base line FG and EM = 3.2 cm.

Work out the slant length of the triangle (marked x in the diagram) to 2 d.p.

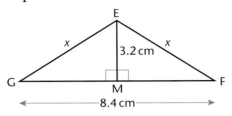

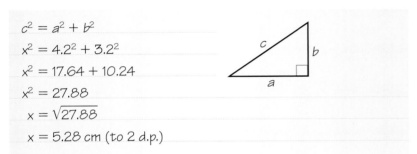

$$c^2 = a^2 + b^2$$
$$x^2 = 4.2^2 + 3.2^2$$
$$x^2 = 17.64 + 10.24$$
$$x^2 = 27.88$$
$$x = \sqrt{27.88}$$
$$x = 5.28 \text{ cm (to 2 d.p.)}$$

Split the triangle EFG into two right-angled triangles:

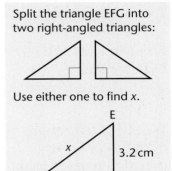

Use either one to find x.

$GM = \frac{1}{2} \times 8.4 \text{ cm} = 4.2 \text{ cm}$

On a calculator
$\sqrt{27.88} = 5.28015 \ldots$
which is 5.28 to 2 d.p.

Exercise 21B

1 Calculate the length marked with a letter in each triangle.

(a)

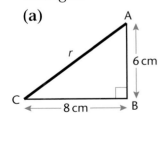

(b)

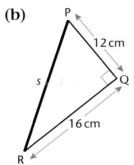

(c)

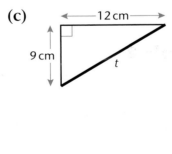

2 Calculate the lengths marked with letters to 1 d.p.

(a)

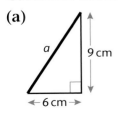

(b)

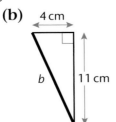

(c)

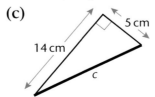

(d)

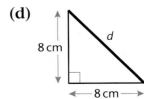

3 A rectangular gate measures 3.5 m long by 1.5 m high. Work out the length of wood needed to make the diagonal.

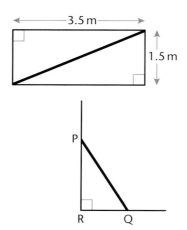

4 An extendable ladder rests against a vertical wall just below a window. PR = 2.6 m and the foot of the ladder is 1.2 m away from the wall (RQ). How long does the ladder need to be to reach the window?

UAM 5 Fiona is flying her kite in a strong breeze. The kite is flying at a height of 18 m and is 20 m away horizontally from Fiona. How long is the kite string?

For questions 5 and 6, sketch a diagram and label it.

UAM 6 A boat sails due east for 24 km then due south for 10 km. How far is the boat from its starting point?

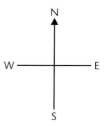

21.3 Finding the length of a shorter side

Key words:
subject

In Pythagoras' theorem: $a^2 + b^2 = c^2$.

To work out the length of the shorter side a, make a the **subject** of the equation.

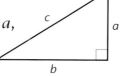

$$a^2 + b^2 = c^2$$
$$a^2 + b^2 - b^2 = c^2 - b^2$$
$$a^2 = c^2 - b^2$$

For help with changing the subject of an equation see Chapter 7.

Subtract b^2 from both sides of the equation.

You can make b the subject in the same way.

$$b^2 = c^2 - a^2$$

You can calculate the lengths of the shorter sides of a right-angled triangle using
- $a^2 = c^2 - b^2$
- $b^2 = c^2 - a^2$

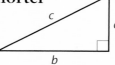

Example 4

In this triangle $b = 4$ cm and $c = 7$ cm.

Calculate the length a (to 1 dp).

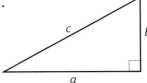

c is the longest side (the hypotenuse).

Using Pythagoras' theorem:

$a^2 = c^2 - b^2$

$a^2 = 7^2 - 4^2$

$a^2 = 49 - 16$

$a^2 = 33$

$a = \sqrt{33}$

$a = 5.744 \approx 5.7$ cm (1 d.p.)

To find a **S**horter side you **S**ubtract squares of sides.

If you add by mistake you get
$a^2 = 49 + 16$
$a^2 = 65$
$a = \sqrt{65} = 8.06 \ldots$
This is impossible because it is longer than the hypotenuse.

Example 5

The end face of a chocolate bar is in the shape of an isosceles triangle EFG. The mid-point of side EG is H. Using the dimensions shown on the diagram work out the base length EG of the chocolate bar (to 1 d.p.).

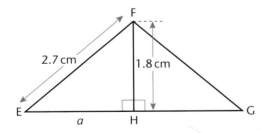

Triangle EHF is a right-angled triangle, so use Pythagoras' theorem.

$a^2 = c^2 - b^2$

$a^2 = 2.7^2 - 1.8^2$

$a^2 = 7.29 - 3.24$

$a^2 = 4.05$

$a = \sqrt{4.05}$

$a = 2.012$

$EG = 2a = 4.024$ cm ≈ 4.0 cm (to 1 d.p.)

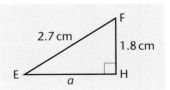

Shorter side → Subtract.

Multiply by 2 on your calculator before rounding your final answer.

Exercise 21C

1 Calculate the lengths marked with letters in these triangles.
Give your answers to 1 d.p.

(a)

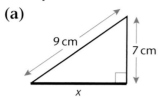

9 cm

7 cm

x

(b)

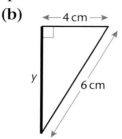

←—4 cm—→

y

6 cm

(c)

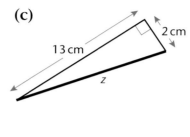

2 cm

13 cm

z

2 Calculate the lengths   **(a)** XY   **(b)** QR.
Give your answers to 1 d.p.

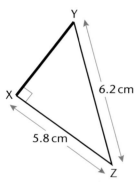

Y

6.2 cm

X

5.8 cm

Z

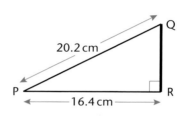

20.2 cm

Q

P

16.4 cm

R

3 A children's slide is 3.6 m long.
The vertical height of the slide above the ground is 2.1 m.
Work out the horizontal distance between each end
of the slide.

For questions 3 to 6, sketch
a diagram and label it.

4 Jotinder walks 120 m up a gentle slope. From the map he
sees he has walked a horizontal distance of only 95 m.
Work out the height he has climbed.

5 A ladder 5.2 metres long is
leaning against a wall.
The top of the ladder is
4.6 metres up the wall.
How far from the bottom of the
wall is the foot of the ladder?

6 A ship sails 46 km North East.
Then it changes course and sails 62 km South East.
How far is the ship from its starting point?

7 Work out the lengths *c* and *d* on this diagram.

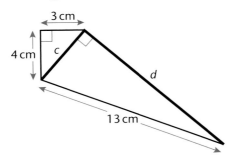

8 The diagram shows the
cross-section of a shed.
Work out the width of
the shed, *w*, to 2 d.p.

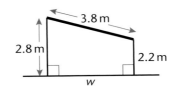

Examination style questions

1 The diagram shows a right-angled triangle *ABC*.
   AB = 10 cm and *AC* = 15 cm.

   Calculate the length of *BC*.
   Leave your answer as a square root.

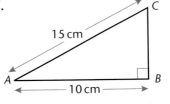

Not drawn
to scale

(3 marks)
AQA, Spec B, 5I, June 2004

2 A support for a flagpole is attached at a height
of 3 m and is fixed to the ground at a distance
of 1.2 m from the base.

   Calculate the length of the support
   (marked *x* on the diagram).

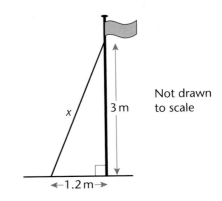

Not drawn
to scale

(3 marks)
AQA, Spec A, I, June 2003

3 A balloon is held to the ground by a
cable at point O.
The balloon is 30 m horizontally
from point O.
The balloon is held at a height of 75 m.

How long is the cable?

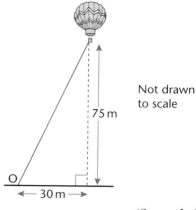

Not drawn
to scale

75 m

O

30 m

(3 marks)

4 A square of side length *a* has a diagonal of 15 cm.
Calculate the value of *a* to 1 d.p.

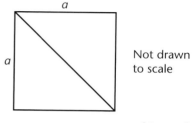

a

a

Not drawn
to scale

(4 marks)

Summary of key points

A right-angled triangle has one angle of 90°.

**In a right-angled triangle the longest side is called the
hypotenuse.
The hypotenuse is always opposite the right angle.**

**For a right-angled triangle with sides of lengths
a, *b* and *c*, where *c* is the hypotenuse,
Pythagoras' theorem states that**

$$a^2 + b^2 = c^2$$

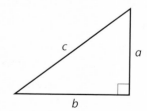

c

a

b

**You can calculate the lengths of the shorter sides
using**

$$a^2 = c^2 - b^2$$
$$b^2 = c^2 - a^2$$

This work is grade C.

22 Proof

This chapter will show you how to:

✔ tell the difference between 'verify' and 'proof'
✔ prove results using simple, step-by-step chains of reasoning
✔ use a counter example to disprove a statement
✔ prove simple geometrical statements

22.1 Proof questions

Proof questions test the 'Using and Applying' part of what you learn in mathematics.

You need to:

- find examples that match a general statement, or give counter examples which disprove a statement,
- show how one statement follows from another,
- explain your reasoning or criticise a piece of faulty reasoning,
- know the difference between a practical demonstration of something and a proof.

Many proof questions are based on the properties of numbers. You need to know about:

- odd and even numbers
- prime numbers
- factors and multiples
- square and cube numbers.

Certain words occur in proof questions:

- consecutive (one after the other)
- integer (whole number)
- product (multiply).

You need to know the rules for addition and multiplication of whole numbers:

For the general statement 'Dogs have two ears'. You can find examples of dogs with two ears to match the statement.

For 'All dogs are black' a brown dog is a counter example that disproves the statement.

Chapter 14 covers most of these. Look back if you need help.

Addition	Multiplication
odd + odd = even	odd × odd = odd
odd + even = odd	odd × even = even
even + odd = odd	even × odd = even
even + even = even	even × even = even

22.2 Proof vs. verify

Key words:
verify
proof

You need to know the difference between 'proof' and 'verify'.

Verify means to check something is true by substituting numbers into an expression or formula.

Proof means to show something is true using logical reasoning.

Example 1 illustrates the difference between 'proof' and 'verify'.

Example 1

integer = whole number

n is any integer.
Explain why $2n + 1$ must be an odd number.

Answer 1

Try $n = 1$ $2n + 1 = 2 \times 1 + 1 = 2 + 1 = 3$
Try $n = 2$ $2n + 1 = 2 \times 2 + 1 = 4 + 1 = 5$
Try $n = 3$ $2n + 1 = 2 \times 3 + 1 = 6 + 1 = 7$
The answer is always an odd number.

Is it *always*?
The answer only shows it is for $n = 1, 2, 3$.

This is not a proof because it does not explain *why* $2n + 1$ must be an odd number.

This answer verifies that $2n + 1$ is an odd number when $n = 1, 2$ and 3.

Will $2n + 1$ be an odd number for $n = 4, 5, 6, 7, ...$?

You cannot test it for all numbers – it would take forever!

Look at Answer 2.

Answer 2

$2n = 2 \times n$

Multiplying any whole number by 2 gives an even number, so $2n$ is an even number.

So $2n + 1 =$ even number $+ 1 =$ odd number

$2n + 1$ must always be an odd number.

Answer 2 is a *proof*. It shows whatever number you start with (*n* could stand for *any* whole number), $2n + 1$ will always be an odd number.

This example shows that there is a huge difference between verifying a result (simply checking with numbers) and proving a result.

> **Verify is a practical demonstration of the result. Proof shows how one statement follows from another using simple chains of reasoning.**

Example 2

Explain why the sum of any three consecutive integers is always a multiple of 3.

'Consecutive' means one after the other.

Verify

Try 1, 2 and 3	$1 + 2 + 3 = 6 = 2 \times 3$
Try 5, 6 and 7	$5 + 6 + 7 = 18 = 6 \times 3$
Try 10, 11 and 12	$10 + 11 + 12 = 33 = 11 \times 3$

The result works for these examples.

Multiple of 3 ... a number in the 3 times table.

Proof

If x is one of the numbers then the others are $(x + 1)$ and $(x + 2)$.

$x + (x + 1) + (x + 2) = 3x + 3 = 3(x + 1)$

$3(x + 1)$ means $3 \times (x + 1)$

So the answer is always a multiple of 3.

For the consecutive numbers you could choose $(x - 1)$, x and $(x + 1)$.

> **Proof questions often have the word 'explain' in the question.**

You only get full marks if you *prove* the result. You may get some marks for verifying (giving numerical examples), but answers which just check the result using numbers often score no marks at all!

Proof usually involves some algebra skills. In Example 1 you needed to know that $2n$ means $2 \times n$, a very basic algebra fact. In Example 2 you had to 'collect like terms' and 'take out a common factor'.

For help with these, look back at Chapter 4.

Exercise 22A

UAM **1** p is an odd number and q is an even number.
    (a) Explain why $p + q - 1$ is always an even number.
    (b) Explain why $pq + 1$ is always an odd number.

UAM **2** n is a positive integer.
    Explain why $n(n + 1)$ must be an even number.

UAM **3** x is an odd number.
    Explain why $x^2 + 1$ is always an even number.

UAM **4** If b is an even number, prove that $(b - 1)(b + 1)$ is an odd number.

UAM **5** x is an odd number and y is an even number.
    Explain why $(x - y)(x + y)$ is an odd number.

UAM **6** Explain why the sum of four consecutive numbers is always an even number.

22.3 Proof by counter example

Key words:
counter example

Proof by counter example asks you to show that a statement is incorrect by finding *one* example where the stated result does not work.

You can substitute numbers into an expression or formula until you find a case where the result is not true.

Example 3

Tony says that when n is an even number, $\frac{1}{2}n + 3$ is always even. Give an example to show that he is wrong.

> Even numbers are 2, 4, 6, ...
> Try $n = 2$ $\frac{1}{2}n + 3 = \frac{1}{2}(2) + 3 = 1 + 3 = 4$ ... even
> Try $n = 4$ $\frac{1}{2}n + 3 = \frac{1}{2}(4) + 3 = 2 + 3 = 5$ ... odd
> When $n = 4$ the result is not true, so Tony is wrong.

The case where $n = 4$ is a counter example.

Sometimes you need to try several values before you find a counter example.

Example 4

Wes says that when you square a number, the answer is always larger than or equal to the original number.

Give a counter example to show that Wes is wrong.

Try 1 $1^2 = 1 \times 1 = 1$ ... *the same*
Try 2 $2^2 = 2 \times 2 = 4$ ... *larger*
Try 3 $3^2 = 3 \times 3 = 9$ ... *larger*
Trying 4, 5, ... will give even larger answers than these.
Try -2 $(-2)^2 = (-2) \times (-2) = 4$ ... *larger*
Try a decimal number smaller than 1, such as 0.5.
$(0.5)^2 = 0.5 \times 0.5 = 0.25$ ... *which is smaller*
$(0.5)^2$ *gives a smaller number, so Wes is wrong.*

Work systematically, trying different values.

Squaring numbers greater than 3 will give larger numbers.

negative × negative = positive

$(0.5)^2$ is a counter example.

Exercise 22B

1 Sam says that when k is an even number, $k^2 + \frac{1}{2}k$ is always odd.
Give an example to show that he is wrong.

2 Heather says that $m^3 + 2$ is never a multiple of 3.
Give a counter example to show that she is wrong.

3 p is an odd number and q is an even number.
Andrew says that $p + q - 1$ cannot be a prime number.
Explain why he is wrong.

4 a and b are both prime numbers. Give an example to show that $a + b$ is not always an even number.

5 Ian says that $n^2 + 3n + 1$ is a prime number for all values of n.
Give a counter example to show that he is wrong.

6 Give a counter example to each of these statements:
   (a) the square root of any number is always smaller than the original number
   (b) the cube of any number is always greater than the square of the same number.

22.4 Proof in geometry

To prove a result in geometry you use step-by-step reasoning, showing clearly how one statement follows from another.

Don't just verify by finding examples that work.

For geometry proofs you need to know:
- angle properties of parallel lines
- angle properties at a point and on straight lines.

Never use a protractor to check angles. The diagrams are not usually accurately drawn.

Example 5

Prove that the sum of the angles of a triangle is 180°.

This result was proved in Chapter 6.

Start by drawing a diagram.

Draw a triangle ABC.

Extend side AC to E.

Draw line CD parallel to AB.

Label the angles as shown on the diagram.

$x = a$ (corresponding angles)

$y = b$ (alternate angles)

$x + y + c = 180°$ (sum of angles on a straight line at point C)

So $a + b + c = 180°$.

The sum of the angles of $\triangle ABC$ is 180°.

Look back at angle properties and parallel line properties in Chapter 6 to remind yourself of these facts.

You cannot simply measure the angles and find that the sum is 180°. This is not a proof. Look back at Section 6.2 to see the proof written out. Note that it uses general angles, not specific ones.

To set out a geometry proof:
- state each step clearly,
- give a reason for each step.

Steps need to follow on from each other.

Exercise 22C

1 Prove that the exterior angle of a triangle is equal to the sum of the two opposite interior angles.

2 Prove that the sum of the interior angles of a quadrilateral is 360°.

3 Prove that the interior angles of a regular pentagon are 108°.

Examination style questions

1 (a) Paul thinks that when an even number is halved the answer is always even.
Give an example to show that Paul is wrong.

(b) p is an odd number.
Is $2p + 1$ an odd number, an even number or could it be either?

(2 marks)
AQA, Spec B, 5F, June 2003

2 (a) k is an even number.
Jo says that $\frac{1}{2}k + 1$ is always even.
Give an example to show that Jo is wrong.

(b) p and q are both odd numbers.
p is greater than q.
Is $p - q$ an odd number, an even number or could it be either?

(2 marks)
AQA, Spec B, 5I, June 2004

3 p and q are odd numbers.

(a) Is $p + q$ an odd number, an even number or could it be either?

(b) Is pq an odd number, an even number or could it be either? *(2 marks)*
AQA, Spec A, F, June 2005

4 P is a prime number.
Q is an odd number.
State whether each of the following is always odd or always even or could be either odd or even.

(a) $P(Q + 1)$ **(b)** $Q - P$ *(2 marks)*
AQA, Spec A, I, November 2004

5 *n* is a positive integer.

(a) Explain why $n(n + 1)$ must be an even number.

(b) Explain why $2n + 1$ must be an odd number. *(2 marks)*

AQA, Spec A, H, June 2004

6 Tom is investigating the equation $y = x^2 - x + 5$.
He starts to complete a table of values of y for some integer values of x.

x			-2	-1	0	1	2	3			
y			11	7	5	5	7	11			

Tom says, "When x is an integer, y is **always** a prime number".
Find a counter example to show that Tom is wrong.
Explain your answer. *(2 marks)*

AQA, Spec A, I, November 2004

7 Andy says that when you work out the value of $2x^2 - 2x + 7$ for any positive integer you always get a prime number.
Give a counter example to show that he is wrong. *(3 marks)*

8 Explain why the sum of three consecutive odd numbers is always three times as big as the middle odd number. *(3 marks)*

9 Explain why the sum of two consecutive square numbers is always an odd number. *(3 marks)*

10 Prove that the value of $n^2 - 3n$ for $n \geqslant 4$ is always an even number. *(4 marks)*

Summary of key points

Proof vs. verify

You need to know the difference between proof and verify.

Verify means to check something is true by substituting numbers into an expression or a formula. It is a practical demonstration of a result.

Proof means to show something is true using logical reasoning.

Proof questions often have the word 'explain' in the question.

Proof by counter example

Proof by counter example asks you to show that a statement is incorrect by finding *one* example where the stated result does not work. You can substitute numbers into an expression or formula until you find a case where the result is not true.

Proof in geometry

To prove a result in geometry you use step-by-step reasoning, showing clearly how one statement follows from another.

Questions on proof can be set at various grade levels depending on the content of the material. They will typically be at grades E, D or C.

Examination practice

Exam paper, non-calculator

1 The attendance at three football matches are

Archester 38 217   Blackborough 29 066   Chelburn 34 829

  (a) Write the number 34 829 to the nearest thousand. *(1 mark)*

  (b) Put the attendances in order with the smallest first. *(1 mark)*

2 (a) Write down two multiples of 6. *(1 mark)*

  (b) Write down two multiples of 8. *(1 mark)*

  (c) Write down a number which is a multiple of both 6 and 8. *(1 mark)*

3 Here is a sequence of numbers,

76, 66, 57, 49, ..., ...,

  (a) Write down the next two numbers in the sequence. *(2 marks)*

  (b) Explain how you worked out your answer. *(2 marks)*

4 Molly scored 7, 4, 5, 9, 3, 7, 2, 7, 5, 8 in ten spelling tests.

  (a) State the mode. *(1 mark)*

  (b) What was her median score? *(2 marks)*

5 Match up a lettered card with a shape drawn on it to a numbered card with the name of a shape on it.

A   B   C   D

1 Parallelogram	2 Kite	3 Trapezium
4 Square	5 Rectangle	6 Rhombus

(4 marks)

6 If $p = 6$ $q = 5$ $r = -2$ work out,

  (a) $4p - 3q$ *(2 marks)*

  (b) $2q + 5r$ *(2 marks)*

7 Bottles of pop are sold in packs of 6. Each pack costs £2.16.
Joe buys 7 packs.

  (a) How many bottles of pop does he buy altogether? *(1 mark)*

  (b) How much does Joe pay for the 7 packs? *(2 marks)*

  (c) Joe pays with a £20 note. How much change should he receive? *(1 mark)*

8 Write down the value of

   (a) 6^2 *(1 mark)*

   (b) $\sqrt{49}$ *(1 mark)*

   (c) 4^3 *(1 mark)*

   (d) $\sqrt[3]{27}$ *(1 mark)*

9 **(a)** In this triangle two of the sides are the same length.

     (i)   What name is given to this special type of triangle?   *(1 mark)*

     (ii) Draw all the lines of symmetry on this triangle.   *(1 mark)*

  (b) Two of these triangles are used to make these different shapes.

 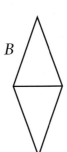

     (i)   How many lines of symmetry has shape A?   *(1 mark)*

     (ii)   How many lines of symmetry has shape B?   *(1 mark)*

     (iii) Write down the order of rotational symmetry of shape A.   *(1 mark)*

     (iv) Write down the order of rotational symmetry of shape B.   *(1 mark)*

10 Which of these amounts of money is the largest?
You **must** show your working.

    $\boxed{A:\ 40\% \text{ of } £80}$ $\boxed{B:\ \frac{2}{3} \text{ of } £45}$ *(4 marks)*

11 The data shows the number of children per family in 40 families.

$\cancel{1}$ $\cancel{4}$ $\cancel{0}$ $\cancel{5}$ $\cancel{1}$ 3 2 3 0 1
2 1 1 3 0 0 1 2 3 1
0 2 4 0 2 1 4 1 2 2
1 2 3 4 0 3 2 1 1 1

(a) Copy and complete this frequency table.

Number of children	Tally	Frequency
0	|	
1	||	
2		
3		
4	|	
5	|	

(3 marks)

(b) Draw a bar chart to show this information. Draw the horizontal axis from 0 to 5 and label it 'Number of children'. Draw the vertical axis from 0 to 13 and label it 'Frequency'. *(3 marks)*

12 Simplify   **(a)** $5x + 4x - 2x$ *(1 mark)*
                        (b) $4x + 6y - x + 5y$ *(2 marks)*
                        (c) $5 \times p \times 3$ *(1 mark)*
                        (d) $3(2m + 5) + 2(m - 4)$ *(2 marks)*

13 The diagram shows a scale drawing of one side of a triangular field, *ABC*.

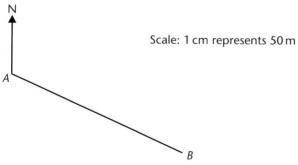

Scale: 1 cm represents 50 m

(a) Use the diagram to work out the actual distance from *A* to *B*. *(2 marks)*

(b) Measure and write down the three figure bearing of *B* from *A*. *(1 mark)*

(c) The bearing of *C* from *A* is 230°.
The actual distance from *A* to *C* is 200 m.
Copy and mark point *C* on the diagram. *(2 marks)*

14 This cuboid has a volume of 90 cm³.
Its base measures 5 cm by 3 cm.
Find the height of the cuboid.

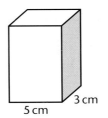

3 cm

5 cm

(2 marks)

15 Two tins contain discs of the same size with numbers on them.

Tin *A*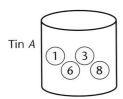

Tin *B*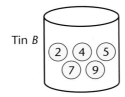

(a) A disc is drawn at random from tin *A* and a disc is drawn at random from tin *B*.
The numbers on the two discs are added to give the total score.
Copy and complete the table to show all the possible scores.

(2 marks)

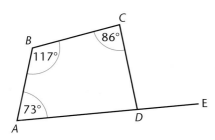

	Tin B				
	2	**4**	**5**	**7**	**9**
Tin A **1**	3	5	6		
3					
6					
8					

(b) Find the probability of scoring more than 11. *(2 marks)*

16 Solve these equations.
(a) $6x + 3 = 15$ *(2 marks)*
(b) $2(y + 3) = 26$ *(3 marks)*
(c) $8z + 5 = 20 - 2z$ *(3 marks)*

17 *ABCD* is a quadrilateral with angles as shown.
Side *AD* is extended to *E*.

Calculate the size of angle *CDE*. *(3 marks)*

18 The cost of hiring a carpet cleaner is £15.50 for the first day then
£2.60 for each additional day.
Jim pays a total of £25.90.
For how many days did he hire the carpet cleaner?   *(3 marks)*

19 Sam travels from Newcastle to Birmingham in his car.
The distance he travels is 210 miles.
The journey takes 3 hours 30 minutes.
Calculate Sam's average speed.   *(3 marks)*

20 Work out
  (a) $9.3 - 2.76$ *(1 mark)*
  (b) $4.81 \times 2 + 15.5$ *(3 marks)*
  (c) $1.7 \times 10 + 0.4 \times 100$ *(2 marks)*

21 A triangle has angles of $x°$, $(x + 20)°$ and $(2x - 60)°$ as shown in the diagram.

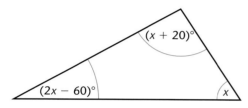

Find the value of x.   *(3 marks)*

22 Simplify   **(a)** $w^5 \times w^4$ *(1 mark)*
                   (b) $x^9 \div x^6$ *(1 mark)*

23 A is the point $(-6, 7)$ and B is the point $(2, 1)$.
Find the mid-point of the line segment AB.   *(2 marks)*

24 Use approximations to estimate the value of

$$\frac{217 \times 2.94}{0.396}$$

You **must** show your working.   *(3 marks)*

25 The diagram shows a shape made up of three sides of a rectangle and
a semi-circle.
The longest side of the rectangle is 20 cm and the shorter sides are of
length 8 cm.

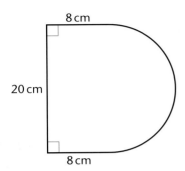

Calculate the perimeter of this shape.
Give your answer in terms of π.   *(3 marks)*

26 The scatter graph shows the results of class tests in Maths and Science.

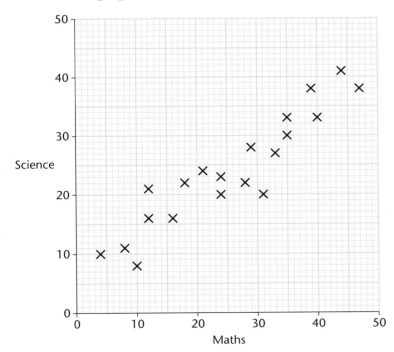

  (a) What was the highest Science score?   *(1 mark)*
  (b) Copy the scatter graph and draw on a line of best fit.   *(1 mark)*
  (c) Describe the relationship between the Maths and Science scores.   *(1 mark)*
  (d) Sarah was absent for the Science test but scored 37 on the Maths test.
      Use your line of best fit to estimate her score on the Science test.   *(1 mark)*

Exam paper, calculator

1 Ron has a mobile phone and pays £6 per month plus 3p per minute for all calls.

    (a) In March he made no calls, how much was his bill? *(1 mark)*

    (b) In April he made 120 minutes of calls, how much was his bill? *(2 marks)*

    (c) In May Ron's bill was £10.50, how many minutes of calls did he make? *(2 marks)*

2 Here is a list of numbers.　1, 8, 15, 24, 29, 42, 50

    From this list write down

    (a) a multiple of 7, *(1 mark)*

    (b) a factor of 45, *(1 mark)*

    (c) a square number, *(1 mark)*

    (d) a prime number. *(1 mark)*

3 State the most appropriate metric unit for these measurements:

    (a) the length of a garden, *(1 mark)*

    (b) the mass of a jar of marmalade, *(1 mark)*

    (c) the distance from London to Manchester, *(1 mark)*

    (d) the volume of the petrol tank of a car. *(1 mark)*

4 Toby went on holiday and recorded the number of hours of sunshine each day. Here are his results.

Monday	◎ ◎	◎ = 2 hours of sunshine
Tuesday	◎ ◎ ⌒	
Wednesday	◎ ⌒	
Thursday	◎ ◎ ◎	
Friday	◎ ◎ ◎ ◎ ⌒	
Saturday		

    (a) How many more hours of sunshine were there on Friday than on Monday? *(1 mark)*

    (b) On Saturday there were 7 hours of sunshine. Copy and complete the pictogram. *(2 marks)*

5 Monica is paid each week using this formula:

total pay = rate per hour × number of hours worked + bonus

Calculate her pay in a week when she works for 37 hours at
£6.24 per hour and she receives a bonus of £25.   *(2 marks)*

6 **(a)** Measure and write down the size of angle *x*.   *(1 mark)*

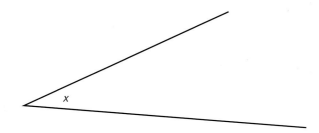

    (b) What type of angle is this?   *(1 mark)*

7 Holly keeps receipts for what she spends.
This table shows the number of each type of receipt she had during
one month.

Meals	Clothes	Groceries	Cinema	Petrol	Other
5	4	10	2	3	6

Holly picks a receipt at random.
What is the probability that the receipt she picks is

    (a) for groceries   *(1 mark)*

    (b) for a meal or going to the cinema   *(1 mark)*

    (c) not for petrol?   *(1 mark)*

8 Here is a number machine.

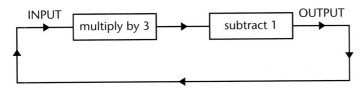

The output is fed back in again to generate the next number.
If you start with an input of 2, write down the first two output
numbers.   *(2 marks)*

9 Here is a sequence of patterns of matchsticks.

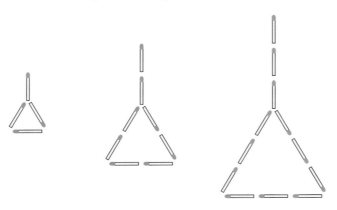

   (a) Draw the next two diagrams in this sequence.   *(2 marks)*
   (b) How many matchsticks will be in the 14th diagram of the
       sequence?   *(2 marks)*

10 Write the number 23.01974
   (a) to 1 decimal place   *(1 mark)*
   (b) to 2 decimal places   *(1 mark)*
   (c) to 3 decimal places.   *(1 mark)*

11 Jon estimates the weight of a cake to be $4\frac{1}{2}$ kg.
    Jo estimates the weight of the same cake to be 13 pounds.
    The actual weight of the cake is 11 pounds.
    Who made the best estimate?
    You **must** show your working.   *(3 marks)*

12 Here are some February temperatures, in °C, of seven European cities.

Glasgow	London	Madrid	Warsaw	Rome	Oslo	Moscow
−8	−1	5	−17	7	−19	−23

   (a) What is the temperature difference between the warmest place
       and the coldest place?   *(2 marks)*
   (b) Calculate the mean temperature.   *(2 marks)*

13 At a garden centre, tomato plants can be bought for 45p each or
    £5.20 for a box of 12.
    Work out the cheapest price for 18 tomato plants.   *(3 marks)*

14 (a) Complete the table for the graph of $y = 2x - 3$ *(2 marks)*

x	0	2	4	6
y		1		

(b) Draw the graph of $y = 2x - 3$ for values of x from 0 to 6.   *(1 mark)*

15 Use the formula $A = \dfrac{(a + b) \times h}{2}$

to calculate A when $a = 2.6$, $b = 7.8$ and $h = 6.4$.   *(3 marks)*

16 Ticket prices for the '*House of Horrors*' are
Adult £8   Child £6   Family £24
A family ticket can be used for up to 2 adults and up to 3 children.
(a) How much does a family of 2 adults and 3 children save by
buying a Family ticket?   *(2 marks)*
(b) A family of 1 adult and 5 children visit the '*House of Horrors*'.
Work out the cheapest price they could pay.   *(2 marks)*

17 Using a scale of 1 cm for 50 m, make an accurate construction of a
triangular field that has sides of 550 m, 250 m and 400 m.   *(3 marks)*

18 Convert a speed of 60 mph to km/h.   *(2 marks)*

19 Which of these fractions is nearest to $\frac{1}{4}$?
(a) $\frac{2}{9}$ **(b)** $\frac{5}{18}$ **(c)** $\frac{7}{27}$ **(d)** $\frac{11}{36}$ *(2 marks)*

20 Calculate   **(a)** $4.32 + \sqrt{(18.4)}$
                    (b) the cube of 2.9
Write down all the figures of your calculator display.   *(2 marks)*

21 Simon takes £350 spending money on holiday.
The exchange rate is £1 = 1.48 euros.
(a) How much is £350 in euros?   *(2 marks)*
(b) He spends 472 euros.
How much money does he bring home?
Give your answer in £ correct to the nearest penny.   *(3 marks)*

22 Fifteen runners in the school cross-country race recorded these times (to the nearest minute).

36, 29, 40, 38, 46, 40, 27, 31, 42, 51, 34, 31, 40, 30, 35

   (a) Draw an ordered stem-and-leaf diagram to show this information.
      Use a key of 3 | 6 to represent 36 minutes.   *(3 marks)*

   (b) Using your stem-and-leaf diagram, or otherwise, write down
      (i) the mode   **(ii)** the median   **(iii)** the range
      of these times.   *(3 marks)*

23 p is an odd number and q is an even number.
For each of these expressions say whether they are odd, even or could be either.

   (a) $p + q$ *(1 mark)*
   (b) $pq + 1$ *(1 mark)*
   (c) $p^2 + 1$ *(1 mark)*
   (d) $p^2 + q^2$ *(1 mark)*

24 Ian buys some building materials and the bill is £460 + VAT.
If VAT is charged at the rate of $17\frac{1}{2}\%$, how much is Ian's total bill?   *(3 marks)*

25 The diagram shows triangle A at points (1, 3), (1, 4) and (3, 4).

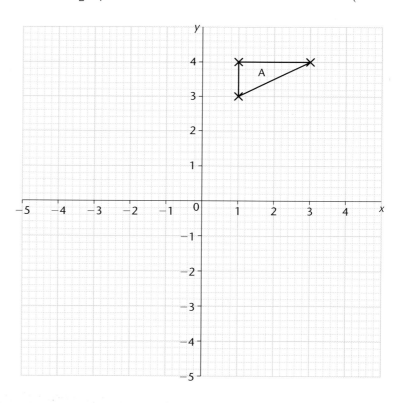

Copy the diagram and:

(a) reflect triangle A in the line $y = 1$. Label the image B.   *(2 marks)*

(b) rotate triangle A through 90° anticlockwise about the origin.
Label the image C.   *(3 marks)*

26 Factorise   (a) $8m + 10$

   (b) $y^2 - 5y$ *(2 marks)*

27 A dog's mass increases from 13 kg to 16 kg.
What percentage increase is this?
Give your answer to 3 significant figures.   *(4 marks)*

28 Draw the points P and Q, 5 cm apart (see diagram).

$P \bullet$ $\bullet Q$

Shade in the region where all points satisfy both of these conditions.

(a) They are less than 3 cm from Q.

(b) They are nearer to P than to Q.   *(3 marks)*

29 This table shows how long people talk on their mobile phone when
making a phone call.

Time, t (minutes)	Frequency (number of people)
$0 < t \leqslant 10$	9
$10 < t \leqslant 20$	25
$20 < t \leqslant 30$	16
$30 < t \leqslant 40$	10
$40 < t \leqslant 50$	5

(a) Draw a frequency polygon to represent this data.
Use a scale of 2 cm for 10 minutes horizontally and 2 cm for
10 people vertically.   *(2 marks)*

(b) Which class interval contains the median?   *(1 mark)*

30 Mavis is using trial and improvement
to solve the equation $x^3 - 2x = 16$.
The table shows her first trial.

x	$x^3 - 2x$	comment
2	4	too small

Continue the table to find a solution to the equation.
Give your answer correct to 1 decimal place.   *(5 marks)*

Answers

Chapter 1 Number 1

Exercise 1A

1 (a) 4 (b) 400 (c) 4000 (d) 40 000
   (e) 400 000
2 300
3 Ten thousand
4 (a) 5000   (b) 9   (c) 800   (d) 60 000
   (e) 500 000 (f) 70   (g) 200   (h) 40 000
   (i) 700 000 (j) 40 000 (k) 6   (l) 3000

Exercise 1B

1 (a) one thousand six hundred and twenty five
   (b) twenty one thousand eight hundred
   (c) four thousand and four
   (d) three million, three hundred and three
   (e) one hundred and one thousand and ten
2 (a) 1145   (b) 7 700 000   (c) 15 000 505
   (d) 60 006   (e) 202 202
3 eight thousand four hundred and fifty
4 34 603
5 four hundred and sixty five thousand pounds,
   two hundred and eighty thousand five hundred
   pounds,
   one hundred and sixty five thousand two hundred and
   fifty pounds,
   seventy nine thousand nine hundred and ninety five
   pounds,
   fifty three thousand seven hundred and fifty pounds
6 23 408, 39 062, 80 304, 100 009

Exercise 1C

1 (a) 1546   (b) 125 462 (c) 5642   (d) 19 350
   (e) 7203   (f) 1 203 000
2 (a) 79, 87, 93, 108, 112, 145, 205
   (b) 203, 213, 230, 231
   (c) 16 324, 16 432, 17 939, 17 940
   (d) 236, 2136, 2345, 2349
   (e) 563 454, 1 567 324, 1 634 324
3 32 598, 32 486, 27 921, 25 645, 25 645
4 £1325, £1327, £2130, £2340, £2360, £3280
5 Sample C 160 983 000
6 Yes Tor 619 m, Sawell 683 m, Scafell Pike 979 m,
   Snowdon 1085 m, Ben Nevis 1344 m

Exercise 1D

1 (a) 6 < 9   (b) 34 < 41 (c) 53 > 19 (d) 159 > 143
2 (a) T   (b) F   (c) T   (d) T
3 (a) 487   (b) 158
   (c) any from 27 369 to 27 385
   (d) any from 5991 to 6000
4 (a) T   (b) T   (c) F   (d) T   (e) F

Exercise 1E

1 (a) 39   (b) 161   (c) 321
2 2435   3 531
4 (a) 679   (b) 19 591   (c) 1883   (d) 1515
5 470

---

6 (a) A = 40, B = 49, C = 42, D = 69, E = 64
   (b) 264
7

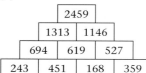

		2459		
	1313		1146	
694		619		527
243	451		168	359

8 391
9 463 miles

Exercise 1F

1 658
2 3465
3

×	8	14	28	123
5	40	70	140	615
9	72	126	252	1107
11	88	154	308	1353

4 8448
5 (a) 312   (b) 2356   (c) 6847   (d) 17 658
6 432
7 3060
8 330
9 227 100
10 (a) £3780   (b) £749   (c) £4529

Exercise 1G

1 55   2 56   3 325
4 (a) 657   (b) 1293   (c) 265   (d) 1438
5 44
6 £4800
7 108
8 90
9 Peter 63, Gill 42, Laura 54
10 2436 miles

Exercise 1H

1 (a) 6   (b) 64   (c) 133
2 14
3 (a) 7   (b) 8   (c) 14
4 £5   5 13   6 372
7 31   8 134   9 7   10 6

Exercise 1I

1 1074   2 16   3 13
4 £9   5 £56   6 96 hours
7 (a) 48   (b) 4
8 (a) 30   (b) 60   (c) 105
9 172
10 360
11 (a) 360   (b) 57
12 (a) 79   (b) £108.60

Exercise 1J

1 (a) 17   (b) 37   (c) 27   (d) 63
   (e) 15   (f) 18
2 (a) 28   (b) 8   (c) 20   (d) 30
   (e) 6   (f) 8
3 (a) 43   (b) 36   (c) 16   (d) 85
   (e) 21   (f) 2
4 (a) $3 \times 2 + 4 = 10$ (b) $3 \times (2 + 4) = 18$
   (c) $(5 + 1) \times (7 + 3) = 60$ (d) $10 \times (6 - 2) + 3 = 43$
   (e) $2 \times 7 - 6 \div 2 = 11$ (f) $9 \times (2 + 3) - 20 = 25$
5 fantastic, E

Exercise 1K

1 10 °C
2 (a) 11 °C   (b) −3 °C
3 (a)

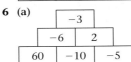

   (b) 7, 1, 0, −2, −6
4 −25 °C, −14 °C, −5 °C, −2 °C, 3 °C
5 (a) bird 22 m, flagpole 14 m, deck chair 5 m,
     shark 0 m, diver −10 m, fish −20 m, star fish −25 m
   (b) 5 m   (c) 15 m   (d) −15 m
6 −25 °C, −22 °C, −18 °C, −15 °C, −8 °C, −6 °C,
   −4 °C, −2 °C

Exercise 1L

1 (a) 5   (b) −2   (c) −11   (d) −11
   (e) −4   (f) −7
2 (a) 3   (b) −11   (c) −6   (d) −18
   (e) −1   (f) 5   (g) 1   (h) 11
3 (a) −4 °C, −3 °C, −1 °C, 1 °C, 2 °C   (b) 6 °C
4 (a) 11 °C   (b) 26 °C
5

+	4	1	−2	−3
5	9	6	3	4
2	6	3	0	−1
−1	3	0	−3	−4
−3	1	−2	−5	−6

−	3	1	−2	−3
4	1	3	6	7
2	−1	1	4	5
−3	−6	−4	−1	0
−5	−8	−6	−3	−2

6 (a) $-4 + 7 = 3$ (b) $8 - -2 = 10$
   (c) $7 + -9 = -2$ (d) $-2 - -3 = 1$

Exercise 1M

1 (a) −27   (b) 20   (c) −12   (d) −10
   (e) 9   (f) −24
2 (a) −5   (b) 9   (c) 4   (d) 4
   (e) −3   (f) 6
3 (a) $-9 \div -3 = +3$ (b) $21 \div -7 = -3$
   (c) $-40 \div 4 = -10$ (d) $36 \div -6 = -6$
   (e) $34 \div 2 = 17$ (f) $-120/-12 = 10$
4 (a) $3 \times -7 = -21$ (b) $-4 \times -6 = 24$
   (c) $-3 \times 3 = -9$ (d) $6 \times -8 = -48$
   (e) $-5 \times -5 = 25$ (f) $3 \times 40 = 120$

5

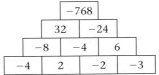

6 (a)

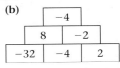

Exercise 1N

1 (a) −2   (b) −4   (c) −5   (d) 2
   (e) 2   (f) −14   (g) −6   (h) 7
   (i) −20   (j) −4   (k) −21   (l) −6
   (m) 30   (n) 0   (o) 3   (p) 11
2 (a) 4 °C   (b) 0 °C   (c) −4 °C   (d) −7 °C
   (e) −15 °C
3 12 °C
4 (a) $-6 + -3 = -9$ (b) $5 + -6 = -1$
   (c) $4 - 3 = 1$ (d) $8 - -2 = 10$
   (e) $5 \times -6 = -30$ (f) $4 \times -3 = -12$
   (g) $12 \div -4 = -3$ (h) $-2 - -8 = 6$
5 −28 °C
6 11 °C
7 −3 °C

Exercise 1O

1 (a) 5 thousandths   (b) 5 hundredths
   (c) 5 tenths
2 Possible answers include:
   (a) eight tenths
   (b) four tenths no hundredths and seven thousandths
   (c) one point two five six
3 (a) 0.4   (b) 1.4   (c) 0.002   (d) 6.103
4 (a) 4 tenths   (b) 4 hundredths
   (c) 4 tenths   (d) 4 thousandths
   (e) 4 units
5 (a) 0.1   (b) 0.03   (c) 0.43   (d) 0.27   (e) 0.304
6 two point four
7 0.001

Exercise 1P

1 (a) 52   (b) 520   (c) 5200   (d) 4.8
   (e) 6400   (f) 38.5   (g) 300   (h) 0.6
   (i) 6.6
2 (a) £99.90   (b) £999
3 (a) 0.64   (b) 0.064   (c) 0.9   (d) 0.085
   (e) 1.32   (f) 0.002   (g) 0.014   (h) 0.004
   (i) 2.15
4 5.34 kg
5 (a) 360   (b) 0.49   (c) 73.2   (d) 0.128
   (e) 0.5   (f) 0.7   (g) 1.53   (h) 602
   (i) 0.05
6 800 g
7 £675
8 £2.35
9 (a) 17 000   (b) 170   (c) 17
10 (a) T   (b) F ($4000 = 100 \times 40$)
   (c) T   (d) T   (e) F ($60 = 6000 \div 600$)

Exercise 1Q

1 (a) 0.05   (b) 1.23   (c) 6.5   (d) 4.023
2 (a) 0.4, 0.41, 0.6   (b) 2.345, 2.349, 2.36
   (c) 0.2, 0.203, 0.23, 0.231   (d) 9.305, 9.5, 9.503, 9.53
3

0.05   0.25   0.65 0.75 0.85 0.95

4 (a) 6.4 > 6.04   (b) 0.305 > 0.3
   (c) 7.23 > 7.203   (d) 11.206 > 11.026
5 Various possible answers
6 10.3, 10.302, 10.32, 10.325, 10.4
7 23.4, 33.04, 33.35, 33.38, 33.4
8 (a) 25.3, 25.4, 25.5   (b) 3.69, 3.68, 3.67
   (c) 4.50, 4.51, 4.52... 4.59, 4.60

Exercise 1R

1 (a) 0.86   (b) 20.835   (c) 6.963   (d) 18.144
2 £5.99
3 4.295 m
4 (a) 6.39 m   (b) 6.47 m   (c) 12.86 m
5 35.05 kg
6 £19.59

Exercise 1S

1 (a) 4.5   (b) 1.74   (c) 4.33   (d) 7.58
2 5.8 kg
3 0.946
4 54.6 cm
5 59.71 seconds
6 £2.36

Exercise 1T

1 (a) 9.6   (b) 26.8   (c) 7.1   (d) 6.36
2 (a) 3.01   (b) 8.06   (c) 1.232   (d) 57.4
3 12.6 m
4 60.8 cm^2
5 22.09 m^2
6 £5.18

Exercise 1U

1 (a) 5.45   (b) 6.28   (c) 14.5   (d) 0.321
   (e) 0.0425
2 (a) $\frac{16.4}{5}$ (b) $\frac{10\,432}{1}$ (c) $\frac{735}{7}$ (d) $\frac{1336}{15}$
3 (a) 21   (b) 670   (c) 145   (d) 30   (e) 250
4 (a) 34   (b) 130   (c) 3.3   (d) 50   (e) 0.15
5 £9.40   **6** £2.90
7 6, 0.4 m   **8** 8.2 m

Exercise 1V

1 £2.25   **2** £3.85   **3** £477.30
4 (a) 14   (b) 10p
5 (a) £68.97   (b) £11.03
6 (a) 9   (b) £1080   (c) £2.70
7 £60
8 £17.73
9 (a) 16 500   (b) £546

10

Exercise 1W

1 (a) 5 thousand   (b) 9 units
   (c) 8 hundred   (d) 6 ten thousand
   (e) 5 hundred thousand   (f) 7 million
   (g) 2 hundred million   (h) 4 ten million
2 (a) 35 030
   (b) one hundred and eight thousand one hundred
     and eight
3 35 654, 35 456, 31 545, 29 248, 27 427
4 Any value in the range:
   (a) 26 to 29   (b) 4290 to 4351
   (c) 89 522 or below   (d) 619 or above
5 746
6 5504
7 1564 miles
8 159 m
9 (a) 13   (b) 9   (c) 36   (d) 17
   (e) 16   (f) 61   (g) 21   (h) 11
10 9 °C
11 (a) 4   (b) −10   (c) 11   (d) −12
12 (a) −32   (b) 15   (c) −7   (d) 8
13 (a) 9 tenths   (b) 9 hundredths   (c) 9 tenths
   (d) 9 thousandths   (e) 9 units
14 (a) 16.2   (b) 0.342   (c) 1460   (d) 0.325
   (e) 580   (f) 0.9   (g) 0.054   (h) 0.6
15 (a) 1.54 kg   (b) 60 g
16 12.3, 12.304, 12.34, 12.4, 12.405
17 £43.53   **18** 39.75 kg
19 62.055 m^2 **20** 6.5 units

Examination style questions

1 (a)

Leeks	3 kg at £1.60 per kg	£4.80
Bananas	2 kg at £1.15 per kg	£2.30
3 bottles of water at £0.75 each		£2.25
	Total	£9.35

   (b) 4   (c) £15.59
2 (a) £24.10
   (b) (i) 8
      (ii) 7 adults, 3 children or 1 adult, 13 children
   (c) (i) Wednesday   (ii) Thursday
3 40 cm
4 (a) 700   (b) 20 000
5 (a) (i) Seven thousand four hundred and eighty
       three
     (ii) 7000
   (b) (i) 2, 58
     (ii) 29, 39

6

BEST BURGERS		
	£	Pence
3 burgers at £1.49	4	47
4 large fries at £1.19	4	76
2 milkshakes at £1.10	2	20
3 colas at 69p	2	07
Total	13	50

7 (a) 2
   (b) (i) $18 - (4 - 2) = 16$
      (ii) $(3 + 4) \times 5 = 35$
      (iii) $20 \div (5 - 3) = 10$
8 (a) (i) three thousand seven hundred and forty eight
      (ii) 3700
      (iii) 3750
   (b) 2193
9 (a) London (b) Moscow (c) 3 °C   (d) 1 °C

Chapter 2   Number 2

Exercise 2A

1 (a) 40   (b) 590   (c) 250
   (d) 3190   (e) 24 390
2 (a) 600   (b) 4300   (c) 6900
   (d) 25 400   (e) 165 400
3 (a) 9000   (b) 15 000   (c) 244 000
   (d) 424 000   (e) 401 000
4 (a) (i) 156 380, 285 920 (ii) 156 400, 285 900
      (iii) 156 000, 286 000 (iv) 160 000, 290 000
      (v) 200 000, 300000
   (b) Not sensible as someone will be born or die so the numbers change constantly.
5 (a) 2 355 000, 15 000, 9000, 12 000
   (b) Not sensible as they need to know exactly how many items they have in stock.
6 (a) T   (b) F   (c) T   (d) F   (e) T

Exercise 2B

1 5   **2** 12   **3** 8
4 (a) 8 litres   (b) 3
5 (a) 208 333 (b) 5208   (c) 289   (d) 6

Exercise 2C

1 (a) 5.8   (b) 7.4   (c) 2.2   (d) 5.7
   (e) 4.3   (f) 15.8   (g) 11.3 (h) 17.2
   (i) 145.1 (j) 522.0
2 (a) 3.26   (b) 6.54   (c) 0.88 (d) 0.03
   (e) 11.06 (f) 4.01   (g) 3.90 (h) 2.31
   (i) 0.00   (j) 5.13
3 (a) 1.255 (b) 5.293 (c) 4.127 (d) 0.001
   (e) 0.000
4 (a) 15.2   (b) 15.15 (c) 15.153 (d) 15
5 (a) £15   (b) £14.80
6 (a) 6 kg   (b) 5.6 kg   (c) 5.63 kg
7 (a) 16 secs   (b) 15.6 secs   (c) 15.63 secs
   (d) 16 secs, 15.6 secs, 15.63 secs. These are the same answers as (a), (b) and (c)
   (e) If you rounded the results you might not be able to tell who had won.

Exercise 2D

1 (a) 400   (b) 3000   (c) 20 000
   (d) 300 000 (e) 80 000   (f) 5
2 (a) 610   (b) 6800   (c) 33 000
   (d) 150 000 (e) 270 000 (f) 5.3
3 (a) 0.04   (b) 0.04   (c) 0.006
   (d) 0.004   (e) 0.5
4 (a) 0.57   (b) 0.000 38   (c) 0.0019
   (d) 0.20   (e) 0.0020
5 (a) 300 000 (b) 290 000
6 (a) 50 000, 20 000, 10 000, 6000, 2000
   (b) 48 400, 21 900, 11 600, 5940, 2050
   (c) Possible answers include: for the first 2 or 3 they will round to 1 s.f. but for those with a smaller attendance they may use 3 s.f.
7 0.039 mm
8 (a) 20   (b) 15   (c) 15.0

Exercise 2E

1 Answers may vary:
   (a) 3500   (b) 4000   (c) 3 or 4 (d) 30
   (e) 10 000 (f) 30
2 3551, 5016, 3.34, 24.97, 10 889.45, 29.05
3 (a) 20   (b) 5   (c) 1000   (d) 3
   (e) 10   (f) 0.5
4 22.43, 4.41, 1423.3, 2.85, 9.43, 0.54
5 900 miles
6 1000
7 80 m, yes they have enough rod.
8 1800
9 4 million
10 120 km

Examination style questions

1 $\dfrac{40 \times 200}{80} = 100$

2 $70p \times 20p = £14$

3 $\dfrac{300 \times 8}{0.4} = 6000$

4 (a) 20 488   (b) (i) 34.2   (ii) 34.250
5 (a) 1479   (b) 9714   (c) 9700
6 (a) (i) 4.1666666
      (ii) 4
   (b) (i) 1.7818181
      (ii) 1.78 (2 d.p.)
7 (a) 16 000   (b) Barton, Carton, Arton
8 14.3

Chapter 3   Number 3

Exercise 3A

1 (a) (i) $\frac{1}{4}$ (ii) $\frac{3}{4}$ (b) (i) $\frac{2}{5}$ (ii) $\frac{3}{5}$
   (c) (i) $\frac{5}{12}$ (ii) $\frac{7}{12}$ (d) (i) $\frac{4}{6}$ (ii) $\frac{2}{6}$
   (e) (i) $\frac{3}{8}$ (ii) $\frac{5}{8}$
2 Various possible answers
3 (a) $\frac{9}{17}$ (b) $\frac{6}{17}$ (c) $\frac{2}{17}$
4 (a) 21   (b) $\frac{12}{21}$ (c) $\frac{9}{21}$

5

Activity	Sleeping	School	Eating
Hours	10	7	1
Fraction	$\frac{10}{24}$	$\frac{7}{24}$	$\frac{1}{24}$

Activity	TV	Football	Homework
Hours	2	3	1
Fraction	$\frac{2}{24}$	$\frac{3}{24}$	$\frac{1}{24}$

6 (a) $\frac{6}{13}$ (b) $\frac{4}{13}$ (c) $\frac{3}{13}$

Exercise 3B

1 Various possible answers
2 $\frac{8}{20}, \frac{18}{45}, \frac{14}{35}$
3 (a) $\frac{3}{5} = \frac{6}{10} = \frac{9}{15} = \frac{30}{50} = \frac{36}{60}$
   (b) $\frac{4}{7} = \frac{8}{14} = \frac{12}{21} = \frac{20}{35} = \frac{40}{70}$
   (c) $\frac{3}{4} = \frac{6}{8} = \frac{15}{20} = \frac{21}{28} = \frac{75}{100}$
4 (a) $\frac{3}{4}$ (b) $\frac{3}{5}$ (c) $\frac{2}{3}$ (d) $\frac{3}{10}$
   (e) $\frac{5}{6}$ (f) $\frac{2}{3}$ (g) $\frac{5}{6}$ (h) $\frac{3}{5}$
5 $\frac{1}{2}, \frac{1}{4}$
6 (a) $\frac{1}{6}, \frac{1}{3}, \frac{1}{2}$
   (b) Answers should show $\frac{1}{6}, \frac{2}{6}$ and $\frac{3}{6}$ labelled

Exercise 3C

1 (a) $\frac{1}{4}, \frac{3}{8}, \frac{1}{2}$ (b) $\frac{4}{10}, \frac{1}{2}, \frac{3}{5}$ (c) $\frac{1}{3}, \frac{3}{4}, \frac{5}{6}$
2 (a) $\frac{24}{40} < \frac{25}{40}$ (b) $\frac{10}{30} > \frac{9}{30}$ (c) $\frac{15}{18} > \frac{8}{18}$
   (d) $\frac{9}{12} < \frac{10}{12}$ (e) $\frac{21}{35} < \frac{25}{35}$ (f) $\frac{5}{20} < \frac{8}{20}$
3 (a) $\frac{1}{4}, \frac{3}{5}, \frac{7}{10}$ (b) $\frac{7}{12}, \frac{2}{3}, \frac{5}{6}$ (c) $\frac{17}{40}, \frac{5}{8}, \frac{7}{10}$
   (d) $\frac{13}{25}, \frac{57}{100}, \frac{14}{20}$
4 $\frac{9}{40}, \frac{7}{30}, \frac{18}{72}$, Meg has the biggest fraction.

Exercise 3D

1 (a) 6 m   (b) £24   (c) 36 kg
   (d) 28 l (e) 28 min   (f) 30 pages
2 (a) 135 km   (b) 77 days   (c) £546   (d) 136 euro
   (e) 337.5 g   (f) 3.33 ℓ
3 (a) $\frac{2}{3}$ of 24   (b) $\frac{7}{8}$ of 96   (c) $\frac{1}{4}$ of 440   (d) $\frac{2}{3}$ of 69
4 15 hours
5 (a) £7.50   (b) £2.50
6 $\frac{2}{3}$ of £75
7 £357.68

Exercise 3E

1 (a) $4\frac{1}{2}$ (b) $2\frac{2}{3}$ (c) $4\frac{4}{5}$ (d) $3\frac{3}{4}$ (e) $2\frac{3}{10}$
   (f) $2\frac{53}{100}$ (g) $3\frac{2}{7}$ (h) $5\frac{2}{9}$ (i) 9
2 (a) $3\frac{2}{5} < 5\frac{1}{2}$ (b) $3\frac{1}{3} > 2\frac{5}{9}$ (c) $3\frac{3}{4} < 4\frac{1}{6}$
   (d) $6\frac{1}{2} > 5\frac{3}{8}$ (e) $8\frac{3}{10} > 7\frac{13}{20}$ (f) $6 > 5$
3 Megan

Exercise 3F

1 (a) $\frac{5}{2}$ (b) $\frac{21}{5}$ (c) $\frac{27}{4}$ (d) $\frac{16}{3}$ (e) $\frac{20}{9}$
   (f) $\frac{29}{8}$ (g) $\frac{29}{9}$ (h) $\frac{37}{10}$ (i) $\frac{16}{7}$

2 Molly $\frac{9}{4}$, Shona $\frac{7}{4}$
3 (a) $3\frac{1}{2} = \frac{7}{2}$ (b) $\frac{17}{5} > 2\frac{4}{5}$ (c) $\frac{11}{4} < 3\frac{1}{4}$
   (d) $1\frac{8}{9} > \frac{16}{9}$ (e) $2\frac{4}{11} < \frac{28}{11}$ (f) $\frac{300}{50} = 6$

Exercise 3G

1 (a) $\frac{6}{8}$ (b) $1\frac{2}{9}$ (c) $1\frac{2}{11}$
2 (a) $1\frac{1}{12}$ (b) $\frac{9}{10}$ (c) $1\frac{4}{15}$
   (d) $1\frac{11}{20}$ (e) $1\frac{1}{6}$ (f) $\frac{13}{24}$
3 (a) $3\frac{11}{12}$ (b) $3\frac{1}{2}$ (c) $2\frac{7}{10}$
   (d) $4\frac{1}{20}$ (e) $4\frac{4}{30}$ (f) $4\frac{1}{24}$
4 $\frac{5}{6}$
5 $5\frac{7}{12}$ metres
6 $6\frac{1}{4}$ hours
7 $2\frac{19}{24}$ kg

Exercise 3H

1 (a) $\frac{2}{8}$ (b) $\frac{4}{6}$ (c) $\frac{4}{5}$
2 (a) $\frac{1}{10}$ (b) $\frac{1}{18}$ (c) $\frac{7}{20}$
3 (a) $1\frac{7}{12}$ (b) $\frac{8}{15}$ (c) $1\frac{11}{24}$
4 $\frac{11}{12}$
5 $1\frac{3}{4}$ metres
6 $\frac{5}{6}$ litres
7 (a) $\frac{11}{15}$ (b) $\frac{4}{15}$
8 $\frac{1}{2}$

Exercise 3I

1 (a) $\frac{6}{35}$ (b) $\frac{5}{36}$ (c) $\frac{1}{4}$
   (d) $\frac{1}{16}$ (e) $\frac{4}{7}$ (f) $\frac{5}{24}$
2 (a) $\frac{1}{4}$ (b) $\frac{2}{3}$ (c) $\frac{3}{5}$
3 (a) 4   (b) $1\frac{1}{2}$ (c) 2
   (d) 3   (e) $2\frac{1}{12}$ (f) $1\frac{5}{16}$
4 $10\frac{2}{5}$ inches^2
5 1 kg
6 $1\frac{1}{4}$ m^2
7 35 min

Exercise 3J

1 (a) $\frac{3}{2}$ (b) $\frac{5}{4}$ (c) $\frac{7}{9}$ (d) $\frac{9}{23}$
2 (a) $\frac{7}{3}$ (b) $\frac{2}{9}$ (c) $\frac{1}{9}$ (d) $\frac{5}{19}$
   (e) $\frac{8}{21}$ (f) $\frac{3}{17}$ (g) $\frac{1}{18}$ (h) $\frac{7}{73}$

Exercise 3K

1 (a) $\frac{1}{12}$ (b) $1\frac{1}{15}$ (c) $17\frac{1}{2}$
   (d) $\frac{1}{16}$ (e) 2   (f) $\frac{5}{6}$
2 (a) 2   (b) 1   (c) $1\frac{1}{2}$
3 $12\frac{1}{2}$
4 $10\frac{1}{2}$
5 6
6 $\frac{3}{8}$

Exercise 3L

1 (a) $\frac{47}{100}$ (b) $\frac{1}{5}$ (c) $\frac{3}{4}$ (d) $\frac{21}{25}$
 (e) $\frac{1}{100}$ (f) $\frac{13}{20}$ (g) $\frac{24}{25}$ (h) $\frac{1}{8}$
2 (a) 0.69 (b) 0.43 (c) 0.4 (d) 0.25
 (e) 0.7 (f) 0.333 (g) 1.84 (h) 0.05
3 (a) (i) $\frac{3}{10}$ (ii) 0.3 (b) (i) $\frac{17}{20}$ (ii) 0.85
 (c) (i) $\frac{1}{50}$ (ii) 0.02 (d) (i) $1\frac{23}{200}$ (ii) 1.23
4 Sam

Exercise 3M

1 (a) 32% (b) 79% (c) 239% (d) 12.5%
2 (a) 0.7, 70% (b) 0.8, 80%
 (c) 0.75 75% (d) 0.375, 37.5%
3 7%, 10%, 32%, 40%
4 (a) 50% (b) 25% (c) 20% (d) 156.25%
5

Percentage	Fraction	Decimal
60%	$\frac{3}{5}$	0.6
48%	$\frac{12}{25}$	0.48
30%	$\frac{3}{10}$	0.3
175%	$\frac{7}{4}$	1.75
5%	$\frac{1}{20}$	0.05

6 English 70%, Maths 75%, History 80%
7 $\frac{1}{4}$ off
8 (a) $\frac{14}{25}$, 0.625, 64%, $\frac{7}{10}$ (b) 0.4, 42%, 0.438, $\frac{8}{20}$

Exercise 3N

1 (a) $\frac{3}{10}$ (b) $2\frac{7}{10}$ (c) $\frac{9}{100}$ (d) $\frac{7}{1000}$
 (e) $\frac{3}{4}$ (f) $\frac{1}{40}$ (g) $\frac{9}{20}$ (h) $\frac{5}{8}$
 (i) $2\frac{1}{20}$ (j) $16\frac{1}{8}$ (k) $4\frac{21}{200}$ (l) $23\frac{6}{25}$

Exercise 3O

1 (a) (i) 0.222222222 (ii) $0.\dot{2}$
 (b) (i) 0.166666666 (ii) $0.1\dot{6}$
 (c) (i) 0.636363636 (ii) $0.\dot{6}\dot{3}$
 (d) (i) 0.4166666 (ii) $0.41\dot{6}$
 (e) (i) 0.2666666 (ii) $0.2\dot{6}$
2 (a) $0.\dot{3}$ (b) $0.\dot{6}$ (c) $0.\dot{7}$ (d) $0.3\dot{6}$
 (e) $0.8\dot{3}$ (f) $0.8\dot{1}$ (g) $0.09\dot{2}\dot{5}$ (h) $1.1\dot{6}$
 (i) $3.58\dot{3}$ (j) $8.1\dot{6}$ (k) $2.\dot{4}$ (l) $3.\dot{1}4285\dot{7}$
3 (a) $\frac{63}{99}$ (b) $\frac{4}{9}$ (c) $\frac{17}{99}$

Examination style questions

1 $\frac{1}{2} + \frac{1}{3} = \frac{3}{6} + \frac{2}{6} = \frac{5}{6}$
2 (a) $\frac{2}{8}, \frac{6}{24}$ (b) 0.25
3 £770
4 $\frac{44}{15}$
5 $\frac{7}{30}$
6 (a) $\frac{3}{5}$ (b) £35 (c) $\frac{7}{10}$
7 (a) $\frac{1}{3}$ (b) Any 9 triangles shaded

Chapter 4 Algebra 1

Exercise 4A

1 (a) $x + 5$ (b) $w - 3$ (c) $m + 8$ (d) $d - 12$
 (e) $x + 6$ (f) $y - 2$ (g) $4 + p$ (h) $a - 1$
 (i) $x + y$ (j) $r - t$
2 (a) $4g$ (b) $5r$ (c) $6h$ (d) $3t$
3 (a) $3y$ (b) $\frac{z}{3}$ (c) $\frac{k}{4}$ (d) $8f$
 (e) $10n$ (f) $\frac{12}{x}$ (g) ab (h) g^2
4 (a) $x + 2$ (b) $7x$ (c) $x - 5$ (d) $2x$
 (e) $\frac{x}{2}$ or $\frac{1}{2}x$ (f) $3x - 2$ (g) $\frac{x}{4} + 5$ (h) x^2
5 (a) a^2 (b) x^2 (c) $4l$ (d) xy
6 (a) $4x$ (b) $6y$ (c) $4d$ (d) $13t$
 (e) $5j + 8k$ (f) $3a + 11b$
7 (a) $4n$ (b) $n + 2$ (c) $4(n + 2)$ (d) $n - 1$

Exercise 4B

1 (a) $3n$ (b) $5d$ (c) $6a$ (d) $2g$ (e) $10f$
 (f) $8h$ (g) $6c$ (h) $12t$ (i) $10x$ (j) $16l$
2 (a) $n + n + n$
 (b) $h + h + h + h + h$
 (c) $y + y + y + y + y + y + y + y$
 (d) $r + r + r + r$
 (e) $w + w$
 (f) $v + v + v + v + v + v$
3 (a) $3b$ (b) $5y$ (c) $4z$ (d) $6t$ (e) $4j$
 (f) $2u$ (g) $3h$ (h) $5t$ (i) $2x$ (j) $6r$
4 (a)

$2x$	$7x$	$6x$
$9x$	$5x$	x
$4x$	$3x$	$8x$

(b)

$7y$	$13y$	$4y$
$5y$	$8y$	$11y$
$12y$	$3y$	$9y$

Exercise 4C

1 (a) $5c + 7d$ (b) $5m + 4r$ (c) $7x + 4y$
 (d) $10a + 8b$ (e) $5q + 8$ (f) $12p + 5$
 (g) $11j + 5$ (h) $8w + 10$
2 (a) $3x + 4y$ (b) $2a + 3b$ (c) $4k + 3m$
 (d) $8h + 7j$ (e) $7q + 3$ (f) $8p + 4$
 (g) $2t + 3$ (h) $z + 10$
3 (a) $12a$ (b) $4x + 4$ (c) $4x + 2y$
 (d) $10x + 2$
4 (a) $10a + 7b$ (b) $6m + 5r$ (c) $5x + 8y$
 (d) $11q + 9r$ (e) $2k + 6l$ (f) $2v - w$
 (g) $y + 3z$ (h) $c - 13d$
5 (a) $10x + 6y$ (b) $11p + 6q$ (c) $4g + h$
 (d) $2t + 6n$ (e) $2a + 8b - c$ (f) $6j + 4k + 7l$
 (g) $5d + 6e + 2f$ (h) $x - 5y + 4z$
6 (a) $7ab + 4a$ (b) $8x^2 + 9x$
 (c) $4t^2 + 5$ (d) $4xy + 8x^2 + 2x$
 (e) $3xy + 6x^2 + 7x$ (f) $3ab + 7a - 7b$
7 (a)

$6a + 7b$	$7a$	$2a + 8b$
$a + 6b$	$5a + 5b$	$9a + 4b$
$8a + 2b$	$3a + 10b$	$4a + 3b$

(b)

3a + 2b	4a + 3b	8a − 2b
10a − 3b	5a + b	5b
2a + 4b	6a − b	7a

(c)

7a + b + 2c	6b − 3c	8a − b + 4c
6a + 3c	5a + 2b + c	4a + 4b − c
2a + 5b − 2c	10a − 2b + 5c	3a + 3b

Exercise 4D

1 10k	**2** 18b	**3** 12x	**4** 20a
5 14h	**6** 12m	**7** 6ab	**8** 12cd
9 42pq	**10** 6gh	**11** 5xy	**12** 56jk
13 $12t^2$	**14** $42x^2$	**15** $30a^2$	**16** $4n^2$
17 24cd	**18** $7x^2$		

Exercise 4E

1 (a) 100 + 14 = 114   (b) 200 + 30 = 230
    (c) 240 + 32 = 272   (d) 420 + 18 = 438
    (e) 120 − 6 = 114   (f) 350 − 28 = 322
    (g) 300 − 6 = 294   (h) 320 − 24 = 296
2 (a) 5(m + 3) = 5m + 15   (b) 4(n + 7) = 4n + 28
3 (a) 5p + 30   (b) 3a + 15   (c) 7k + 14
    (d) 4m + 36b   (e) 35 + 5f   (f) 16 + 2q
    (g) 2a + 2b   (h) 5x + 5y   (i) 8g + 8h + 8i
    (j) 4u + 4v + 4w
4 (a) 2y − 16   (b) 3x − 15   (c) 6b − 24
    (d) 7d − 56   (e) 14 − 2x   (f) 32 − 4n
    (g) 5a − 5b   (h) 2x − 2y   (i) 28 + 7p − 7q
    (j) 8a − 8b + 48
5 (a) 6c + 18   (b) 12m + 8
    (c) 20t + 15   (d) 24y + 36
    (e) 12e + 4f   (f) 10p + 2q
    (g) 6a − 3b   (h) 18c − 12d
    (i) 2m − 8n   (j) 14x + 7y − 21
    (k) 18a − 24b + 6c   (l) 8u − 20v − 12w
6 (a) $2x^2 + 6x + 4$ (b) $3x^2 + 15x − 18$
    (c) $2a^2 − 2a + 4$ (d) $4y^2 − 12y − 40$
7 A and F, B and K, C and I, D and L, E and G, H and J

Exercise 4F

1 $b^2 + 4b$	**2** $a^2 + 5a$	**3** $k^2 − 6k$
4 $m^2 − 9m$	**5** $2a^2 + 3a$	**6** $4g^2 + g$
7 $2p^2 + pq$	**8** $t^2 + 5tw$	**9** $m^2 + 3mn$
10 $2x^2 − xy$	**11** $4r^2 − rt$	**12** $a^2 − 4ab$
13 $2t^2 + 10t$	**14** $3x^2 − 24x$	**15** $5k^2 + 5kl$
16 $6a^2 + 12a$	**17** $8g^2 + 2gh$	**18** $15p^2 − 10pq$
19 6xy + 15xz	**20** $12p^2 + 8pq$	

Exercise 4G

1 −6k − 8   **2** −6x − 18
3 −15n − 5   **4** −12t − 20
5 −12p + 3 = 3 − 12p   **6** −6x + 14 = 14 − 6x
7 −6x + 18 = 18 − 6x   **8** −10x + 15 = 15 − 10x

Exercise 4H

1 5y + 22   **2** 5k + 21   **3** 2a + 18
4 7t − 16   **5** 8y + 19   **6** 7x + 40

7 8x + 7   **8** 13n − 5   **9** 5x − 21
10 14x − 8   **11** 2b − 5   **12** 4m + 2
13 2k − 14   **14** 4p + 14   **15** 2g − 8
16 −4w − 5

Exercise 4I

1 A 2×5 C $2 \times 2t$ D $2 \times 4x$
2 B 3×3 E $3 \times y^2$
3 B $x \times 5$ C $x \times x$ E $x \times w$
4 A Yes $3 \times 4y$ and 3×2 B No   C Yes $3 \times 2q$ and 3×7
5 A Yes $x \times 4x$ and $x \times 2$ B No   C Yes $x \times y$ and $x \times t$

Exercise 4J

1 (a) 3(x + 5)   (b) 5(a + 2)
    (c) 2(x − 6)   (d) 4(m − 4)
    (e) 4(t + 3)   (f) 3(n + 6)
    (g) 2(b − 7)   (h) 4(t − 5)
2 (a) 5(p + 4)   (b) 2(a + 6)
    (c) 3(y + 5)   (d) 7(b + 3)
    (e) 4(q + 3)   (f) 6(k + 4)
    (g) 5(a + 1)   (h) 4(g + 2)
3 (a) 4(t − 3)   (b) 3(x − 3)
    (c) 5(n − 4)   (d) 2(b − 4)
    (e) 6(a − 3)   (f) 7(k − 1)
    (g) 4(r − 4)   (h) 6(g − 2)
4 (a) y(y + 7)   (b) x(x + 5)
    (c) t(t + 2)   (d) n(n + 1)
    (e) x(x − 7)   (f) z(z − 2)
    (g) p(p − 8)   (h) a(a − 1)
5 (a) 2(3p + 2)   (b) 2(2a + 5)
    (c) 2(2t − 3)   (d) 4(2m − 3)
    (e) 5(2x + 3)   (f) 3(2y − 3)
    (g) 4(a + 2b)   (h) 5(2p + q)
6 A and G, B and L, C and J, D and I, E and H, F and K

Exercise 4K

1 $a^2 + 9a + 14$	**2** $x^2 + 4x + 3$	**3** $x^2 + 10x + 25$
4 $t^2 + 3t − 10$	**5** $x^2 + 3x − 28$	**6** $n^2 + 3n − 40$
7 $x^2 + x − 20$	**8** $p^2 − 16$	**9** $x^2 − 13x + 36$

Examination style questions

1 7x grams
2 500 − 22x
3 (a) 7(x + 2)   (b) 10m − 3
4 7x + 17
5 (a) 5(2a + 1)   (b) c(c − 4)
6 (x + 2)(x + 3) = $x \times x + x \times 3 + 2 \times x + 2 \times 3$
               = $x^2 + 3x + 2x + 6$
               = $x^2 + 5x + 6$
7 (a)

	n − 9
n	n + 1
	n + 11

    (b) 4n + 3

Chapter 5 Algebra 2

Exercise 5A

1 (a) h = 2   (b) b = 11   (c) x = 1
    (d) k = 7   (e) y = 20   (f) a = 50
2 (a) s = 1   (b) d = 33   (c) r = 2
3 (a) f = 27   (b) w = 80   (c) n = 6
4 t + 25 = 43, t = 18

Exercise 5B

1 (a) $b = 3$ (b) $t = 5$ (c) $p = 3$
    (d) $g = 7$ (e) $x = 4$ (f) $n = 3$
    (g) $m = 13$ (h) $y = 60$ (i) $k = 7$
    (j) $h = 4$ (k) $r = 8$ (l) $a = 8$
2 (a) $t = 24$ (b) $y = 18$ (c) $f = 36$
    (d) $r = 33$ (e) $q = 36$ (f) $g = 72$
    (g) $s = 112$ (h) $b = 150$ (i) $n = 8$
    (j) $m = 20$ (k) $v = 48$ (l) $u = 900$
3 (a) $t = 6$ (b) $p = 3$ (c) $r = 56$
    (d) $m = 60$ (e) $x = 80$ (f) $g = 650$
    (g) $s = 60$ (h) $d = 11$ (i) $j = 6$
4 $\frac{t}{4} = 20, t = 80$

Exercise 5C

1 (a) $w = 4$ (b) $u = 5$ (c) $t = 8$
    (d) $m = 3$ (e) $d = 2$ (f) $g = 9$
    (g) $s = 3$ (h) $b = 0$ (i) $a = 5$
2 (a) $r = 9$ (b) $h = 28$ (c) $f = 30$
    (d) $x = 2$ (e) $d = 110$ (f) $a = 9$
3 (a) $x = 8$ (b) $t = 5$ (c) $w = 10$
    (d) $a = 2$ (e) $n = 4$ (f) $q = 4$

Exercise 5D

1 (a) $b = 4\frac{1}{2}$ (b) $c = 1\frac{2}{7}$ (c) $y = 3\frac{1}{3}$
    (d) $j = 2\frac{5}{7}$ (e) $g = 1\frac{2}{3}$ (f) $p = 1\frac{1}{2}$
2 (a) $k = 1.25$ (b) $e = 1.5$ (c) $i = 2.2$
    (d) $s = 6.2$ (e) $m = 1.75$ (f) $a = 7.5$

Exercise 5E

1 (a) $k = -2$ (b) $f = -1$ (c) $m = -3$
    (d) $g = -5$ (e) $w = -2$ (f) $s = -2$
    (g) $u = -2$ (h) $f = -2$ (i) $v = -5$
    (j) $x = -6$ (k) $a = -1$ (l) $n = -5$
2 (a) $m = -4$ (b) $q = -3$ (c) $x = -10$
    (d) $b = -24$ (e) $e = -12$ (f) $t = -36$
3 $\frac{w}{20} + 2 = 18, w = £320$

Exercise 5F

1 (a) $x = 2$ (b) $u = 4$ (c) $r = 3$
    (d) $p = 5$ (e) $w = 7$ (f) $b = 4$
2 (a) $a = -2$ (b) $m = -3$
    (c) $d = -2$ (d) $y = -4$

Exercise 5G

1 (a) $g = 2$ (b) $k = 2$ (c) $s = 3$
    (d) $n = 7$ (e) $f = 10$ (f) $v = 9$
    (g) $m = -1$ (h) $w = -3$
2 (a) $b = 3$ (b) $r = 1$ (c) $t = 2$
    (d) $v = 6$ (e) $k = 1$ (f) $y = 3$
    (g) $x = -3$ (h) $f = -2$
3 (a) $b = 3$ (b) $a = 2$ (c) $x = 4$
    (d) $p = 5$ (e) $s = 3$ (f) $t = 1$
    (g) $w = -3$ (h) $t = -4$

Exercise 5H

1 (a) $5 < 8$ (b) $10 > 5$ (c) $-7 < 6$
    (d) $20 < 80$ (e) $2 > -4$ (f) $23 > 21$
    (g) $-5 > -8$ (h) $5.5 < 6.4$ (i) $12.7 > 12.6$
    (j) $0 < 0.01$ (k) $0.01 < 10.01$ (l) $1112 < 1121$
2 (a) F   (b) T   (c) T
    (d) F   (e) T   (f) T
    (g) F   (h) F   (i) T
3 (a) 5, 6, 7, 8   (b) 8, 9, 10, 11, 12
    (c) 1, 2, 3, 4, 5   (d) 95, 96, 97, 98, 99, 100
    (e) $-1, 0, 1, 2, 3$ (f) $-5, -6, -7, -8, -9$
4 (a) 3.3, 3.4, 3.5, 3.6, 3.7, 3.8, 3.9, 4.0
    (b) 20.8, 20.9, 21.0, 21.1, 21.2
    (c) 0.3, 0.4, 0.5, 0.6, 0.7, 0.8, 0.9
    (d) 78.9, 79.0, 79.1, 79.2, 79.3, 79.4, 79.5, 79.6, 79.7,
       79.8, 79.9, 80.0
5 (a)

    (b)

    (c)

    (d)

    (e)

    (f)

    (g)

    (h)

6 3 (a) $4 < b < 9$ 3 (b) $7 < r \leqslant 12$
    3 (c) $0 < p \leqslant 5$ 3 (d) $94 < k \leqslant 100$
    3 (e) $-2 < x \leqslant 3$ 3 (f) $-9 \leqslant q < -4$
    4 (a) $3.2 < h < 4.1$ 4 (b) $20.7 < t \leqslant 21.2$
    4 (c) $0.3 \leqslant y \leqslant 0.9$ 4 (d) $78.9 \leqslant m < 80.1$

Exercise 5I

1 (a) $k > 4$ (b) $h < 9$ (c) $g \geqslant 2$
    (d) $m \leqslant 3$ (e) $v > 3$ (f) $x \leqslant 3$
    (g) $w \leqslant -4$ (h) $a > -9$ (i) $t \leqslant -3$
2 (a) $x > 2$ (b) $s < 2$ (c) $u \leqslant 3$
    (d) $a \geqslant 10$ (e) $q > 5$ (f) $r \leqslant 1$
    (g) $n \geqslant -2$ (h) $p \leqslant -3$ (i) $y > -4$
3 (a) $q > 4$ (b) $c \leqslant 10$ (c) $v \leqslant 20$
    (d) $n \leqslant 2$ (e) $x \geqslant 7$ (f) $g \geqslant -2$
    (g) $m \leqslant -5$ (h) $w > -4$
4 (a) $3 < z < 10$ (b) $3 < c < 4$ (c) $2 < x \leqslant 6$
    (d) $20 > b \geqslant 15$ (e) $0 < h < 9$ (f) $0.75 < t \leqslant 2.5$
    (g) $-2 < p \leqslant 1$ (h) $-2.5 \leqslant q \leqslant -0.5$

Examination style questions

1 (a) $d = 3$ (b) $k = 12$ (c) $y = 4$
(d) $n = 24$
2 (a) $x = 4$ (b) $y = 6$ (c) $c = 2$
(d) $w = 5$
3 (a) $r = 1$ (b) $s = 0.5$ (c) $y = -3$
4 (a) $x = 4$ (b) $x < 7$
5 (a) $x \geqslant -1$ (b) $x < 2$ (c) $-1, 0, 1$

Chapter 6 Shape 1

Exercise 6A

1 (a) $\frac{1}{2}$ turn clockwise (b) $\frac{1}{4}$ turn clockwise
(c) $\frac{3}{4}$ turn clockwise (d) $1\frac{1}{2}$ turns clockwise
2 (a) E (b) NE (c) NE (d) S
(e) SE (f) E (g) SW (h) E

Exercise 6B

1 (a) reflex (b) acute (c) obtuse
(d) right angle (e) acute (f) reflex
2 (a) BAC or CAB (b) XZY or YZX
(c) LMN or NML (d) DFE or EFD
(e) QRP or PRQ (f) WUV or VUW

Exercise 6C

1 (a) $30°$ (b) $70°$ (c) $45°$
(d) $130°$ (e) $25°$ (f) $155°$

Exercise 6D

1 $a = 38°$ **2** $b = 59°$
3 $c = 56°$ **4** $d = 39°$
5 $e = 24°$ **6** $f = 128°, g = 15°$
7 $h = 22.5°$ **8** $j = 60°$
9 $k = 18°$

Exercise 6E

1 $a = 62°$
2 $b = 109°$
3 $c = 65°$
4 $d = 110°, e = 110°$
5 $f = 105°$
6 $g = 63°, h = 117°$
7 $i = 74°, j = 106°, k = 106°$
8 $l = 53°, m = 127°, n = 53°, p = 53°$
9 $q = 82°, r = 75°$

Exercise 6F

1 (a) (i) $065°$ (ii) $245°$ (b) (i) $115°$ (ii) $295°$
(c) (i) $225°$ (ii) $045°$ (d) (i) $325°$ (ii) $145°$

2 (a) 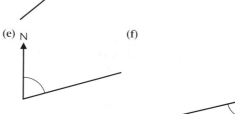 (b)

(c) 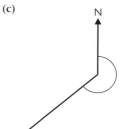 (d)

(e) (f)

(g)  (h)

3 (a) $057°$ (b) $112°$ (c) $018°$
(d) $236°$ (e) $293°$ (f) $197°$

Exercise 6G

1 $a = 70°$
2 $b = 37°$
3 $c = 75°$
4 $d = 62°, e = 138°$
5 $f = 53°, g = 127°$
6 $h = 60°$
7 $i = 36°, j = 108°, k = 72°$
8 $l = 75°, m = 30°$
9 $n = 134°, p = 118°, q = 108°$

Exercise 6H

1 $a = 96°, b = 40°$
2 $c = 78°, d = 78°$
3 $e = 47°, f = 80°$
4 $g = 38°, h = 66°$

5 $i = 264°$
6 $j = 78°, k = 51°$
7 $l = 32°, m = 32°, n = 62°$
8 $p = 27°$

Exercise 6I

1 (a) rectangle   (b) trapezium
   (c) rhombus   (d) square
   (e) parallelogram   (f) kite
2 (a) square, rhombus, kite
   (b) square, rhombus
   (c) trapezium
   (d) parallelogram, rhombus
   (e) square, rectangle
   (f) square, rectangle, parallelogram, rhombus
   (g) kite
   (h) rectangle, parallelogram, kite
   (i) square, rectangle, parallelogram, rhombus
   (j) square, rectangle, parallelogram, rhombus, trapezium
   (k) square, rectangle
   (l) square, rhombus, kite

Exercise 6J

1 $a = 92°$ **2** $b = 55°, c = 125°$
3 $d = 69°$ **4** $e = 135°$
5 $f = 64°$ **6** $g = 137°$
7 $h = 100°, i = 105°$ **8** $k = 45°, j = 153°$
9 $l = 116°, m = 64°$ **10** $n = 120°, p = 40°$
11 $q = 77°, r = 75°$ **12** $t = 72°$
13 $u = 105°, w = 75°$ **14** $x = 64°, y = 116°$
15 $160°$
16 No, $\dfrac{360}{180 - 130}$ is not a whole number of sides

Exercise 6K

1 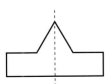   **2**

3   **4** No lines of symmetry

5   **6** No lines of symmetry

7 (a)   (b)

(c)

Exercise 6L

1   1 1   2 4   3 1
   4 2   5 2   6 2
2 Students' own answers
3 (a) (i)   (ii)

   (b) (i)   (ii)

Examination style questions

1 (a) (i) 3   (ii)

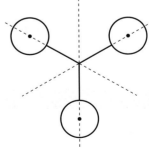

   (b) Students' own answers
2 $95°$
3 (a) $a = 45°$ (b) $b = 115°, c = 50°$
4 (a) A: parallelogram, B: rhombus
   (b) (i)

      (ii) trapezium
5 $x = 122°$
6 12 sides
7 $a = 135°, b = 67.5°, c = 45°$

Chapter 7 Algebra 3

Exercise 7A

1 (a) d^4 (b) a^6 (c) x^5 (d) y^8
     (e) b^2 (f) m^4 (g) p^3 (h) r^7
2 (a) $5p^4$ (b) $4a^3$ (c) $8x^4$ (d) $15f^5$
     (e) $8b^3$ (f) $60h^5$ (g) $27z^3$ (h) $25r^4$
3 (a) b^2c^3 (b) f^4g^2 (c) $12x^3y$ (d) $10d^4e^2$
     (e) $8s^3t^2$ (f) $30j^2k^3$ (g) $4p^2q^3$ (h) $16a^2b^2$
4 (a) $6x^2$ (b) $12y^3$ (c) $6a^4$ (d) $24b^3$
     (e) $40n^5$ (f) $126k^4$ (g) $27x^3$ (h) $36z^6$
5 (a) $6a^2b$ (b) $8p^3q^2$ (c) $18xy^2$ (d) $60r^2s^2$
     (e) $42c^2d^2$ (f) $120x^3y$ (g) $16p^2r^2$ (h) $49r^4s^2$
6 1g   2c   3j   4a   5h
     6e   7i   8d   9b   10f

Exercise 7B

1 (a) a^5 (b) b^7 (c) c^6 (d) d^{10}
2 (a) x^4 (b) y^3 (c) z^5 (d) w^4
3 (a) $6a^7$ (b) $15b^9$ (c) $4c^8$ (d) $9d^8$
4 Add the powers
5 (a) p^3 (b) q^4 (c) $r^1 = r$ (d) s^3
6 (a) $\dfrac{1}{e^3}$ (b) $\dfrac{1}{f^4}$ (c) $\dfrac{1}{g}$ (d) $\dfrac{1}{h^5}$
7 (a) $5x^3$ (b) $2y^4$ (c) $4r$ (d) $\dfrac{1}{5s^3}$
8 Subtract the powers

Exercise 7C

1 (a) m^5 (b) a^7 (c) n^6 (d) u^{10} (e) t^9
2 (a) d^4 (b) a^3 (c) r^5 (d) e^4 (e) t^7
3 (a) $6h^7$ (b) $15e^9$ (c) $4g^8$ (d) $9r^5$ (e) $24e^4$
4 (a) a^3 (b) t^4 (c) $e^1 = e$ (d) s^3 (e) $t^0 = 1$
5 (a) $e^{-3} = \dfrac{1}{e^3}$ (b) $f^{-4} = \dfrac{1}{f^4}$ (c) $g^{-1} = \dfrac{1}{g}$
     (d) $h^{-3} = \dfrac{1}{h^3}$ (e) $w^{-5} = \dfrac{1}{w^5}$
6 (a) $5x^3$ (b) $3y^4$ (c) $3r^1 = 3r$
     (d) $2t^3$ (e) $2s^{-3} = \dfrac{2}{s^3}$
7 (a) h^9 (b) e^8 (c) $24c^9$
8 (a) d^2 (b) $3a^3$ (c) $6r^2$
     (d) $2e^1 = 2e$ (e) $\dfrac{4}{b^2}$ (f) $3x^4$

Exercise 7D

1 (a) £60   (b) £110
2 (a) £6   (b) £5.25   (c) 24 stamps
3 (a) £72.50   (b) £98
4 (a) 50 mph   (b) 60 mph   (c) 3 mph   (d) 6.5 mph

Exercise 7E

1 (a) 40   (b) 42   (c) 60   (d) 63
2 (a) 12   (b) 24   (c) 13.5   (d) 240
3 (a) 48   (b) 100   (c) 14   (d) 20.8
4 (a) 24   (b) 25   (c) 36   (d) 17.5
5 (a) 26   (b) 34   (c) 23   (d) 22

6 (a) $2 \times 13 = 26$ (b) $2 \times 17 = 34$
     (c) $2 \times 11.5 = 23$ (d) $2 \times 11 = 22$

Exercise 7F

1 $C = 500 - 48x$ **2** $t = 7r + 5s$
3 $t = 45w + 20$ **4** $t = 65w + 30$
5 $p = 8x + 6$

Exercise 7G

1 8   **2** 10   **3** 18   **4** 7   **5** 9
6 1   **7** 7   **8** 16   **9** 5   **10** 28
11 27   **12** 47   **13** 21   **14** 4   **15** 11
16 11   **17** 10   **18** 20   **19** 2   **20** 24

Exercise 7H

1 30   **2** 1   **3** -10 **4** 19   **5** 9
6 8   **7** 20   **8** $17\frac{1}{2}$ **9** 14   **10** 2
11 23   **12** $13\frac{1}{2}$ **13** 42   **14** 6   **15** -18
16 7.5   **17** 3   **18** 3   **19** 4.5   **20** 12

Exercise 7I

1 4   **2** 3   **3** 5   **4** 76   **5** 30
6 37   **7** 84   **8** 27   **9** -5 **10** 4
11 6   **12** 3   **13** 76.5   **14** 9   **15** -6.5
16 65   **17** 12.5   **18** 5.375
19

x	1	2	3	4	5
$x^2 + 2x$	3	8	15	24	35

20

x	3	4	3.5	3.7	3.8
$x^3 - x$	24	60	39.375	46.953	51.072

21 90   **22** -18 **23** 31   **24** 7   **25** 8

Exercise 7J

1 (a) 6   (b) 4   (c) 6   (d) $12\frac{1}{2} = 12.5$
2 (a) 32   (b) 88   (c) 30   (d) 75
3 (a) 26   (b) 110   (c) 50   (d) 24
4 (a) $21.604... \to 22$ (b) $19.387... \to 19$
     (c) $23.545... \to 24$ (d) $20.415... \to 20$
5 (a) 7   (b) 10   (c) 11   (d) 13
6 (a) £16.25   (b) £21.65   (c) £34.55
7 (a) £98.05   (b) 33.12 (...) so 34 or more

Exercise 7K

1 (a) $l = 10$ (b) $l = 5$ (c) $l = 20$
2 (a) $w = 3$ (b) $w = 4$ (c) $w = 6$ (d) $w = 4$
3 (a) $l = 5$ (b) $l = 9$ (c) $w = 13$ (d) $w = 11.5$
4 (a) $u = 6$ (b) $u = 11$ (c) $a = 5$
     (d) $t = 9$ (e) $u = -8$
5 (a) $a = d - 8$ (b) $a = t - 12$
     (c) $a = k + 6$ (d) $a = w + 7$
6 (a) $w = \dfrac{P}{4}$ (b) $w = \dfrac{a}{3}$ (c) $w = \dfrac{A}{l}$ (d) $w = \dfrac{h}{k}$

7 (a) $x = \dfrac{y + 6}{5}$ (b) $x = \dfrac{y + 7}{4}$

(c) $x = \dfrac{y - 1}{2}$ (d) $x = \dfrac{y - 5}{6}$

8 (a) $59\,°F$ (b) $C = \dfrac{F - 32}{1.8}$

(c) $20\,°C$ (d) $\dfrac{82 - 32}{1.8} = \dfrac{50}{1.8}$

$= 27.77\,(\ldots)$

$= 28\,°C$

9 (a) $r = \dfrac{p - 2t}{4}$ (b) $r = \dfrac{v - 4h}{7}$

(c) $r = \dfrac{w + 2s}{3}$ (d) $r = \dfrac{y + 5p}{6}$

10 (a) $a = \dfrac{u - v}{t}$ (b) $t = \dfrac{u - v}{a}$

11 (a) $a = 2b - 12$ (b) $a = 2b - 14$ (c) $a = 3b + 3$

(d) $a = 4b + 12$ (e) $a = \tfrac{1}{2}b - 1$ (f) $a = \tfrac{1}{3}b + 5$

or $a = \dfrac{b}{2} - 1$ or $a = \dfrac{b}{3} + 5$

Exercise 7L

1 $x = 4.6$

x	$x^3 + x$
5	130
4.5	95.625
4.6	101.936
4.55	98.746375

2 $x = 2.8$

x	$x^3 + x$
3	30
2	10
2.8	24.752
2.9	27.289
2.85	25.999125

3 $x = 3.2$

x	$x^3 - x$
3	24
4	60
3.2	29.568
3.3	32.637
3.25	31.078125

4 $x = 6.1$

x	$x^3 + 2x$
6	228
7	357
6.2	250.728
6.1	239.181
6.15	244.908375

5 $x = 9.7$

x	$x^3 + x^2$
10	1100
9	810
9.7	1006.763
9.6	976.896
9.65	991.754625

Examination style questions

1 (a) (i) £70 (ii) £195 (b) 10 days
2 (a) £1050 (b) 100 people
3 6
4 (a) $x + 5$ (b) $x - 2$ (c) $2x$
5 (a) 26 (b) 33
6 $-4\,°C$
7 (a) $a = 7$ (b) $a = 2$
8 -3.34

9 (a) n^2 (b) $\dfrac{3}{n}$

10 (a) $2\tfrac{3}{8}$ (b) $\tfrac{5}{8}$

11 (a) 64 (b) 2 (c) $5p + 2q$

12 $r = \dfrac{p - 3}{2}$

Chapter 8 Shape 2

Exercise 8A

1 Possible approximate answers include:
  (a) 75 cm–1 m , 2–3 ft
  (b) 5–10 cm, 2–4 inches
  (c) 1–2 cm, $\tfrac{1}{2}$–1 inch
  (d) 140–190 cm, 5 ft 2 inches–6 ft 6 inches
  (e) 22–30 cm, 8–12 inches
  (f) 6–8 m, 18–25 feet
  (g) 3–6 m, 12–20 feet
  (h) 2.5–4 m, 8–12 feet
  (i) 600–800 km, 400–500 miles
  (j) 1–2 mm, $\tfrac{1}{8}$ inch

Exercise 8B

1 Possible approximate answers include:
  Cola bottle: 500 ml, 17.5 fl. ounces
  Milk bottle: 1 litre, $1\tfrac{3}{4}$ pints
  Measuring jug: $\tfrac{1}{2}$ litre, pint
  Mug: 300 ml, $\tfrac{1}{2}$ pint
  Bucket: 10 litres, 2–3 gallons
  Bath: 60–100 litres, 15–20 gallons

Exercise 8C

1 Possible approximate answers include:
  (a) 1 kg (b) 15 kg (c) 800 g
  (d) 5 kg (e) 1 tonne (f) 15 g
  (g) 63 kg (h) 500 g (i) 1 kg
  (j) 20 kg
2 Possible approximate answers include:
  (a) 2 lb (b) 30 lb (c) $1\tfrac{1}{2}$ lbs
  (d) 10 lb (e) 1 ton (f) $\tfrac{1}{2}$ oz
  (g) 10 st (h) 1 lb (i) 2 lbs
  (j) 4 lb

Exercise 8D

1 (c) and (d) are sensible estimates
  (a) 140–180 cm (b) 5–30 kg (e) 800 km

2 (a) and (b) are sensible estimates
  (c) $\frac{1}{16}$ ounce   (d) 1 fl. ounce   (e) 500 miles
3 (a) kg, stones   (b) m, feet or yards
  (c) ml, fluid ounces   (d) kg, pounds
  (e) cm, inches   (f) tonnes, tons
  (g) km, miles   (h) l, gallons
4 (c) and (e) are appropriate
  (a) metres   (b) kg   (d) ml
  (f) yards   (g) pints   (h) km

Exercise 8E

1 (a) 140 g   (b) 40 ml   (c) 15 cm   (d) 5.25 m
  (e) 58 mph   (f) 66 kg   (g) 200 ml   (h) $\frac{1}{4}$ full
2 (a)   (b)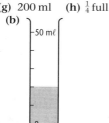

  (c) 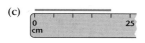   (d)

  (e) 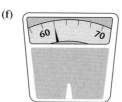   (f)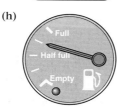

  (g)    (h)

Exercise 8F

1 (a) 8.2 cm   (b) 370 g   (c) 14 ml   (d) 1.5 l
  (e) 5.75 m   (f) 63 mph   (g) 270 g   (h) 0.3 units

Exercise 8G

1 (a) (i) 6.20 am   (ii) 0620
  (b) (i) 10.45 am   (ii) 1045
  (c) (i) 3.05 pm   (ii) 1505
  (d) (i) 6.50 pm   (ii) 1850
  (e) (i) 10.30 pm   (ii) 2230
2 (a) quarter past nine in the evening
  (b) twenty to eight in the morning
  (c) noon
  (d) half past 2 in the afternoon
  (e) midnight

3 (a) 1900   (b) 0600   (c) 0830   (d) 1245
  (e) 1415   (f) 0215   (g) 2323   (h) 1655
  (i) 1147   (j) 0015   (k) 2145   (l) 0220
4 (a) 11 am   (b) 1 pm   (c) 6 am
  (d) 4.30 pm   (e) 9.20 am   (f) 12.30 am
  (g) 11.15 pm   (h) 10.05 am   (i) 3.45 pm
  (j) 4.20 am   (k) 6.22 pm   (l) 12 midnight

Exercise 8H

1 2 hrs 5 min
2 2 hrs 50 min
3

Time	Add on 45 minutes	Take off 55 minutes
0920	1005	0825
1135	1220	1040
1317	1402	1222
1648	1733	1553
2121	2206	2026
2352	0037	2257

4 7.42 am
5 (a) 21 days   (b) 730 days   (c) 14 400 sec
  (d) 168 hours   (e) 200 min   (f) 150 sec
  (g) 4hrs 10min   (h) 5 hrs 43 min   (i) 13 weeks
  (j) 527 040 min

Exercise 8I

1 (a) 20th April   (b) 14th April
  (c) 30th April   (d) 29th April
2 (a) 9   (b) 21   (c) 71   (d) 89   (e) 92
3 76
4 11.3.97

Exercise 8J

1 (a) 2   (b) 1310   (c) 1233   (d) 1455
  (e) 2 hr 41 min
2 (a) 1233   (b) 2 hrs 17min
  (c) 1157 from Newcastle to make sure to get to the
      interview on time.
3 (a) 1000   (b) 1738
  (c) (i) 46 min   (ii) 14 min   (d) 1631
  (e) 1014   (f) 3 min
  (g) No, the buses run 14, 34 and 54 minutes past the
      hour.
4 (a) 1634   (b) 40 min   (c) 1720
  (d) 1746–1749   (e) 15 min
5 (a) 2   (b) 1800
  (c) 1800, 1830   (d) 1606
  (e) 1500   (f) 1630, 1722 & 1806
6 (a) Bristol   (b) Bristol   (c) 7 min   (d) 1845

Exercise 8K

1 (a) 600 cm   (b) 8000 m
2 (a) 3000 mm   (b) 500 cm   (c) 120 mm
  (d) 4 km   (e) 6 m   (f) 30 cm
3 (a) 5000 kg   (b) 8 kg   (c) 60 000 g
  (d) 16 t
4 (a) 7 ℓ   (b) 8 ℓ   (c) 300 mℓ
  (d) 500 cℓ

5 (a) 7 m   (b) 5 cℓ (c) 4000 g
    (d) 15 000 m   (e) 5 t   (f) 2 ℓ
6 1200 g   **7** 20

Exercise 8L

1 (a) 3.2 m   (b) 4.5 m   (c) 4200 m
   (d) 485 m   (e) 0.87 m   (f) 0.75 m
   c is the longest, **f** is the shortest
2 (a) 5.64 kg   (b) 2600 kg   (c) 0.8 kg
   c is the smallest
3 (a) 16.3 cℓ (b) 930 cℓ (c) 70 cℓ
   a is the largest
4 450 cm, 4620 mm, 5.1 m , 0.05 km
5 2450 mm
6 0.22 ℓ, 230 mℓ, 25 cℓ, 0.3ℓ

Exercise 8M

1 36.8
2 (a) inch   (b) kg   (c) mile
   (d) litre   (e) metre   (f) gallon
3 (a) 5 cm   (b) 3 inches   (c) 60 cm
   (d) 14 inches   (e) 8 km   (f) 15 miles
   (g) 4 km   (h) 1.25 miles
4 (a) 2 ℓ (b) $8\frac{3}{4}$ pints   (c) $22\frac{1}{2}$ ℓ
   (d) 3 gallons   (e) $\frac{1}{2}$ gallon   (f) 18.9 ℓ
5 (a) 6.6 pounds   (b) 14.3 pounds   (c) 5 kg
   (d) 20 kg   (e) 200 g   (f) 15 ounces
6 20 cm by 15 cm
7 No (the tank is 11.1 ℓ)
8 200 g flour, 125 g fat, 50 g cocoa and 100 g sugar
9 Lynn (43.18 kg)
10 The 200 ℓ butt (350 pints)
11 Yes
12 John's car which does 36 m.p.g

Exercise 8N

1 (a) 40 000 cm^2 (b) 0.19 m^2
2 (a) 2 500 000 cm^3 (b) 0.3 m^3
3 (a) 3 m^2 = 30 000 cm^2 (b) 5 m^3 = 5 000 000 cm^3
   (c) 70 000 cm^2 = 7 m^2 (d) $\frac{1}{2}$ m^3 = 500 000 cm^3
   (e) $\frac{1}{2}$ m^2 = 5000 cm^2 (f) 3 000 000 cm^3 = 3 m^3
4 (a) 100   (b) 10 000 000 000   (c) 9
5 (a) 1 000 000 000   (b) 1 000 000 000   (c) 1728
6 3.25 km^2
7 (a) 159 600 cm^2 (b) 15.96 m^2
   (c) 16   (d) £168
8 (a) 248 625 cm^3 (b) Yes (250 000 cm^3)

Exercise 8O

1 (a) 8.5 cm ≤ 9 cm < 9.5 cm
   (b) 11.5 kg ≤ 12 kg < 12.5 kg
   (c) 64.5 cℓ ≤ 65 cℓ < 65.5 cℓ
   (d) 9.5 sec ≤ 9 sec < 9.5 sec
2 (a) 13.5 mm ≤ 14 mm < 14.5 mm
   (b) 1.35 cm ≤ 1.4 cm < 1.45 cm
   (c) 3.75 cm ≤ 3.8 cm < 3.85 cm
   (d) 7.45 cm ≤ 7.5 cm < 7.55 cm
3 (a) 124.5 cm ≤ 125 cm < 125.5 cm
   (b) 1.245 m ≤ 1.25 m < 1.255 m
   (c) 4.065 m ≤ 4.07 m < 4.075 m
   (d) 6.195 m ≤ 6.20 m < 6.205 m

4 (a) continuous   (b) discrete   (c) continuous
   (d) discrete   (e) continuous
5 (a) 16.5 cm ≤ 17 cm < 17.5 cm,
      10.5 cm ≤ 11 cm < 11.5 cm,
      4.5 cm ≤ 5 cm < 5.5 cm
   (b) 779.625 cm^2
6 (a) 1.5525 m^2 (b) 1.8125 m^2

Exercise 8P

1 15 mph   **2** 8 m/sec
3 260 miles   **4** $3\frac{1}{2}$ hrs
5 37.5 mph   **6** 55.7 mph
7 5 hrs 24 min   **8** 42 miles
9 2.5 m/sec
10 (a) 4 hrs 15 min   (b) 1292 miles   (c) 304 mph

Exercise 8Q

1 4 g/cm^3
2 4750 g or 4.75 kg
3 550 cm^3
4

	Mass	Density	Volume
(a)	24 g	1.6 g/cm^3	15 cm^3
(b)	82 g	2.05 g/cm^3	40 cm^3
(c)	28 g	2.1 g/cm^3	13.3 cm^3
(d)	4.8 g	3.2 g/cm^3	1.5 litres
(e)	5.7 kg	1.24 g/cm^3	4596.8 cm^3

5 (a) steel   (b) steel
6 0.03 m^3
7 34 200 kg

Examination style questions

1 25 km/h
2 2 hours 5 minutes
3 45 miles per hour
4 (a) (i)   (ii) 8 pounds

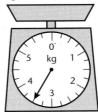

   (b) 2 hours 10 minutes
5 74.5 m, 75.49 m
6 (a)

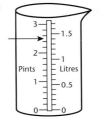

   (b) Approx. 1.75 pints
   (c) 4.6 litres
7 80p
8 2 hours 24 minutes
9 3 hours 32 minutes

Chapter 9 Shape 3

Exercise 9A

1 (a) P = 22 cm, A = 24 cm² (b) P = 29 cm, A = 25 cm²
　(c) P = 27 cm, A = 35 cm² (d) P = 44 cm, A = 96 cm²
　(e) P = 12 cm, A = 6 cm² (f) P = 24 cm, A = 24 cm²
　(g) P = 32 cm, A = 36 cm² (h) P = 30 cm, A = 30 cm²
2 (a) 8 cm (b) 4 cm (c) 4 cm

Exercise 9B

1 (a) 51 m² (b) 96 cm²
2 (a) 2.5 cm (b) 2 cm
3 (a) 30 cm² (b) 20 cm² (c) 70 cm² (d) 40 cm²
　(e) 30 cm² (f) 48 cm² (g) 33 cm² (h) 49.5 cm²
　(i) 96 cm² (j) 48 cm² (k) 45 cm²

Exercise 9C

1 (a) P = 36 cm, A = 56 cm²
　(b) P = 30 cm, A = 46 cm²
　(c) P = 42 cm, A = 81 cm²
　(d) P = 38 cm, A = 48 cm²
　(e) P = 50 cm, A = 72 cm²
　(f) P = 42 cm, A = 66 cm²
2 (a) 48 cm² (b) 84 cm² (c) 66 cm²
　(d) 74.5 cm² (e) 47 cm²

Exercise 9D

1 (a) 12π cm (b) 8π cm (c) 30π m
2 (a) 73.8 cm (b) 42.1 cm (c) 5.3 m
3 (a) 15.3 cm (b) 5.8 cm (c) 18.1 mm
4 (a) 3.3 cm (b) 11.3 mm (c) 0.99 m
5 (a) 185.7 cm (b) 582.8 cm (c) 441.9 m
6 348.5 cm
7 392
8 (a) 961.3 m (b) 729

Exercise 9E

1 (a) 36π (b) 16π (c) 225π
2 (a) 433.7 cm² (b) 141 cm² (c) 2.22 cm²
3 (a) 2123.7 cm² (b) 21251.5 cm² (c) 12356.6 cm²
4 14.5 cm
5 (a) 8.14 cm² (b) 51 cm
6 (a) 417.8 cm² (b) 55 cm² (c) 134 cm²
7 7.9 cm
8 575 cm²

Exercise 9F

1

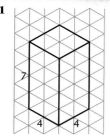

2

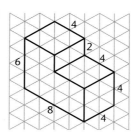

3

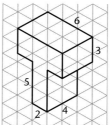

4

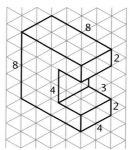

Exercise 9G

1 (a)

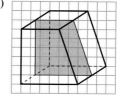

(b)

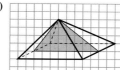

(c)

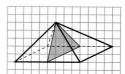
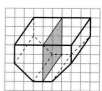

(d)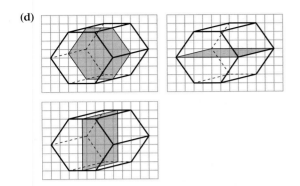

Exercise 9H

1 (a)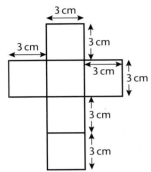

(b)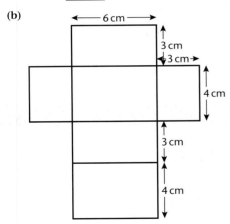

(c)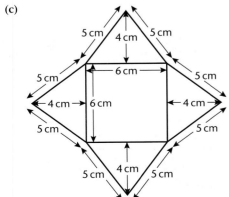

(d)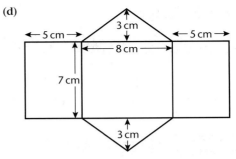

2 (a) 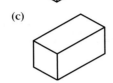 **(b)**

(c) **(d)**

Exercise 9I

1

2

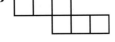

3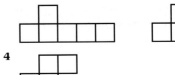

4

5

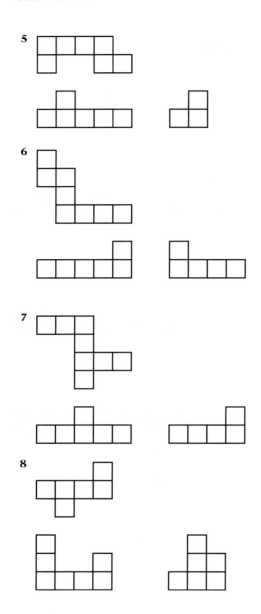

6

7

8

Exercise 9J

1 $V = 4\,cm^3$, $SA = 18\,cm^2$
2 $V = 7\,cm^3$, $SA = 30\,cm^2$
3 $V = 7\,cm^3$, $SA = 30\,cm^2$
4 $V = 6\,cm^3$, $SA = 26\,cm^2$
5 $V = 8\,cm^3$, $SA = 34\,cm^2$
6 $V = 9\,cm^3$, $SA = 38\,cm^2$
7 $V = 9\,cm^3$, $SA = 38\,cm^2$
8 $V = 9\,cm^3$, $SA = 38\,cm^2$

Exercise 9K

1 (a) $135\,cm^3$ (b) $154\,cm^3$ (c) $108\,cm^3$
2 (a) $85\,cm^3$ (b) $231\,cm^3$
3 (a) $160\,cm^3$ (b) $264\,cm^3$ (c) $290\,cm^3$

Exercise 9L

1 (a) $124\,cm^2$ (b) $214.5\,cm^2$ (c) $171\,cm^2$
2 (a) $840\,cm^2$ (b) $336\,cm^2$

Exercise 9M

1 (a) $32\pi\,cm^3$ (b) $45\pi\,cm^3$ (c) $980\pi\,cm^3$
   (d) $32\pi\,cm^3$
2 (a) $72\pi\ell$
3 $5\,cm$
4 (a) $196.3\,cm^3$ (b) $77.8\,cm^3$
5 Frisk
6 $6.5\,cm$
7 (a) $6126\,cm^3$ (b) $20.83\,kg$
8 $6.76\,g/cm^3$
9 (a) 16 (b) $942.5\,cm^3$ (c) $15\,080\,cm^3$
   (d) $19\,200\,cm^3$ (e) $4120\,cm^3$

Exercise 9N

1 (a) $40\pi\,cm^2$ (b) $48\pi\,cm^2$ (c) $378\pi\,cm^2$
   (d) $1632\pi\,cm^2$
2 $2513\,cm^2$
3 $266\,cm^2$
4 $1440\pi\,cm^3$
5 (a) $276.7\,cm^2$ (b) $132\,cm^2$

Examination style questions

1 (a) $5.7\,cm$ (b) $53.58\,cm^2$
2 (a) $25.5\,cm^2$ (b) $24.5\,cm^2$
3   A   B

4 $1648\,cm^2$
5 (a) (i) $1440\,m^3$
     (ii) $336\,m^3$
   (b) 93%

Chapter 10 Handling data 1

Exercise 10A

1 (a) quantitative   (b) quantitative   (c) qualitative
   (d) qualitative   (e) quantitative   (f) quantitative
2 (a) discrete   (b) discrete   (c) continuous
   (d) continuous   (e) continuous   (f) discrete
   (g) continuous
3 (a) primary   (b) secondary   (c) secondary
   (d) secondary   (e) primary
4 (i)   Internet/magazines, secondary, quantitative
   (ii)   Internet, secondary, quantitative and qualitative
   (iii)   Questionnaire/survey, primary, quantitative
   (iv)   Internet, secondary, quantitative

Exercise 10B

1

Pet	Tally	Frequency
Dog	⫴⫴ ||	7
Cat	⫴⫴ ||||	9
Bird	⫴⫴		6
Other	⫴⫴ ⫴⫴	10
	Total	32

2

Mark	Tally	Frequency
1–10		0
11–20	⫴⫴ ⫴⫴ ⫴⫴		16
21–30	⫴⫴ ⫴⫴ |||	13
	Total	29

3

Mass m (kg)	Tally	Frequency
$40 \leqslant m < 50$	||	2
$50 \leqslant m < 60$	||||	4
$60 \leqslant m < 70$	⫴⫴ ||	7
$70 \leqslant m < 80$	⫴⫴	5
$80 \leqslant m < 90$	||	2
	Total	20

4

Time t (min)	Tally	Frequency
10–14		0
15–19	⫴⫴ ⫴⫴ ⫴⫴	15
20–24	⫴⫴ ⫴⫴ ||||	14
25–29	⫴⫴ ⫴⫴ ⫴⫴ ||	17
30–34	|||	3
35–40			1
	Total	50

5

Amount m (£)	Tally	Frequency
$0 \leqslant m < 20$	⫴⫴	5
$20 \leqslant m < 40$	⫴⫴	5
$40 \leqslant m < 60$	⫴⫴	5
$60 \leqslant m < 80$	⫴⫴	5
$80 \leqslant m < 100$	⫴⫴ |||	8
	Total	28

6

Number on dice	Tally	Frequency
1		
2		
3		
4		
5		
6		
	Total	

7

Colour of car	Tally	Frequency
Black		
Silver		
Red		
Blue		
Green		
Other		
	Total	

Exercise 10C

1 (a) yes
   (b) no, need a third option such as 'don't know'
   (c) no, personal questions not allowed
   (d) yes
   (e) no, too vague
2 Students' own questionnaire

Exercise 10D

1 (a) random   (b) biased   (c) random
   (d) random   (e) random   (f) biased
2 Students' own questionnaire
3 Students' own questionnaire

Exercise 10E

1

	Bicycle	No bicycle	Total
Men	6	7	13
Women	8	9	17
Total	14	16	30

2 , 15%

	Left-handed	Right-handed	Total
Girls	6	44	50
Boys	9	41	50
Total	15	85	100

3 (a) and (b)

	Population Poynton (Central)	Population Poynton (West)	Total
Male	3522	2743	6265
Female	3270	3898	7168
Total	6792	6641	13 433

(c) 53%

4 (a)

		Level						Total
		3	4	5	6	7	8	
English	Boys	11	28	34	31	15	1	120
	Girls	4	20	36	43	22	5	130
Total		15	48	70	74	37	6	250

(b) 250   (c) 68%   (d) 17%

Exercise 10F

1 (a) 158   (b) 69   (c) 357   (d) 56%
2 (a) 153   (b) 103   (c) 39   (d) 9%

Exercise 10G

1 (a) 649 km   (b) 536 km   (c) 1529 km
2 (a) £1231   (b) £1326

Examination style questions

1

Berry	Tally	Frequency
Strawberry	ⅢⅡ	7
Blackberry	‖	2
Raspberry	‖‖	4
None	‖	2

2 (a)

Distances (miles)	Tally	Frequency
1–10	Ⅲ Ⅲ Ⅲ Ⅲ Ⅰ	21
11–20	Ⅲ Ⅲ ‖	12
21–30	‖‖	4
31–40	‖	2
41–50	Ⅰ	1

(b) 12
3 (a) 7   (b) 9
4 (a)

		Number of children				
		1	2	3	4	5+
Number of pets	1 2 3 4 5+					

(b) Students' own data
5 (a) The categories used mean different things to different people, there is no 'never' response, it doesn't state whether it is asking per week/per month etc
(b) Number each pupil and pick 50 numbers from a hat, or use systematic sampling with a random starting point.

Chapter 11 Handling data 2

Exercise 11A

1 (a)

Tutor group	
A	📱📱📱
B	📱📱📱📱
C	📱📱📱📱📱📱
D	📱📱📱📱

(b) 36

2

Sport	
Football	☺
Netball	☺ ☹
Riding	☺ ☺
Other	☺ ☹

3

Drink	
Tea	☕☕☕☕☕☕☕
Coffee	☕☕☕☕☕☕☕☕☕
Chocolate	☕☕☕☕☕
Soup	☕☕☕☕
Fruit juice	☕☕☕

Exercise 11B

1 (a)

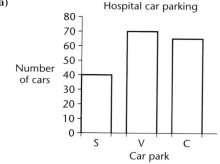

Hospital car parking

(b) 30

2

TV programme survey

3 (a)

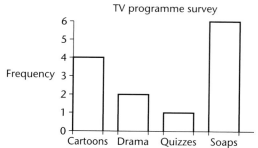

Popularity of crisps

(b) 100

4 (a) 180   (b) Audi   (c) Ford
   (d) Volvo   (e) Renault and Citroen

5

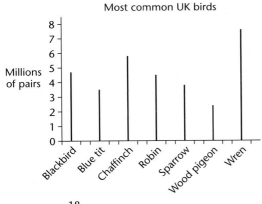

Most common UK birds

6

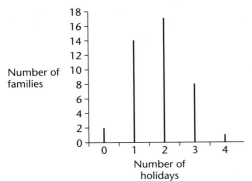

Exercise 11C

1

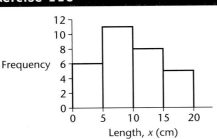

2

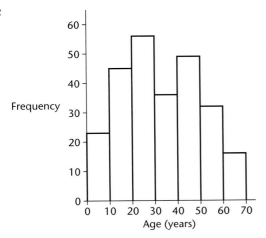

Exercise 11D

1 (a) 90   (b) 4°
   (c)

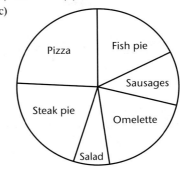

2

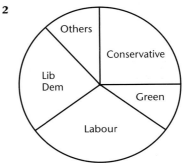

3 Art 96, English 45, French 42, Geography 63, History 57, Maths 108, PE 78, Science 51

4 Stayed on in 6th form 41, went to College 20, got a full-time job 31, on a training scheme 18, out of work 14, other 11

5 (a) 3.6°
   (b) Protein = 54°, Carbohydrate = 225°, Fat = 36°, Fibre = 45°
   (c)

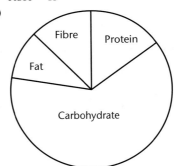

6 (a) $\dfrac{360}{720} g = 0.5°$

Ingredient	Amount in g	Sector angle calculation	Angle
Carbohydrate	400 g	400 × 0.5° =	200°
Protein	150 g	150 × 0.5° =	75°
Fibre	120 g	120 × 0.5° =	60°
Fat	50 g	50 × 0.5° =	25°
Total		Total angle	360°

(b)

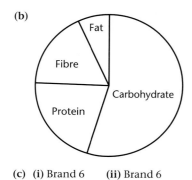

(c) (i) Brand 6 **(ii)** Brand 6

Exercise 11E

1 (a)

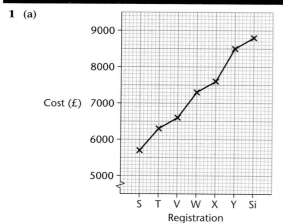

(b) Older second-hand cars cost less

2

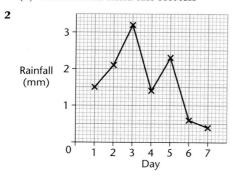

3

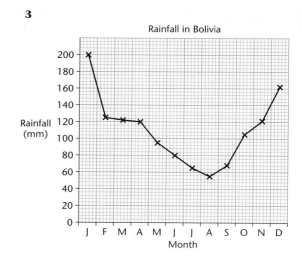

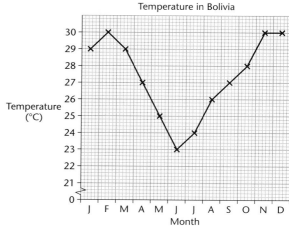

Exercise 11F

1 0 9
  1 1 2 4 5 6 7 7 8 8 9 9 9
  2 0 1 1 2 3 3 4

2 (a) 0 9
  1 2 4 7 8 9
  2 5 6 9 9
  3 1 2 3 3 7 7
  4 0 3 3

  (b) 2

3 (a) 18

  (b) 0 8
  1 2 3 4 4 7 8 9
  2 0 1 1 5 9
  3 2 4 5 8
  4 1

4 (a) 1 2 5 8
  2 0 0 3 4 5 6 7 8 9
  3 0 0 0 0 1 2 3 4 5 5 5 6 6 7 8 8 9
  4 0 0 0 2 2 3 4 4 5 6
  5 0 1

  (b) 9

Exercise 11G

1 (a) 3   **(b)** 1, 4   **(c)** 2
2 (a) positive correlation   **(b)** no correlation
   (c) positive correlation   **(d)** positive correlation
   (e) positive correlation
3 (a) and **(b)**

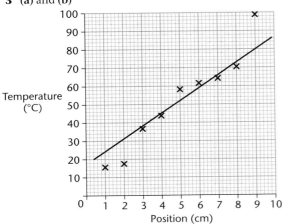

  (c) positive correlation
4 (a) and **(b)**

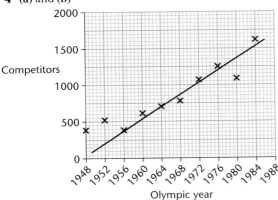

  (c) positive correlation
  (d) trend shows an increasing number of competitors every games
5 (a)

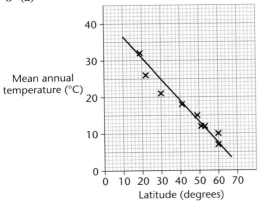

(b) negative correlation
(c) temperature becomes colder
6 (a) and **(b)**

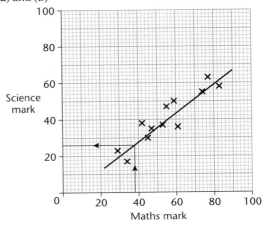

  (c) positive correlation   **(d)** 26 marks

Exercise 11H

1 (a) 165 cm   **(b)** 17 kg   **(c)** 5 s
  (d) 17.5 m^2 **(e)** 150 mℓ
2 (a)

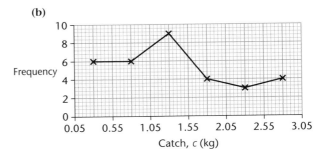

(b) girls' peak height is 165–170 cm whilst the boys' peak height is between 170 and 175 cm

3 (a)

Catch, c (kg)	Frequency	Mid-point values
$0.05 < c \leqslant 0.55$	6	0.3
$0.55 < c \leqslant 1.05$	6	0.8
$1.05 < c \leqslant 1.55$	9	1.3
$1.55 < c \leqslant 2.05$	4	1.8
$2.05 < c \leqslant 2.55$	3	2.3
$2.55 < c \leqslant 3.05$	4	2.8

(b)

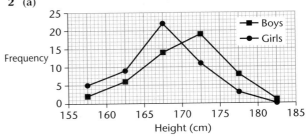

4 (a)

Temperature, t (°C)	Frequency	Mid-point values
$0 < t \leqslant 4$	6	2
$4 < t \leqslant 8$	14	6
$8 < t \leqslant 12$	11	10
$12 < t \leqslant 16$	15	14
$16 < t \leqslant 20$	8	18
$20 < t \leqslant 24$	3	22
$24 < t \leqslant 28$	3	26

(b)

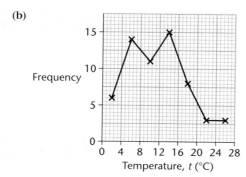

5

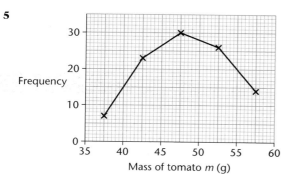

Examination style questions

1 (a) (i)

Colour	Tally	Frequency
Blue	||||| ||||| ||||	14
Red	||||	4
Green	||||| |||||	10
Black	||||| |||	8

(ii) 36

(b)

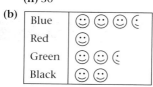

2 (a)

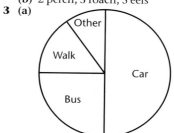

(b) 2 perch, 3 roach, 5 eels

3 (a)

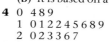

(b) It is based on a sample

4 0 4 8 9
  1 0 1 2 2 4 5 6 8 9   1 | 2 represents 12
  2 0 2 3 3 6 7

5 (a) and (b)

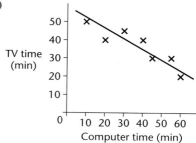

(c) As the time spent on home computer increases, the time spent watching television decreases (or vice versa)

6

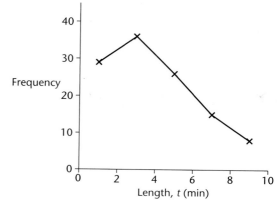

Chapter 12 Number 4

Exercise 12A

1 (a) 400 g flour, 200 g margarine, 160 g dried fruit
  (b) 100 g, 50 g, 40 g
  (c) 250 g, 125 g, 100 g
2 (a) 12   (b) 35
3 (a) 125 mℓ cranberry and 75 mℓ orange juice
  (b) 225 mℓ
4 (a) 27   (b) 20
5 (a) 144   (b) 70
6 (a) 7.5 kg   (b) 0.8 kg
7 (a) 7.5 kg   (b) 24 kg peat, 9 kg sand
8 (a) 120 g meat, 60 g mushrooms
  (b) 380 mℓ

Exercise 12B

1 (a) 1:2   (b) 3:2   (c) 1:5   (d) 3:5
  (e) 8:5   (f) 4:3   (g) 12:7   (h) 4:7
  (i) 4:1
2 (a) 5:1   (b) 1:500   (c) 20:1   (d) 3:20
  (e) 3:20   (f) 7:40   (g) 14:5   (h) 9:40
  (i) 1:20
3 (a) 2:1   (b) 8:9   (c) 6:7   (d) 2:1
  (e) 3:1   (f) 4:1
4 7:6
5 40:8:1
6 6:1
7 16:6:3

Exercise 12C

1 94
2 £360
3 1680
4 (a) 245 cm   (b) 385 cm
5 (a) 5:4   (b) 240 g   (c) 540 g
6 (a) 96   (b) 114
7 50
8 (a) 7.5   (b) 133.5 kg

Exercise 12D

1 (a) 5:1   (b) 3.5:1   (c) 2.25:1   (d) 2.4:1
  (e) 15:1   (f) 1.5:1   (g) 12.5:1   (h) 2.4:1
  (i) 5:1
2 (a) 1:1.5   (b) 1:1.2   (c) 1:2.67   (d) 1:1.6
  (e) 1:5.6   (f) 1:4.5   (g) 1:0.25   (h) 1:0.3

Exercise 12E

1 (a) 30:20   (b) 9:21   (c) 25:20   (d) 24:40
2 (a) £180:£20   (b) 50 g:75 g
  (c) 350 m:250 m   (d) 160 ℓ:90 ℓ
3 (a) £84:£36   (b) £30:£90
  (c) £72:£48   (d) £50:£70
4 18
5 24
6 400 mℓ orange juice, 480 mℓ wine, 80 mℓ lime juice.
7 (a) 5:3:2   (b) £65, £39, £26
8 Maths 112 min, Geography 80 min, French 48 min
9 cement 250 kg, sand 750 kg, gravel 1000 kg
10 Paul £102, John £85, Sarah £68

Exercise 12F

1 (a) 45p   (b) £1.12   (c) 45p
2 £1.11   3 75 miles
4 1.62 kg   5 £5.25
6 £12.30   7 £95.20
8 225 g
9 (a) 3.75 km   (b) 36 min
10 4.67 tins

Exercise 12G

1 12 days   2 4 hrs
3 14 horses   4 10 days
5 5 pumps   6 500 people
7 3 days   8 40 books
9 (a) 6 hrs   (b) 200 mph
10 (a) 12 days   (b) 16 men

Exercise 12H

1 £16.45   2 108 min or 1 hr 48 min
3 288   4 £75
5 2 hrs   6 £1.02
7 (a) 72   (b) 6 days
8 (a) 128 min or 2 hours 8 min   (b) 30

Exercise 12I

1 (a) 1 km   (b) 1.75 km
  (c) 200 m   (d) 3.125 km
2 (a) 2 cm   (b) 10 cm
  (c) 7.5 cm   (d) 6.25 cm
3 (a) 75 km   (b) 30 km
4 (a) 80 cm   (b) 64 cm   (c) 32 cm
5 (a) 325 km   (b) 6 cm
6 (a) 300 m   (b) 1.25 km   (c) 650 m
  (d) North pier 750 m, South pier 1.3 km
  (e) 1.75 km
7 (a) 8 m   (b) 18.5 cm
8 (a) 11 cm by 8.4 cm
  (b)

	Shed	Water feature	Barbeque
Scale measurement	4 cm × 2 cm	3 cm × 1 cm	1.5 cm × 1 cm
Real measurement	2 m × 1 m	1.5 m × 0.5 m	0.75 m × 0.5 m

Examination style questions

1 (a) 16   (b) 12
2 £191.25
3 £6.60
4 2.5 cm
5 £18
6 (a) €340   (b) €85   (c) £10
7 (a) (i) 1600 m   (ii) 1.6 km
  (b) 20
8 41.85 francs

Chapter 13 Handling data 3

Exercise 13A

1 (a) certain   (b) even chance/unlikely
  (c) likely   (d) unlikely/impossible
  (e) even chance

2
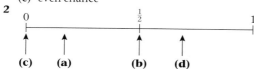

3
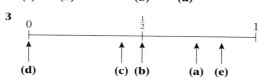

4 (a) P(homework) = 0.4   (b) P(snow) = 1%
  (c) P(two sixes) = 0.028
5 (a) $P(2) = \frac{1}{6}$ (b) $P(\text{even}) = \frac{3}{6} = \frac{1}{2}$
  (c) $P(<4) = \frac{3}{6} = \frac{1}{2}$ (d) $P(3 \text{ or } 5) = \frac{2}{6} = \frac{1}{3}$
  (e) $P(\text{not } 6) = \frac{5}{6}$
6 (a) P(winning) = 0.03 or $\frac{3}{100}$ or 3%
  (b) P(winning) = 0.11 or $\frac{11}{100}$ or 11%
7 (a) $P(\text{blue}) = \frac{11}{16}$ (b) $P(\text{red}) = \frac{5}{16}$
  (c) P(green) = 0   (d) $P(\text{blue or red}) = \frac{16}{16} = 1$

Exercise 13B

1 (a), (b) and (c) Table of your results
  (d) $P(H) = \frac{1}{2}$, $P(T) = \frac{1}{2}$, toss the coin many more times.
2 (a) and (b) Table of your results
  (c) $P(\text{even}) = \frac{1}{2}$, $P(\text{square}) = \frac{2}{6} = \frac{1}{3}$
     $P(\text{prime}) = \frac{3}{6} = \frac{1}{2}$, $P(>4) = \frac{2}{6} = \frac{1}{3}$
3 (a) and (b) Your tally chart and relative frequencies
  (c) $P(R) = \frac{7}{10} = 0.7$ and $P(B) = \frac{3}{10} = 0.3$

Exercise 13C

1 $P(\text{vowel}) = \frac{5}{26}$, $P(\text{consonant}) = \frac{21}{26}$
2 (a) $P(I) = \frac{4}{11}$
  (b) $P(S) = \frac{4}{11}$
  (c) $P(M \text{ or } I \text{ or } S \text{ or } P) = \frac{11}{11} = 1$
3 (a) $P(3) = \frac{1}{6}$
  (b) $P(1) = P(2) = P(4) = P(5) = P(6) = \frac{1}{6}$
  (c) $P(3 \text{ or } 4) = \frac{2}{6} = \frac{1}{3}$
  (d) $P(\text{not } 2) = \frac{5}{6}$
  (e) $P(\text{not } 2 \text{ or } 3) = \frac{4}{6} = \frac{2}{3}$
4 (a) $P(\text{square}) = \frac{5}{30} = \frac{1}{6}$
  (b) $P(\text{prime}) = \frac{10}{30} = \frac{1}{3}$
  (c) $P(>10) = \frac{20}{30} = \frac{2}{3}$
  (d) $P(\text{multiple of } 3) = \frac{10}{30} = \frac{1}{3}$
  (e) $P(\text{factor of } 24) = \frac{8}{30} = \frac{4}{15}$
  (f) $P(\text{containing a } 2) = \frac{12}{30} = \frac{2}{5}$

5 (a) $P(\text{red}) = \frac{2}{14} = \frac{1}{7}$
  (b) $P(\text{blue or green}) = \frac{9}{14}$
  (c) $P(\text{not yellow}) = \frac{11}{14}$
  (d) $P(\text{not green or red}) = \frac{7}{14} = \frac{1}{2}$
6 (a) $P(Y) = \frac{3}{8}$
  (b) $P(\text{not R or Y}) = \frac{2}{8} = \frac{1}{4}$
  (c) 4 green, 6 red, 6 yellow

Exercise 13D

1 Boy, Boy, Boy
  Boy, Boy, Girl
  Boy, Girl, Girl
  Girl, Girl, Girl
2 (a) HHH, HHT, HTH, THH, HTT, THT, TTH, TTT
  (b)

		2 coins			
		HH	HT	TH	TT
1 coin	Heads (H)	HHH	HHT	HTH	HTT
	Tails (T)	THH	THT	TTH	TTT

  (c) Sample Space Diagram

3

		Six-sided dice					
		1	2	3	4	5	6
Four-sided dice	1	1, 1	1, 2	1, 3	1, 4	1, 5	1, 6
	2	2, 1	2, 2	2, 3	2, 4	2, 5	2, 6
	3	3, 1	3, 2	3, 3	3, 4	3, 5	3, 6
	4	4, 1	4, 2	4, 3	4, 4	4, 5	4, 6

4

		Spinner 1		
		0	1	2
Spinner 2	1	1	2	3
	−1	−1	0	1
	−2	−2	−1	0

$P(\text{negative}) = \frac{3}{9} = \frac{1}{3}$

Exercise 13E

1 (a) Head, Red   Head, Blue   Head, Green
     Tail, Red   Tail, Blue   Tail, Green
  (b) $P(\text{Red, Tail}) = \frac{1}{6}$
2 (a) 6 ways (TM, TH, TP, MH, MP and HP)
3 Apple, Apple   Apple, Strawberry
  Apple, Banana   Strawberry, Apple
  Strawberry, Strawberry   Strawberry, Banana
  Banana, Apple   Banana, Strawberry
  Banana, Banana
  (a) $P(\text{No Banana}) = \frac{4}{12} = \frac{1}{3}$
  (b) $P(\text{at least one Apple}) = \frac{5}{12}$
  (c) $P(\text{two identical fruits}) = \frac{3}{12} = \frac{1}{4}$

4 (a)

		Spinner		
		Red (R)	White (W)	Blue (B)
Four-sided dice	1	1, R	1, W	1, B
	2	2, R	2, W	2, B
	3	3, R	3, W	3, B
	4	4, R	4, W	4, B

(b) P(Red or Blue, Even Number) $= \frac{4}{12} = \frac{1}{3}$

5

		1st throw					
		1	2	3	4	5	6
2nd throw	1	2	3	4	5	6	7
	2	3	4	5	6	7	8
	3	4	5	6	7	8	9
	4	5	6	7	8	9	10
	5	6	7	8	9	10	11
	6	7	8	9	10	11	12

(a) P(3, 3) $= \frac{1}{36}$ (b) P(sum < 5) $= \frac{6}{36} = \frac{1}{6}$
(c) P(2nd throw $= 2 \times$ 1st throw) $= \frac{3}{36} = \frac{1}{12}$

Examination style questions

1 (a) red
(b) spinner A, A '1 out of 4' is more than B '1 out of 6'
(c) $\frac{1}{4}$
(d)

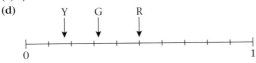

2

3 (a)

		Coin	
		Heads Disc + 1	Tails Disc − 1
Disc	2	3	1
	4	5	3
	7	8	6

(b) $\frac{1}{6}$
4 (a) Mr Key, he bought the most tickets
(b) $\frac{6}{200} = \frac{3}{100}$ (c) $\frac{180}{200} = \frac{9}{10}$
5 (a) $\frac{6}{20} = \frac{3}{10}$ (b) 68

Chapter 14 Number 5

Exercise 14A

1 (a) 2, 8, 12, 86
(b) 1, 5, 7, 9, 11, 13, 27, 33, 51, 69, 81
(c) 2, 5, 7, 11, 13
(d) 13
(e) 1
(f) 2
(g) 5, 7, 11, 13
2 (a) 3, 6, 9, 12, 15, 18 (b) 9, 18, 27, 36, 45, 54
(c) 11, 22, 33, 44, 55, 66 (d) 15, 30, 45, 60, 75, 90
(e) 22, 42, 63, 84, 105, 126
3 (a) 1, 2, 7, 14 (b) 1, 2, 4, 5, 10, 20
(c) 1, 2, 3, 4, 6, 9, 12, 18, 36 (d) 1, 2, 4, 11, 22, 44
4 (a) 44, 48, 54, 60 (b) 48, 54, 57, 60, 75
(c) 60, 75 (d) 54
(e) 48, 54, 60 (f) 60
5 (a) 2, 12 (b) 4, 12 (c) 3, 9
6 (a) (i) 52, 84, 88, 92, 104 (ii) 84, 91, 203
(iii) 52, 91, 104, 299 (iv) 92, 138, 299
(b) 91 (c) none
7 (a) 1, 2, 5, 10 (b) 1, 5, 25 (c) 1, 5
8 (a) 52 (b) 51 (c) 55 (d) 54
9 (a) 1, 2, 3, 4, 6, 8, 12, 24 (b) 1, 2, 4, 8, 16
(c) 1, 2, 3, 5, 6, 10, 15, 30 (d) 1, 2, 4, 8
(e) 1, 2 (f) 1, 2
10 (a) 96 (b) 96 (c) 98 (d) 96
11 1

Exercise 14B

1 (a) 25 (b) 121 (c) 27
(d) 225 (e) 1 000 000 (f) 8
(g) 81 (h) 125
2 (a) 289 (b) 40 000 (c) 1728
(d) 625 (e) 117.649 (f) 0.04
(g) 15.625 (h) 1.5625
3 (a) 52 (b) 20 (c) 875
4 (a) 4^3 (b) 3^2 (c) 6^3
5 72.25 cm^2

Exercise 14C

1 (a) 10 (b) 5 (c) 8 (d) 9
(e) 2 (f) 3 (g) 4 (h) 10
2 (a) 2.8 (b) 3.9 (c) 9.2 (d) 1.9
(e) 2.8 (f) 6.6 (g) 2.1 (h) 2.5
3 (a) 1.41 (b) 2.08 (c) 3.74 (d) 10.5
(e) 7.94 (f) 15.9 (g) 2.88 (h) 1.09

Exercise 14D

1 (a) 32 (b) 78125 (c) 50625
(d) 531441 (e) 39.0625 (f) 0.00032
(g) 0.0256 (h) 0.000001
2 (a) $x = 8$ (b) $x = 5$ (c) $x = 3$
(d) $x = 7$ (e) $x = 15$

Exercise 14E

1 (a) 3^8 (b) 5^{10} (c) 4^9 (d) 9^{14}
(e) 6^{12} (f) $7^7 \times 5^6$

2 (a) 7^2 (b) 9^3 (c) 11   (d) 8^5
   (e) 5^5 (f) 10^3

3 (a) 8^6 (b) 10^4 (c) y^7 (d) 10^{10} (e) 3
   (f) m   (g) t^7 (h) 3   (i) 9^5

4 $7^3 \times 7^5 = 7^8$
   $3^4 \times 3^3 = 3^7$
   $7^{10} \div 7^7 = 7^3$
   $7^2 \times 7^3 \times 7^2 = 7^7$
   $3^5 \div 3^3 = 3^2$
   $7^{10} \div 7^5 \times 7 = 7^6$
   $7^4 \times 7^2 \div 7^1 = 7^5$
   $3^3 \times 3^4 \times 3 = 3^8$
   $7^2 \times 7^3 \times 7^4 = 7^9$
   $3^5 \times 3^3 \div 3^2 = 3^6$

5 (a) 4^2, 16   (b) 7, 7   (c) 2^6, 64
   (d) 8^2, 64   (e) 6^3, 216   (f) 2^4, 16
   (g) 3^5, 243   (h) 5^{-3}, 0.008

Exercise 14F

1 (a) $2^2 \times 5$ (b) $2^2 \times 3$ (c) $3^2 \times 5$
   (d) $2^3 \times 3$ (e) $2^2 \times 5^2$ (f) $2^3 \times 3^2$
   (g) $2 \times 3 \times 5 \times 7$ (h) $2^2 \times 3^3$

2 2

3 (a) $2^4 \times 3$ (b) $2 \times 3 \times 11$ (c) 2, 3

4 (a) $2^2 \times 3 \times 5$, $2^3 \times 3^2$, $2^3 \times 3 \times 5$ (b) 2, 3

Exercise 14G

1 (a) $2^2 \times 7$, 3×7 (b) 7

2 (a) 5   (b) 8   (c) 9   (d) 4   (e) 5   (f) 12

3 (a) 2   (b) 5   (c) 2   (d) 9   (e) 8   (f) 25

4 20

5 (a) Various possible answers ending in 0 or 5
   (b) two from 16, 32, 48
   (c) 28 and 56

6 (a) $\frac{2}{7}$ (b) $\frac{3}{4}$

Exercise 14H

1 (a) 40   (b) 72   (c) 36   (d) 48
   (e) 180   (f) 180   (g) 270   (h) 132
   (i) 264

2 (a) 48   (b) 120   (c) 150   (d) 144
   (e) 240   (f) 360   (g) 450   (h) 630
   (i) 2520

3 12 seconds

4 40 minutes

5 (a) $\frac{17}{18}$ (b) $\frac{41}{60}$ (c) $\frac{1}{40}$ (d) $\frac{31}{300}$
   (e) $\frac{55}{144}$ (f) $\frac{191}{240}$

6 90 seconds or $1\frac{1}{2}$ minutes

Examination style questions

1 (a) 2, 4   (b) 5, 7, 15, 19

2 (a) 64   (b) 0.04   (c) (i) 27   (ii) 19

3 (a) 125   (b) 10 000

4 (a) $2^3 \times 3$ (b) 120

5 (a) $3^2 \times 5$ (b) 3^2, 9   (c) $2^2 \times 3^2 \times 5$, 180

6 40

7 (a) e.g. 8, 12   (b) e.g. 14, 21   (c) e.g. 28

Chapter 15 Number 6

Exercise 15A

1 (a) £35   (b) 10 g   (c) 16 ℓ (d) 16p

2 (a) 25.5 km   (b) 283.5 kg
   (c) £1126.40   (d) 3.52 mℓ

3 (a) £4   (b) 6.9 kg   (c) 1363 g   (d) £258.10

4 112   **5** 185   **6** 225

7 60   **8** £10440   **9** 306

10 122 miles

Exercise 15B

1 (a) £478.80   (b) 684 g   (c) 95.76 ℓ

2 (a) 480 miles   (b) £61.44   (c) 4080 kg

3 £41.40

4 (a) £31.50   (b) 64.96   (c) 2720 mℓ
   (d) 326.4   (e) £145.13   (f) 815.24 mℓ

5 £436.80

6 Burger £1.89, Pasta £2.73, Coffee 95p, Tea 84p Salad £1

7 235   **8** £30.10   **9** £4816.50

10 99   **11** 383

Exercise 15C

1 (a) 60%   (b) 85%   (c) 60%   (d) 85.7%

2 (a) 25%   (b) 20%   (c) 36.67%   (d) 92.86%

3 Maths (80%)

4 (a) 50.2%   (b) 9.8%   (c) 17.2%   (d) 42.5%

Exercise 15D

1 (a) 33.3%   (b) 25%   (c) 50%   (d) 40%

2 5%   **3** 6.25%

4 11.1%   **5** 10%

6 12.5%   **7** 59.5%

8 43.4%   **9** 40%

10 63.9%   **11** 97.6%

12 700%

Exercise 15E

1 (a) £2   (b) 25%

2 (a) £10   (b) 40%

3 30% profit, 29.4% loss, 25% profit, 5.4% loss, 32.8% loss, 3.86% profit

4 5.95%   **5** 14.3%

6 87.5%   **7** (a) £31   (b) 16.1%

8 59%   **9** 38.5%

10 66.6%

Exercise 15F

1 (a) £22.75   (b) £78.75   (c) £7   (d) £14.70

2 (a) £9.40   (b) £86.95   (c) £14.10   (d) £2914

3 £3800 + VAT (£4465)

4 (a) £12.88   (b) £86.46   (c) £21.62

5 £468.83

6 (a) £12.88   (b) £86.46   (c) £21.62

7 £21.74

8 £106.61

Exercise 15G

1 (a) £285.50   (b) £3300
   (c) £3585.50   (d) £730.50
2 (a) £10 060   (b) £3260
3 Paying cash saving £44.80
4 £74.40
5 £29.18
6 (a) £4450   (b) £124 450

Exercise 15H

1 (a) £228   (b) £321.10   (c) £353.40
2 (a) £13.20   (b) £9.90   (c) £8.25
3 £363.60
4 £470.45, £402.55, £363.75
5 £316.20
6 3 hours
7 £12.75
8 6 hours

Exercise 15I

1 (a) 4 g   (b) 18 g   (c) 33.3 g
2 (a) 8p   (b) 0.25p   (c) 0.5p
3 (a) large 1.799, small 1.724
   (b) large box
4 (a) A 5.2p, B 5.15p, C 5.16p
   (b) B
5 545 g jar
6 large block

Exercise 15J

1 (a) £10.50   (b) £37.05   (c) £47.55
2 (a) 775   (b) £54.38
3 (a) £137.87   (b) £25.44   (c) £112.08
4 (a) 13061   (b) £314.26   (c) £327.01
5 (a) £171.20   (b) £3.20   (c) £168   (d) 7000
6 (a) £5.04   (b) £7.35   (c) £30.89   (d) £36.30

Exercise 15K

1 (a) £80   (b) £50   (c) £192.50   (d) £679.69
2 (a) £36   (b) £186
3 Panda Bank (£552)
4 £120
5 3 years
6 3.5%

Exercise 15L

1 (a) £42   (b) £76.88   (c) £181.80
2 £134.98
3 £416.86
4 £9344.26
5 Paul (£231.53)
6 2250
7 1842
8 (a) £13 781.25   (b) £13 659.98
9 (a) 10.5 kg   (b) 11.58 kg   (c) 14.07 kg
10 3 years

Exercise 15M

1 (a) £37.50   (b) 36 g
2 16 100   **3** 287.5 g
4 768 people   **5** French (84%)
6 41.67%   **7** 33.3%
8 55.4%   **9** 16%
10 30%   **11** £258.50
12 (a) £61.10   (b) £12.22
13 (a) £48   (b) £348
14 £2220   **15** £349.20
16 4 hours   **17** 375 g box
18 50 mℓ tube   **19** £35.25
20 £53.31   **21** £123.75
22 4 years   **23** £449.95
24 1 842 375 litres

Examination style questions

1 (a) 180   (b) 75   (c) 8%
2 (a) £152   (b) 3 hours and 55 minutes
3 (a) 20%   (b) $33\frac{1}{3}$%
4 4 years
5 (a) £539   (b) £404.25
6 £147   **7** £23.58
8 £30.59   **9** Large

Chapter 16 Algebra 4

Exercise 16A

1 (a) 4   (b) 10   (c) add 2   (d) 14, 16, 18
2 (a) 15   (b) 23   (c) add 4   (d) 27, 31, 35
3 (a) add 3; 18, 21   (b) add 5; 24, 29
   (c) add 10; 62, 72   (d) add 7; 34, 41
   (e) add 2; 4, 6   (f) subtract 3; $-2, -5$
4 (a) 21   (b) 13
   (c) subtract 4   (d) 9, 5, 1
5 (a) 40   (b) 28
   (c) subtract 6   (d) 16, 10, 4
6 (a) subtract 10; 36, 26   (b) subtract 6; 36, 30
   (c) subtract 4; 16, 12   (d) subtract 9; 51, 42
   (e) subtract 15; 45, 30   (f) subtract 7; 8, 1
7 (a) 0, -5 (b) 1, -2 (c) $-5, -15$ (d) $-2, -9$

Exercise 16B

1 (a) $+2, +3, +4$ (b) go up by 1 each time
   (c) 16, 22
2 (a) $+3, +4, +5$ (b) go up by 1 each time
   (c) 24, 31
3 (a) 20, 25   (b) 20, 29   (c) 70, 75   (d) 25, 36
   (e) 13, 21   (f) 44, 64   (g) 9, 4   (h) 3, 2
   (i) 21, 15   (j) 3, 3
4 (a) (i) $+3, +6, +12$ (ii) 48, 96
   (b) (i) $+2, +6, +18$ (ii) 81, 243
   (c) (i) $-20, -10$ (ii) 5, 2.5
   (d) (i) $-9000, -900$ (ii) 10, 1

Exercise 16C

1 (a) 5, 8, 11   (b) 47
2 (a) 9, 11, 13   (b) 27

3 (a) 3, 7, 11   (b) 199
4 (a) 5, 7, 9, 11   (b) +2   (c) 43
5 (a) 8, 11, 14, 17; +3; 95
     (b) 1, 3, 5, 7; +2; 59
     (c) 1, 5, 9, 13; +4; 117
     (d) 10, 13, 16, 19; +3; 97
     (e) 6, 11, 16, 21; +5; 151
     (f) 4, 10, 16, 22; +6; 178
6 The difference between consecutive tems is the same number that n is multiplied by.
7 (a) 20   (b) 21st term
     (c) $2n + 4 = 35$ gives $n = 15\frac{1}{2}$ which is not a whole number
8 (a) 17   (b) 20th term
     (c) $3n - 1 = 90$ gives $n = 30\frac{1}{3}$ which is not a whole number
9 (a) 57   (b) 15th term
     (c) $5n + 7 = 110$ gives $5n = 103$ which is not in 5 times table
10 (a) 2, 5, 10, 17   (b) 145
11 (a) 5, 8, 13, 20   (b) 85
12 (a) −2, 1, 6, 13   (b) 166
13 (a) 5th term
     (b) $n^2 + 3 = 50$ gives $n^2 = 47$ but 47 is not a square number ($\sqrt{47}$ is not a whole number)

Exercise 16D

1 $3n + 1$; 151   **2** $2n + 3$; 103
3 $4n + 3$; 203   **4** $5n - 2$; 248
5 $3n + 3$; 153   **6** $7n - 1$; 349
7 $10n - 8$; 492   **8** $9n + 7$; 457
9 $n + 4$; 54   **10** $3n - 7$; 143

Exercise 16E

1 (a)   (b) 21

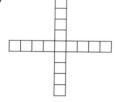

     (c) add 4 tiles each time   (d) $4n + 1$
2 (a)   (b) 11

     (c) add 2 matchsticks each time
     (d) $2n + 1$
     (e) 1 lot of 2 matches + 1
         2 lots of 2 matches + 1
         3 lots of 2 matches + 1
         n lots of 2 matches + 1
     (f) 25 triangles
3 (a)    (b) 21

     (c) add 4 matchsticks each time
     (d) $4n + 1$
     (e) n pentagons need n lots of 4 matches + 1
     (f) 21 pentagons

4 (a) 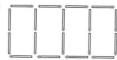   (b) 22

     (c) $4n + 2$ (d) 16
5 (a)   (b) 12

     (c) $2n + 2$ (d) 11
6 (a) ● ● ● ●   (b) 25
         ● ● ● ●
         ● ● ● ●
         ● ● ● ●

     (c) +3, +5, +7, ... add the next odd number (the differences are going up by 2 each time)
     (d) square numbers   (e) n^2 (f) 12
7 (a)   (b) 21

         ●
        ● ●
       ● ● ●
      ● ● ● ●
     ● ● ● ● ●

     (c) +2, +3, +4, +5, ... add the next whole number (to find the no. of dots in the 5th triangle $1 + 2 + 3 + 4 + 5 = 15$)
     (d) $\frac{1}{2} \times 10 \times (10 + 1) = 55$

Examination style questions

1 (a)

Pattern number	1	2	3	4
Number of sticks	5	7	9	11

     (b) 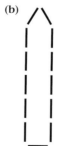   (c) 11   (d) 203

2 14
3 (a) 5, 0   (b) 40, 5
4 11, −11
5 (a) 30, 37   (b) 30   (c) subtract 3
6 (a)

     (b) 2   (c) 20
7 (a) (i) 5, 9, 13   (ii) No, 121 ÷ 4 not an integer
     (b) $3n + 1$
8 (a) 21   (b) $4n + 1$ (c) 50
9 (a) $2n$ (b) $n + 1$

Chapter 17 Shape 4

Exercise 17A–F

Students' own constructions

Exercise 17G

1

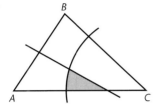

2

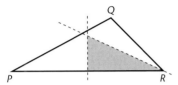

3

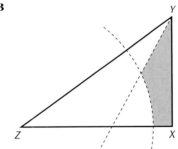

Exercise 17H

1

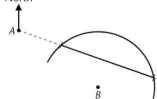

2

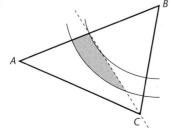

3

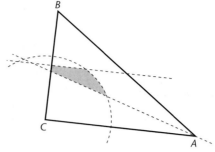

Examination style questions

1 Students' own constructions
2 (a) 290° (b) 31.5 km
(c)

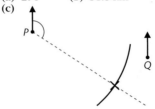

3 Students' own constructions
4 Students' own constructions
5 (a) Students' own constructions
(b) The perpendicular bisector of *LM* is the locus of points which are equidistant from 2 fixed points.
6 (a) Students' own constructions
(b) (i) Students' own constructions
(ii) Students' own constructions

7

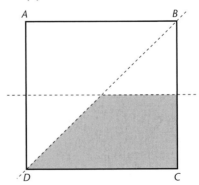

8

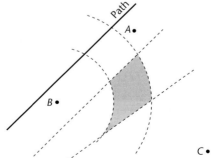

Chapter 18 Algebra 5

Exercise 18A

1 P(5, 2), Q(2, −2), R(−4, 0), S(−2, 2), T(0, 4)

2 (a)

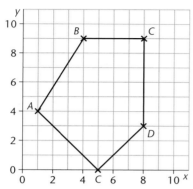

(b) pentagon

3 A(1, 9), B(4, 6)   mid-point (2.5, 7.5)
C(9, 7), D(1, −4)   mid-point (5, 1.5)
E(−3, 9), F(−3, 4)   mid-point (−3, 6.5)
G(−3, 1), H(4, 4)   mid-point (0.5, 2.5)
I(6, 10), J(10, 10)   mid-point (8, 10)

4 (a) A(4, 0, 0), B(4, 5, 0), C(0, 5, 0), D(4, 0, 2), E(4, 5, 2),
F(0, 5, 2), G(0, 0, 2)
  (b) (i) (0, 2.5, 0)   (ii) (0, 5, 1)   (iii) (2, 5, 0)
    (iv) (4, 5, 1)
  (c) (i) (2, 2.5, 0)   (ii) (2, 2.5, 2)   (iii) (2, 5, 1)

5 (a) (4.5, 1)   (b) (7, 5.5)   (c) (3.5, 6)
  (d) (−1, 5)   (e) (0.5, −0.5)

Exercise 18B

1 (a) $y = 3$ (line b), $y = -6$ (line d)
  (b) $x = 2$ (line a), $x = -4$ (line c) and $x = 5$ (line e)

2

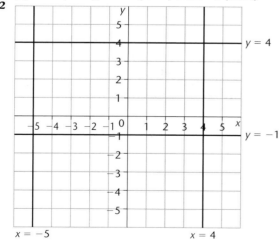

Exercise 18C

1 (a)

x	−2	0	3
$y = x - 3$	−5	−3	0

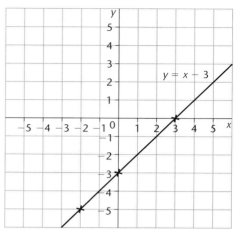

(b)

x	−3	0	1
$y = 2x + 3$	−3	3	5

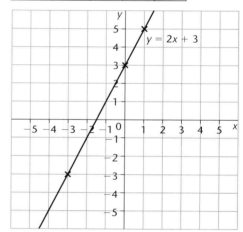

(c)

x	−1	0	3
$y = 4 - x$	5	4	1

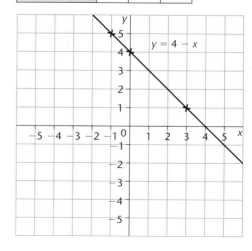

(d)

x	-1	0	2
$y = 3x - 1$	-4	-1	5

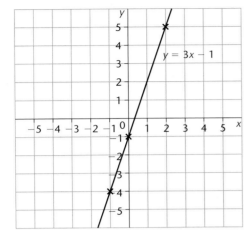

(e)

x	-1	0	3
$y = 2 - 2x$	4	2	-4

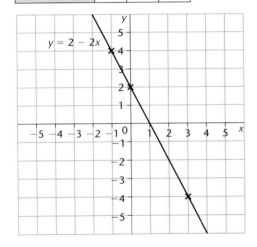

(f)

x	-4	0	4
$y = 1 - \frac{1}{2}x$	3	1	-1

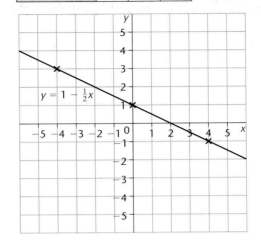

2

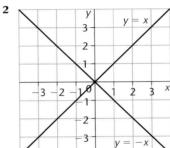

They cross over at right−angles; cross at the origin.

; (2,3)

3

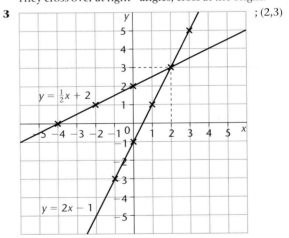

Exercise 18D

1 (a) $m = 1$ (b) $m = 2$ (c) $m = -1$
   (d) $m = 3$ (e) $m = -2$ (f) $m = -\frac{1}{2}$
2 (a) $m = 3$ (b) $m = -1$ (c) $m = 4$
   (d) $m = -3$ (e) $m = \frac{2}{3}$ (f) $m = 0.8$
3 (a), (b), (e) and (f)
4 (b), (c) and (d)

Exercise 18E

1 (a) $c = -3$ (b) $c = 3$ (c) $c = 4$
   (d) $c = -1$ (e) $c = 2$ (f) $c = 1$
2 (a) $c = 2$ (b) $c = 3$ (c) $c = -7$
   (d) $c = -4$ (e) $c = 8$ (f) $c = -0.3$
3 (a) $c = 1$ (b) $c = -7$ (c) $c = \frac{1}{2}$
   (d) $c = -\frac{1}{2}$ (e) $c = -\frac{1}{3}$ (f) $c = \frac{3}{2}$
   (g) $c = -2$ (h) $c = 2$ (i) $c = \frac{2}{3}$
   (j) $c = -\frac{2}{3}$ (k) $c = 1$ (l) $c = -\frac{1}{2}$

Exercise 18F

1 (a) $y = 3x + 6; m = 3, c = 6$
   (b) $y = -4x - 2; m = -4, c = -2$
   (c) $y = -3x + 1; m = -3, c = 1$
   (d) $y = -2x + 7; m = -2, c = 7$
   (e) $y = x + 3; m = 1, c = 3$
   (f) $y = -3x - 4; m = -3, c = -4$
   (g) $y = 3x - 7; m = 3, c = -7$
   (h) $y = \frac{1}{2}x + 5; m = \frac{1}{2}, c = 5$
   (i) $y = -2x + 3; m = -2, c = 3$
   (j) $y = -3x + \frac{7}{2}; m = -3, c = \frac{7}{2}$

Exercise 18G

1

x	-3	-2	-1	0	1	2	3
x^2	9	4	1	0	1	4	9
$+2$	2	2	2	2	2	2	2
$y = x^2 + 2$	11	6	3	2	3	6	11

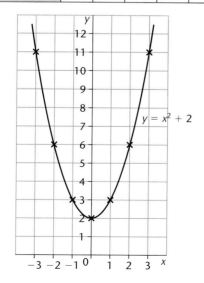

2

x	-2	-1	0	1	2
$2x^2$	8	2	0	2	8
-5	-5	-5	-5	-5	-5
$y = 2x^2 - 5$	3	-3	-5	-3	3

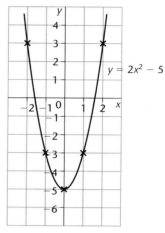

3

x	-2	-1	0	1	2
$3x^2$	12	3	0	3	12
$+1$	1	1	1	1	1
$y = 3x^2 + 1$	13	4	1	4	13

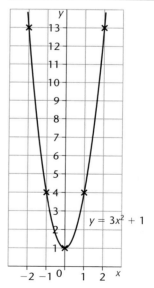

4

x	-3	-2	-1	0	1	2
x^2	9	4	1	0	1	4
$+x$	-3	-2	-1	0	1	2
$y = x^2 + x$	6	2	0	0	2	6

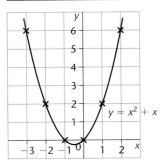

5

x	0	1	2	3	4	5
x^2	0	1	4	9	16	25
$-5x$	0	-5	-10	-15	-20	-25
$y = x^2 - 5x$	0	-4	-6	-6	-4	0

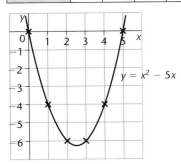

6

x	-4	-3	-2	-1	0	1
x^2	16	9	4	1	0	1
$+3x$	-12	-9	-6	-3	0	3
$y = x^2 + 3x$	4	0	-2	-2	0	4

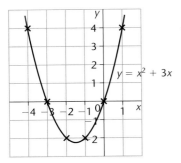

7

x	-4	-3	-2	-1	0	1	2
$2x^2$	32	18	8	2	0	2	8
$+4x$	-16	-12	-8	-4	0	4	8
$y = 2x^2 + 4x$	16	6	0	-2	0	6	16

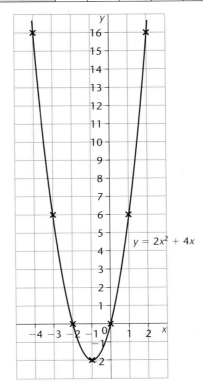

8

x	-1	0	1	2	3	4
$2x^2$	2	0	2	8	18	32
$-6x$	6	0	-6	-12	-18	-24
$y = 2x^2 - 6x$	8	0	-4	-4	0	8

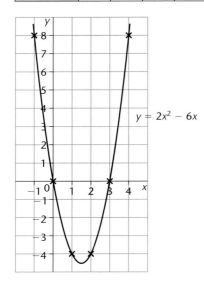

$y = 2x^2 - 6x$

(b)

x	-5	-4	-3	-2	-1	0	1
x^2	25	16	9	4	1	0	1
$+4x$	-20	-16	-12	-8	-4	0	4
-1	-1	-1	-1	-1	-1	-1	-1
$y = x^2 + 4x - 1$	4	-1	-4	-5	-4	-1	4

$; x = -2$

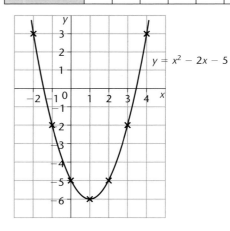

$y = x^2 + 4x - 1$

(c)

x	-2	-1	0	1	2	3	4
x^2	4	1	0	1	4	9	16
$-2x$	4	2	0	-2	-4	-6	-8
-5	-5	-5	-5	-5	-5	-5	-5
$y = x^2 - 2x - 5$	3	-2	-5	-6	-5	-2	3

$; x = 1$

$y = x^2 - 2x - 5$

Exercise 18H

1 (1) $x = 0$ (2) $x = 0$ (3) $x = 0$
  (4) $x = -0.5$ (5) $x = 2.5$ (6) $x = -1.5$
  (7) $x = -1$ (8) $x = 1.5$

2 (a)

x	-3	-2	-1	0	1	2
x^2	9	4	1	0	1	4
$+x$	-3	-2	-1	0	1	2
-2	-2	-2	-2	-2	-2	-2
$y = x^2 + x - 2$	4	0	-2	-2	0	4

$; x = -0.5$

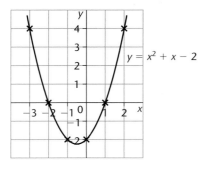

$y = x^2 + x - 2$

(d)

x	-5	-4	-3	-2	-1	0	1	2
x^2	25	16	9	4	1	0	1	4
$+3x$	-15	-12	-9	-6	-3	0	3	6
$+1$	1	1	1	1	1	1	1	1
$y = x^2 + 3x + 1$	11	5	1	-1	-1	1	5	11

; $x = -1.5$

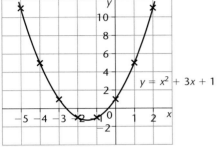

(e)

x	-1	0	1	2	3	4	5	6
x^2	1	0	1	4	9	16	25	36
$-5x$	5	0	-5	-10	-15	-20	-25	-30
-4	-4	-4	-4	-4	-4	-4	-4	-4
$y = x^2 - 5x - 4$	2	-4	-8	-10	-10	-8	-4	2

; $x = 2.5$

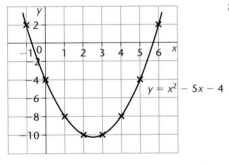

(f)

x	-3	-2	-1	0	1	2	3	4
$2x^2$	18	8	2	0	2	8	18	32
$-2x$	6	4	2	0	-2	-4	-6	-8
$+3$	3	3	3	3	3	3	3	3
$y = 2x^2 - 2x + 3$	27	15	7	3	3	7	15	27

; $x = 0.5$

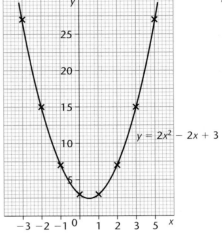

(g)

x	-4	-3	-2	-1	0	1	2
$2x^2$	32	18	8	2	0	2	8
$+4x$	-16	-12	-8	-4	0	4	8
-5	-5	-5	-5	-5	-5	-5	-5
$y = 2x^2 + 4x - 5$	11	1	-5	-7	-5	1	11

; $x = -1$

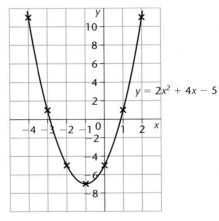

(h)

x	-2	-1	0	1	2	3	4
$2x^2$	8	2	0	2	8	18	32
$-4x$	8	4	0	-4	-8	-12	-16
-9	-9	-9	-9	-9	-9	-9	-9
$y = 2x^2 - 4x - 9$	7	-3	-9	-11	-9	-3	7

$; x = 1$

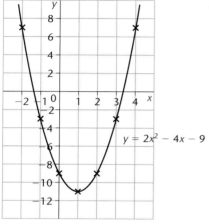

3 (a)

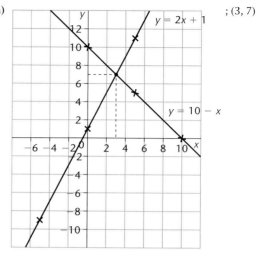

$; (3, 7)$

Exercise 18I

1

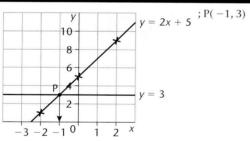

$; P(-1, 3)$

2

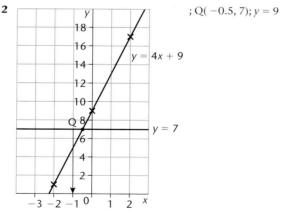

$; Q(-0.5, 7); y = 9$

(b)

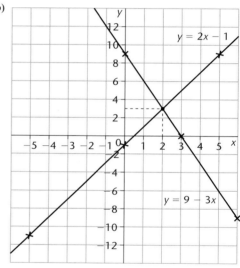

$(2, 3)$

(c)

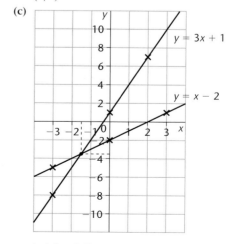

$(-1.5, -3.5)$

Exercise 18J

1 (a) and (b)

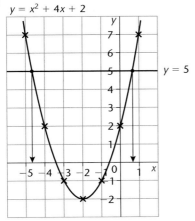

$y = x^2 + 4x + 2$

(c) $x = 0.6$, $x = -4.7$ **d)** $x^2 + 4x - 3 = 0$
(e) $x = -0.7$, $x = -3.4$

2 (a) $x = 1.0$ and $x = -2.0$; $x = 1.5$ and $x = -2.6$
(b) $x = 0.2$ and $x = -4.3$; $x = -0.6$ and $x = 3.4$
(c) $x = 3.4$ and $x = -1.4$; $x = 2.4$ and $x = -0.4$
(d) $x = -0.5$ and $x = -2.7$; $x = 1.5$ and $x = -4.6$
(e) $x = 5.7$ and $x = -0.8$; $x = 4.5$ and $x = 0.4$
(f) $x = 1.6$ and $x = -0.6$; $x = 3.4$ and $x = -2.5$
(g) $x = 0.9$ and $x = -2.9$; $x = 1.4$ and $x = -3.4$
(h) $x = 3.3$ and $x = -1.4$; $x = 2.2$ and $x = -0.3$

Exercise 18K

1 (a)

miles (x)	0	5	10	20	50	100
km (y)	0	8	16	32	80	160

(b)

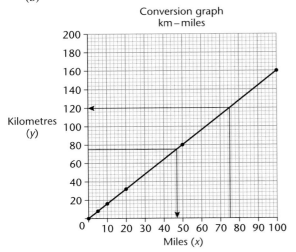

(c) approximately 47 miles
(d) approximately 120 km

2 (a)

cm	0	2.5	5	7.5	12.5	15	25	30
inches	0	1	2	3	5	6	10	12

(b)

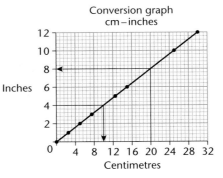

(c) 10 cm   (d) 8 inches

3 (a)

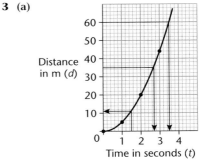

(b) 11 m   (c) 2.7 s   (d) 3.5 s

Exercise 18L

1 (a) 0830   (b) 15 min   (c) 1 hr 15 min
2 (a) 5 hr   (b) On her return; steeper line
(c) 20 km/hr

3

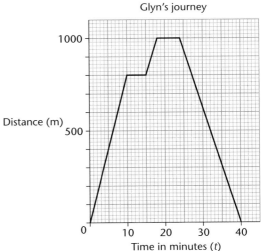

(a) 80 m/min (1.3 m/s)   (b) 50 m/min (0.83 m/s)

4 (a)

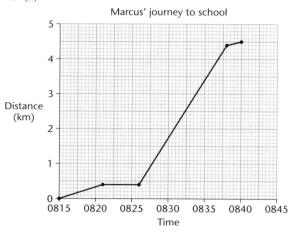

Marcus' journey to school

(b) 333.3 m/min or 20 km/h

5 (a)

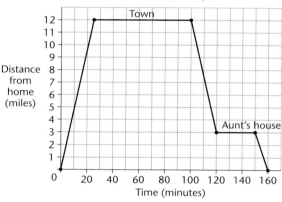

Alison's journey

(b) 28.8 mph

3

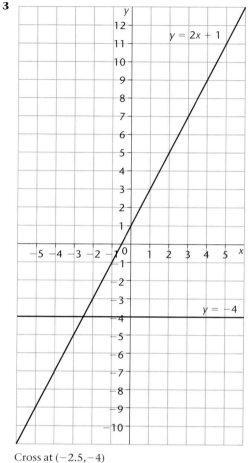

Cross at $(-2.5, -4)$

4 (a) He is not moving   **(b)** 16 miles
  (c) 8 miles per hour   **(d)** 1330

Examination style questions

1 B(2, 2), C(1, −3), D(−3, −3)

2

x	−1	0	1	2	3
y	−3	−1	1	3	5

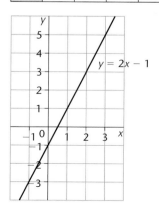

5

x	-2	-1	0	1	2	3
y	15	5	-1	-3	-1	5

(a)

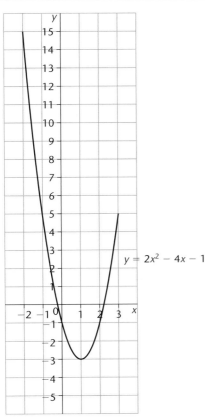

$y = 2x^2 - 4x - 1$

(b) (i) Check where the graph crosses the $x-$axis
(ii) -0.2

Chapter 19 Shape 5

Exercise 19A

1 (a)

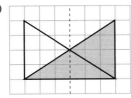

(b)

(c)

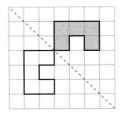

(d)

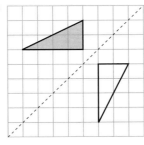

2 (a)

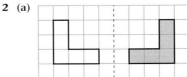

(b)

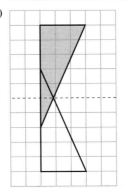

(c)

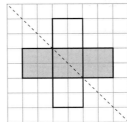

3 (a) and (b)

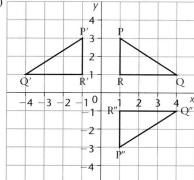

4

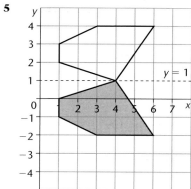

5

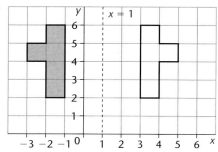

6

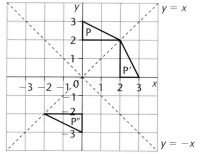

Exercise 19B

1 (a)

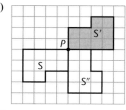

(b)

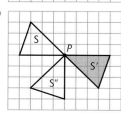

(c)

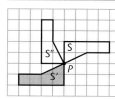

2 (a), (b) and (c)

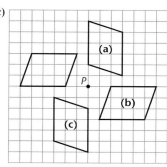

3 (a), (b) and (c)

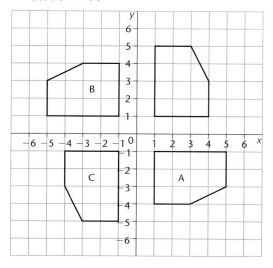

4 90° clockwise about the point (−3, 4)

Exercise 19C

1 (i) (a)

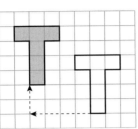

(b)

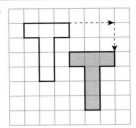

(c)

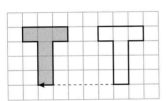

(d)

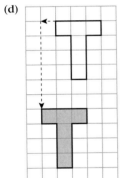

(e)

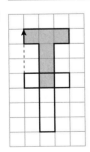

(ii) (a)

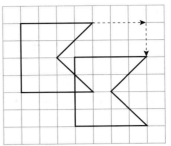

(b)

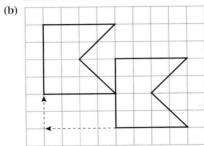

(c)

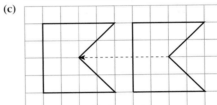

(d)

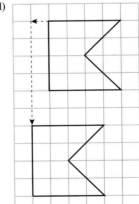

(e)

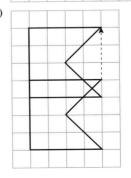

2

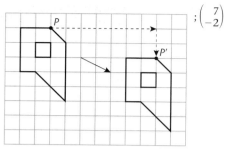

$;\begin{pmatrix} 7 \\ -2 \end{pmatrix}$

3 A to B $\begin{pmatrix} -7 \\ 5 \end{pmatrix}$ A to C $\begin{pmatrix} -2 \\ 6 \end{pmatrix}$ A to D $\begin{pmatrix} 1 \\ 5 \end{pmatrix}$ A to E $\begin{pmatrix} -6 \\ -1 \end{pmatrix}$

Exercise 19D

1 **(a)** scale factor 2   **(b)** scale factor 3
   (c) scale factor 2   **(d)** scale factor 4

2 **(a)**

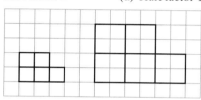

(b)

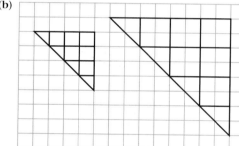

(c)

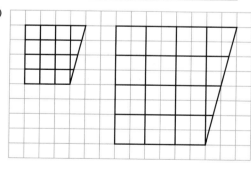

3 **(a)**

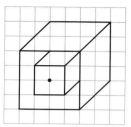

(b)

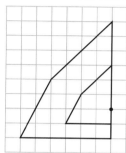

4

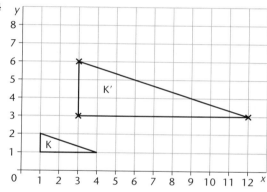

(3, 3), (3, 6) and (12, 3)

5

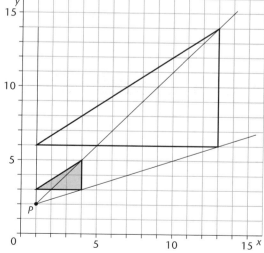

Exercise 19E

1 (a) scale factor $\frac{1}{3}$

(b)

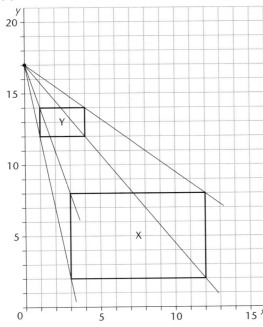

(c) $(0, 17)$

2

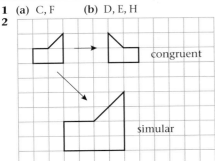

3 (a) $(7, 7)$ **(b)** scale factor $\frac{1}{4}$

Exercise 19F

1 (a) C, F   **(b)** D, E, H

2

congruent

simular

3 (a) A, B   **(b)** A, B   **(c)** A, B, C   **(d)** A, B

4 (a) SSS   **(b)** SAS   **(c)** Not congruent

     (d) ASA   **(e)** RHS   **(f)** RHS

Examination style questions

1

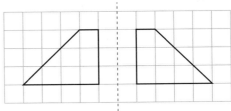

2 (a) 2   **(b)** $(-2, -1)$ **(c)** 4

3 (a)

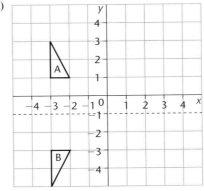

(b)

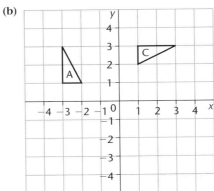

4 (a) B and D   **(b)** A and E

5 (a) (i) A   **(ii)** E   **(iii)** B

   (b) Reflection in the line $x = -2$

6

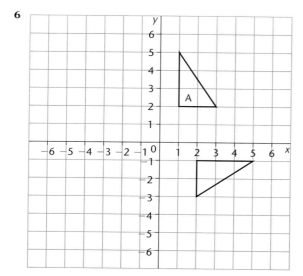

Chapter 20 Handling data 4

Exercise 20A

1 (a) mean = 5, median = 4, mode = 4, range = 9
  (b) mean = 58, median = 61, mode = 80, range = 56
  (c) mean = 23, median = 22, mode = none, range = 17
  (d) mean = 0.91, median = 0.63, mode = 0.25,
    range = 2.25

Exercise 20B

1 (a) mean = 6, median = 6, mode = 6, range = 11
  (b) mean = 22.9, median = 22.0, mode = none, range
    = 9.5
  (c) mean = $\frac{29}{32}$, median = $\frac{20}{32} = \frac{5}{8}$, mode = $\frac{8}{32} = \frac{1}{4}$,
    range = $\frac{9}{4} = 2\frac{1}{4}$
  (d) mean = 158.3, median = 161, mode = 180,
    range = 56
2 mean = 65.2, median = 64 and mode = 64
3 mean score = 6, mode score = 8
4 mean price = £9948.33
5 (a) mean amount = £87   (b) amount added = £59
6 mean height = 173 cm
7 (a) mean score = 22.2   (b) median score = 23
  (c) mode score = 24   (d) range = 12
8 Possible answers include: 4, 5, 6, 6, 7, 8 and median = 6
9 mean temp = 16.75 °C, median temp = 17.5 °C, modal
  temp = 21 °C, range = 30 °C
10 (a) mean = 148 cm, mode = 137 cm,
    median = 152 cm, range = 46 cm
  (b) mean = 155.7 cm, mode = 161 cm,
    median = 158.5 cm, range = 49 cm
  (c) mean = 151.6 cm, mode = 155 cm,
    median = 155 cm, range = 53 cm

Exercise 20C

1 mode = 38, median = 33.5, mean = 30.9, range = 32
  Median better represents the average. Ignore the value
  9 (extreme value).
2 mode = 70, median = 54.5 and mean = 50.2
  Median is a better average. 9 and 12 are extreme values.

Exercise 20D

1 (a)

Number on dice (x)	Frequency (f)	fx
1	7	7
2	9	18
3	6	18
4	11	44
5	10	50
6	7	42
Totals	$\Sigma f = 50$	$\Sigma fx = 179$

  (b) mean $\bar{x}$ = 3.58 ≈ 4, median = 4, mode = 4

2 (a)

Speed mph (x)	Frequency (f)	fx
10	0	0
20	23	460
30	48	1440
40	16	640
50	2	100
60	1	60
Totals	$\Sigma f = 90$	$\Sigma fx = 2700$

  (b) mean $\bar{x}$ = 30 mph, median = 30 mph,
    mode = 30 mph.
  (c) Speed limit is 30 mph.

3 mean $\bar{x}$ = 2.2 goals, median = 2 goals, mode = 1 goal

4 (a)

Test score	Frequency (f)	fx
1	2	2
2	7	14
3	6	18
4	7	28
5	9	45
6	10	60
7	9	63
8	13	104
9	7	63
10	5	50
Totals	75	447

  (b) mean $\bar{x}$ = 5.96 ≈ 6, median = 6, mode = 8,
    range = 9

Exercise 20E

1 (a)

Calls made	Frequency (f)	Mid-interval value (x)	fx
0–2	12	1	12
3–5	18	4	72
6–8	31	7	217
9–11	11	10	110
Totals	$\Sigma f = 72$		$\Sigma fx = 411$

(b) estimated mean = 5.7
(c) mode class interval = 6–8,
median class interval = 6–8

2 (a)

Shoe size	Frequency (f)	Mid-interval value (x)	fx
3–4	5	3.5	17.5
6–6	8	5.5	44
7–8	12	7.5	90
9–10	5	9.5	47.5
11–12	2	11.5	23
Totals	$\Sigma f = 32$		$\Sigma fx = 222$

(b) estimated mean = 6.9375 ≈ 7
(c) mode class interval = 7–8,
median class interval = 7–8

3 (a)

Number of CD's	Frequency (f)	Mid-interval value (x)	fx
$1 \leqslant x < 25$	12	12.5	150
$25 \leqslant x < 50$	0	37	0
$50 \leqslant x < 75$	11	62	682
$75 \leqslant x < 100$	9	87	783
$100 \leqslant x < 125$	15	112	1680
$125 \leqslant x < 150$	33	137	4521
$150 \leqslant x < 175$	28	162	4536
$175 \leqslant x < 200$	41	187	7667
$200 \leqslant x < 225$	16	212	3392
Totals	$\Sigma f = 165$		$\Sigma fx = 23411$

(b) estimated mean = 141.9 ≈ 142
(c) mode class interval = $175 \leqslant x < 200$,
median class interval = $125 \leqslant x < 150$

4 Estimated mean = 10.4 ≈ 10
Mode class interval = 15–19 and median class
interval = 5–9
Mean compares better with the median class interval.

Examination style questions

1 (a) 7   (b) 4
2 (a) 5   (b) 7   (c) 8
3 (a) 4   (b) 20   (c) 21.5
4 1.7

5 (a)

Score	Frequency
0–9	0
10–19	4
20–29	7
30–39	5
40–49	1
50–59	6
60–69	5
70–79	2
80–89	1

(b) 20–29   (c) 43

6

Length x (cm)	Frequency	Mid-point value
$1 < x \leqslant 3$	5	2
$3 < x \leqslant 5$	2	4
$5 < x \leqslant 7$	2	6
$7 < x \leqslant 9$	1	8

Mean length = 3.8 cm

Chapter 21   Shape 6

Exercise 21A

1 (a) b (b) g (c) j
2 (a) $b^2 = a^2 + c^2$ (b) $g^2 = f^2 + h^2$ (c) $j^2 = i^2 + k^2$

Exercise 21B

1 (a) $r = 10$ cm   (b) $s = 20$ cm   (c) $t = 15$ cm
2 (a) $a = 10.8$ cm   (b) $b = 11.7$ cm
    (c) $c = 14.9$ cm   (d) $d = 11.3$ cm
3 3.8 m   **4** 2.86 m
5 26.9 m   **6** 26 km

Exercise 21C

1 (a) $x = 5.7$ cm   (b) $y = 4.5$ cm   (c) $z = 13.2$ cm
2 (a) $XY = 2.2$ cm   (b) $QR = 11.8$ cm
3 2.9 m   **4** 73.3 m
5 2.42 m   **6** 77.2 km
7 $c = 5$ cm, $d = 12$ cm   **8** $w = 3.75$ m

Examination style questions

1 $\sqrt{125}$ **2** 3.23 m
3 80.78 m   **4** 10.6 cm

Chapter 22

Exercise 22A

1 (a) $p + q = O + E = O$, so $p + q - 1 = O - 1 = E$
    (b) $pq = O \times E = E$, so $pq + 1 = E + 1 = O$

2 If n is O then $n + 1 = O + 1 = E$,
so $n \times (n + 1) = O \times E = E$
If n is E then $n + 1 = E + 1 = O$,
so $n \times (n + 1) = E \times O = E$
So $n(n + 1)$ is always E.

3 If x is O then $x^2 = x \times x = O \times O = O$
So $x^2 + 1 = O + 1 = E$

4 If b is E then $b - 1 = E - 1 = O$ and $b + 1 = E + 1 = O$
So $(b - 1) \times (b + 1) = O \times O = O$

5 $x - y = O - E = O$ and $x + y = O + E = O$
So $(x - y) \times (x + y) = O \times O = O$

6 The four consecutive numbers can be $x, x + 1, x + 2, x + 3$.
$x + (x + 1) + (x + 2) + (x + 3) = 4x + 6 = 2(2x + 3)$
$2(2x + 3)$ means $2 \times (2x + 3)$
$2 \times$ 'anything' is always even.

Exercise 22B

1 $k =$ any multiple of 4
e.g. $k = 4$, $k^2 + \frac{1}{2}k = 16 + 2 = 18$ which is even.

2 $m = 1$, $m^3 + 2 = 1^3 + 2 = 1 + 2 = 3$, which is a multiple of 3.

3 When $p = 1$ and $q = 2$, $p + q - 1 = 1 + 2 - 1 = 2$, which is prime.

4 $a = 7$, $b = 2$, $a + b = 7 + 2 = 9$, which is odd.

5 When $n = 6$, $n^2 + 3n + 1 = 6^2 + 3(6) + 1 = 36 + 18 + 1 = 55$
55 is divisible by 5 and 11 and is not a prime number.

6 **(a)** $\sqrt{0.25} = 0.5$ (because $0.5 \times 0.5 = 0.25$)
So the square root (0.5) is greater than the original number (0.25).
(b) Try 0.2 $(0.2)^3 = 0.2 \times 0.2 \times 0.2 = 0.008$
$(0.2)^2 = 0.2 \times 0.2 = 0.04$
0.008 is smaller than 0.04
So the cube of 0.2 is smaller than the square of 0.2.

Exercise 22C

1

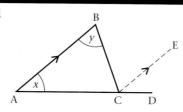

Draw CE parallel to AB.
$\hat{\text{ECD}} = x$ (corresponding angles)
$\hat{\text{ECB}} = y$ (alternate angles)
Exterior angle, $\hat{\text{BCD}} = \hat{\text{ECD}} + \hat{\text{ECB}}$
$= x + y$
$=$ sum of opposite interior angles

2

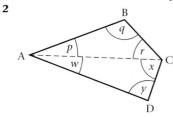

Join AC
In $\triangle$ABC, $p + q + r = 180°$ (sum of angles of $\triangle$ABC)
In $\triangle$ADC, $w + x + y = 180°$ (sum of angles of $\triangle$ABC)
Sum of angles of quadrilateral ABCD
$= p + w + q + r + x + y$
$= (p + q + r) + (w + x + y)$
$= 180° + 180°$
$= 360°$

3

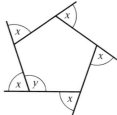

Regular pentagon, all sides equal, all interior angles equal (y), so all exterior angles equal (x).
5 exterior angles $= 360°$
$5x = 360°$
$x = 72°$
$x + y = 180°$ (angles on a straight line)
$y = 180° - 72° = 108°$

Examination style questions

1 **(a)** e.g. $10 \div 2 = 5$ **(b)** odd
2 **(a)** e.g. $k = 4$ **(b)** even
3 **(a)** even **(b)** odd
4 **(a)** Always even **(b)** Could be either odd or even
5 **(a)** even $\times$ odd so even product
(b) $2 \times n$ always even so $2n + 1$ is odd
6 e.g. $x = -4$, $y = 25$ which is not prime ($5 \times 5 = 25$)
7 When $x = 7$, $2x^2 - 2x + 7 = 2 \times 49 - 2 \times 7 + 7$
$= 98 - 14 + 7$
$= 91$
91 is not prime ($7 \times 13 = 91$)
8 An odd number is always of the form $2n + 1$ so three consecutive odd numbers are $2n + 1$, $2n + 3$ and $2n + 5$. Their sum is
$2n + 1 + 2n + 3 + 2n + 5 = 6n + 9 = 3(2n + 3)$
which is $3 \times (2n + 3)$.
9 n^2 is a square number. The next one in the sequence is $(n + 1)^2$. Their sum is:
$n^2 + (n + 1)^2 = n^2 + n^2 + 2n + 1$
$= 2n^2 + 2n + 1$
$= 2(n^2 + n) + 1$
$2(n^2 + n)$ must be even because it is a multiple of 2, so $2(n^2 + n) + 1$ must be odd.
10 $n^2 - 3n = n(n - 3)$.
If n is even then $(n - 3)$ is odd.
If n is odd the $(n - 3)$ is even.
The product is always O $\times$ E, which is even.

Exam practice paper, non-calculator

1 **(a)** 35 000 **(b)** 29 066, 34 829, 38 217
2 **(a)** e.g. 12, 18 **(b)** e.g. 16, 24 **(c)** e.g. 24
3 **(a)** 42, 36
(b) The difference is one less each time.
4 **(a)** 7 **(b)** 6
5 A5, B6, C1, D3
6 **(a)** 9 **(b)** 0
7 **(a)** 42 **(b)** £15.12 **(c)** £4.88

8 (a) 36 (b) 7 (c) 64 (d) 3
9 (a) (i) Isosceles (ii)

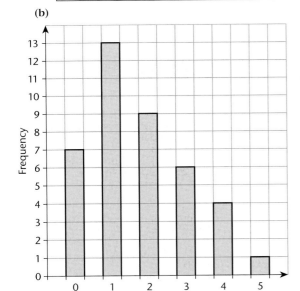

(b) (i) 0 (ii) 2 (iii) 2 (iv) 2
10 A(£32), B is only £30
11 (a)

Number of children	Tally	Frequency
0	JHT II	7
1	JHT JHT III	13
2	JHT IIII	9
3	JHT I	6
4	IIII	4
5	I	1

(b)

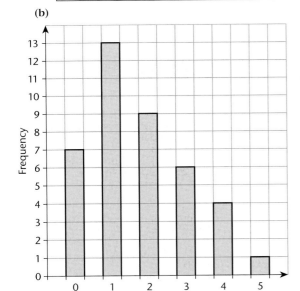

12 (a) $7x$ (b) $3x + 1y$ (c) $15p$ (d) $8m + 7$
13 (a) 250 m (b) 75°
  (c) AC = 4 cm and correct bearing
14 6 cm
15 (a)

		Tin B			
	2	4	5	7	9
1	3	5	6	8	10
Tin A 3	5	7	8	10	12
6	8	10	11	13	15
8	10	12	13	15	17

(b) $\frac{7}{20}$
16 (a) $x = 2$ (b) $y = 10$ (c) $z = 15$
17 96°
18 5 days
19 60 mph
20 (a) 6.54 (b) 25.12 (c) 57
21 $x = 55$
22 (a) w^9 (b) x^3

23 $(-2, 4)$
24 1500
25 $36 + 10\pi$ cm
26 (a) 41 (b) sensible line of best fit
  (c) positive correlation, as maths score increases, so
    does science score
  (d) 32 approx.

Exam practice paper, calculator

1 (a) £6 (b) £9.60 (c) 150 minutes
2 (a) 42 (b) 15 (c) 1 (d) 29
3 (a) m (b) g (c) km (d) 1
4 (a) 5 (b)

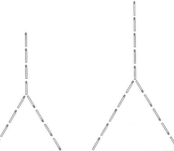

5 £255.88
6 (a) 30° (b) acute
7 (a) $\frac{10}{30}$ (b) $\frac{7}{30}$ (c) $\frac{27}{30}$
8 5, 14
9 (a)

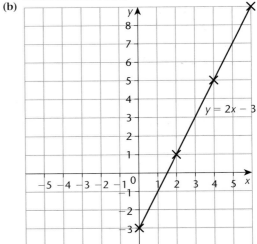

(b) 56
10 (a) 23.0 (b) 23.02 (c) 23.020
11 $4\frac{1}{2}$ kg = 4.5 × 2.2 pounds = 9.9 pounds,
  Jon made the best estimate.
12 (a) 30 °C (b) −8 °C
13 £7.90
14 (a)

x	0	2	4	6
y	−3	1	5	9

(b)

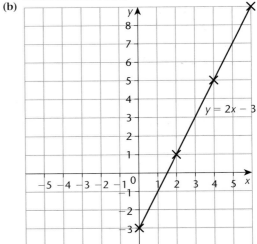

15 33.28
16 (a) £10 (b) £36

17 Triangle with sides of 11 cm, 8 cm and 5 cm; arcs must be shown.

18 96 km/h (multiply by $\frac{8}{5}$)

19 $\frac{7}{27}$ (convert to decimals)

20 **(a)** 8.609522118   **(b)** 24.389

21 **(a)** 518 euros
   (b) 46 euros → £31.08 (to nearest penny)

22 **(a)**

2	7	9				
3	0	1	1	4	5	6	8
4	0	0	0	2	6	
5	1					

   (b) (i) 40 **(ii)** 36 **(iii)** 24

23 **(a)** odd   **(b)** odd   **(c)** even   **(d)** odd

24 £540.50

25

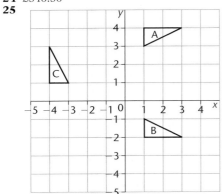

26 **(a)** $2(4m + 5)$ **(b)** $y(y - 5)$

27 23.1%

28 **(a)**

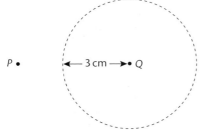

(b)

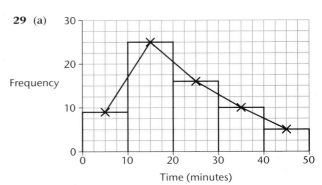

29 **(a)**

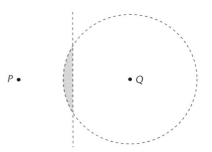

(b) $10 < t < 20$

30

x	$x^3 - 2x$	comment
2	4	too small
3	21	too big
2.5	10.625	too small
2.6	12.376	too small
2.7	14.283	too small
2.8	16.352	too big
2.75	15.296875	too small

Solution is $x = 2.8$ (1 d.p.)

Index